FORD FALCON STRAIGHT SIX OWNERS WORKSHOP MANUAL 1964-1970

Models covered:
Engines: 170 - 200 - 250 CID
Ford Falcon 170 CID 1964-1969
Ford Falcon Futura 200 CID 1965-1970
Ford Falcon Station Wagon 200 CID 1965-1970
Ford Falcon 200 CID 1969-1970
Ford Falcon 250 CID 1970

ISBN 9780851473161

Brooklands Books
BROOKLANDS BOOKS LTD.
P.O. BOX 904, AMERSHAM,
BUCKS, HP6 9JA. UK

FO64WH
www.brooklandsbooks.com
3W5/2930

INTRODUCTION

This do-it-yourself Workshop Manual has been specially written for the owner who wishes to maintain his car in first class condition and to carry out his own servicing and repairs. Considerable savings on garage charges can be made, and one can drive in safety and confidence knowing the work has been done properly.

Comprehensive step-by-step instructions and illustrations are given on all dismantling, overhauling and assembling operations. Certain assemblies require the use of expensive special tools, the purchase of which would be unjustified. In these cases information is included but the reader is recommended to hand the unit to the agent for attention.

Throughout the Manual hints and tips are included which will be found to be invaluable, and there is an easy to follow fault diagnosis at the end of each chapter.

Whilst every care has been taken to ensure correctness of information it is obviously not possible to guarantee complete freedom from errors or to accept liability arising from such errors or omissions.

Instructions may refer to the right-hand or left-hand sides of the vehicle or the components. These are the same as the right-hand or left-hand of an observer standing behind the car and looking forward.

By the Autobooks Team of Technical Writers

Brooklands Books Ltd., PO Box 904, Amersham, Bucks, England. HP6 9JA

brooklandsbooks.com

www.brooklandsbooks.com

CONTENTS

INDEX

HINTS ON MAINTENANCE AND OVERHAUL

There are few things more rewarding than the restoration of a vehicle's original peak of efficiency and smooth performance.

The following notes are intended to help the owner to reach that state of perfection. Providing that he possesses the basic manual skills he should have no difficulty in performing most of the operations detailed in this manual. It must be stressed, however, that where recommended in the manual, highly-skilled operations ought to be entrusted to experts, who have the necessary equipment, to carry out the work satisfactorily.

Quality of workmanship:
The hazardous driving conditions on the roads today demand that vehicles should be as nearly perfect, mechanically, as possible. It is therefore most important that amateur work be carried out with care, bearing in mind the often inadequate working conditions, and also the inferior tools which may have to be used. It is easy to counsel perfection in all things, and we recognize that it may be setting an impossibly high standard. We do, however, suggest that every care should be taken to ensure that a vehicle is as safe to take on the road as it is humanly possible to make it.

Safe working conditions:
Even though a vehicle may be stationary, it is still potentially dangerous if certain sensible precautions are not taken when working on it while it is supported on jacks or blocks. It is indeed preferable not to use jacks alone, but to supplement them with carefully placed blocks, so that there will be plenty of support if the car rolls off the jacks during a strenuous manoeuvres. Axle stands are an excellent way of providing a rigid base which is not readily disturbed. Piles of bricks are a dangerous substitute. Be careful not to get under heavy loads on lifting tackle, the load could fall. It is preferable not to work alone when lifting an engine, or when working underneath a vehicle which is supported well off the ground. To be trapped, particularly under the vehicle, may have unpleasant results if help is not quickly forthcoming. Make some provision, however humble, to deal with fires. Always disconnect a battery if there is a likelihood of electrical shorts. These may start a fire if there is leaking fuel about. This applies particularly to leads which can carry a heavy current, like those in the starter circuit. While on the subject of electricity, we must also stress the danger of using equipment which is run off the mains and which has no earth or has faulty wiring or connections. So many workshops have damp floors, and electrical shocks are of such a nature that it is sometimes impossible to let go of a live lead or piece of equipment due to the muscular spasms which take place.

Work demanding special care:
This involves the servicing of braking, steering and suspension systems. On the road, failure of the braking system may be disastrous. Make quite sure that there can be no possibility of failure through the bursting of rusty brake pipes or rotten hoses, nor to a sudden loss of pressure due to defective seals or valves.

Problems:
The chief problems which may face an operator are:

1 External dirt.
2 Difficulty in undoing tight fixings
3 Dismantling unfamiliar mechanisms.
4 Deciding in what respect parts are defective.
5 Confusion about the correct order for reassembly.
6 Adjusting running clearances.
7 Road testing.
8 Final tuning.

Practical suggestions to solve the problems:

1. Preliminary cleaning of large parts - engines, transmissions, steering, suspensions, etc., should be carried out before removal from the car. Where road dirt and mud alone are present, wash clean with a high pressure water jet, brushing to remove stubborn adhesions, and allow to drain and dry. Where oil or grease is also present, wash down with a proprietary compound (Gunk, Teepol etc.,) applying with a stiff brush - an old paint brush is suitable, into all crevices. Cover the distributor and ignition coils with a polythene bag and then apply a strong water jet to clear the loosened deposits. Allow to drain and dry. The assemblies will then be sufficiently clean to remove and transfer to the bench for the next stage.

   On the bench, further cleaning can be carried out, first wiping the parts as free as possible from grease with old newspaper. Avoid using rag or cotton waste which can leave clogging fibres behind. Any remaining grease can be removed with a brush dipped in paraffin. Avoid using paraffin or petrol in large quantities for cleaning in enclosed areas, such as garages, on account of the high fire risk.

   When all exteriors have been cleaned, and not before, dismantling can be commenced. This ensures that dirt will not enter into interiors and orifices revealed by dismantling. In the next phases, where components have to be cleaned and keep the containers covered except when in use. After the components have been cleaned, plug small holes with tapered hard wood plugs cut to size and blank off larger orifices with greaseproof paper and masking tape. Do not use soft wood plugs or matchsticks as they may break.

2. It is not advisable to hammer on the end of a screw thread, but if it must be done, first screw on a nut to protect the thread, and use a lead hammer. This applies particularly to the removal of tapered cotters. Nuts and bolts seem to 'grow' together, especially in exhaust systems. If penetrating oil does not work, try the judicious application of heat, but be careful of starting a fire.

   Tight bushes or pieces of tail-pipe rusted into a silencer can be removed by splitting them with an open-ended hacksaw. Tight screws can sometimes be started by a tap from a hammer on the end of a suitable screwdriver. Many tight fittings will yield to the judicious use of a hammer, but it must be a soft-faced hammer if damage is to be avoided, use a heavy block on the opposite side to absorb shock. Any parts of the steering system which have been damaged should be renewed, as attempts to repair them may lead to cracking and subsequent failure, and steering ball joints should be disconnected using a recommended tool to prevent damage.

3. It often happens that an owner is baffled when trying to dismantle an unfamiliar piece of equipment. So many modern devices are pressed together or assembled by spinning-over flanges, that they must be sawn apart. The intention is that the whole assembly must be renewed. However, parts which appear to be in one piece to the naked eye, may reveal close-fitting joint lines when inspected with a magnifying glass, and, this may provide the necessary clue to dismantling. Left-handed screw threads are used where rotational forces would tend to unscrew a right handed screw thread.

   Be very careful when dismantling mechanisms which may come apart suddenly. Work in an enclosed space where the parts will be contained, and drape a piece of cloth over the device if springs are likely to fly in all directions. Mark everything which might be reassembled in the wrong position, scratched symbols may be used on unstressed parts, or a sequence of tiny dots from a centre punch can be useful. Stressed parts should never be scratched or centre-popped as this may lead to cracking under working conditions. Store parts which look alike in the correct order for reassembly. Never rely upon memory to assist in the assembly of complicated mechanisms, especially when they will be dismantled for a long time, but make notes, and drawings to supplement the diagrams in the manual, and put labels on detached wires. Rust stains may indicate unlubricated wear. This can sometimes be seen round the outside edge of a bearing cup in a universal joint. Look for bright rubbing marks on parts which normally should not make heavy contact. These might prove that something is bent or running out of truth. For example, there might be bright marks on one side of a piston, at the top near the ring grooves, and others at the bottom of the skirt on the other side. This could well be the clue to a bent connecting rod. Suspected cracks can be proved by heating the component in a light oil to approximately 100°C, removing, drying off, and dusting with French chalk, if a crack is present the oil retained in the crack will stain the French chalk.

4. In determining wear, and the degree, against the permissible limits set in the manual, accurate measurement can only be achieved by the use of a micrometer. In many cases the wear is given to the fourth place of decimals; that is in ten-thousandths of an inch. This can be read by the vernier scale on the barrel of a good micrometer. Bore diameters are more difficult to determine. If, however, the matching shaft is accurately measured, the degree of play in the bore can be felt as a guide to its suitability. In other cases, the shank of a twist drill of known diameter is a handy check.

   Many methods have been devised for determining the clearance between bearing surfaces. Today the best and simplest is by the use of Plastigage, obtainable from most garages. A thin plastic thread is laid between the two surfaces and the bearing is tightened, flattening the thread. On removal, the width of the thread is compared with a scale supplied with the thread and the clearance is read off directly. Sometimes joint faces leak persistently, even after gasket renewal. The fault will then be traceable to distortion, dirt or burrs. Studs which are screwed into soft metal frequently raise burrs at the point of entry. A quick cure for this is to chamfer the edge of the hole in the part which fits over the stud.

5. Always check a replacement part with the original one before it is fitted.

   If parts are not marked, and the order for reassembly is not known, a little detective work will help. Look for marks which are due to wear to see if they can be mated. Joint faces may not be identical due to manufacturing errors, and parts which overlap may be stained, giving a clue to the correct position. Most fixings leave identifying marks especially if they were painted over on assembly. It is then easier to decide whether a nut, for instance, has a plain, a spring, or a shake-proof washer under it. All running surfaces become 'bedded' together after long spells of work and tiny imperfections on one part will be found to have left corresponding marks on the other. This is particularly true of shafts and bearings and even a score on a cylinder wall will show on the piston.

6. Checking end float or rocker clearances by feeler gauge may not always give accurate results because of wear. For instance, the rocker tip which bears on a valve stem may be deeply pitted, in which case the feeler will simply be bridging a depression. Thrust washers may also wear depressions in opposing faces to make accurate measurement difficult. End float is then easier to check by using a dial gauge. It is common practice to adjust end play in bearing assemblies, like front hubs with taper rollers, by doing up the axle nut until the hub becomes stiff to turn and then backing it off a little. Do not use this method with ball-bearing hubs as the assembly is often pre-loaded by tightening the axle nut to its fullest extent. If the split-pin hole will not line up, file the base of the nut a little.

   Steering assemblies often wear in the straight-ahead position. If any part is adjusted, make sure that it remains free when moved from lock to lock. Do not be surprised if an assembly like a steering gearbox, which is known to be carefully adjusted outside the car, becomes stiff when it is bolted in place. This will be due to distortion of the case by the pull of the mounting bolts, particularly if the mounting points are not all touching together. This problem may be met in other equipment and is cured by careful attention to the alignment of mounting points.

   When a spanner is stamped with a size and A/F it means that the dimension is the width between the jaws and has no connection with ANF, which is the designation for the American National Fine thread. Coarse threads like Whitworth are rarely used on cars to-day except for studs which screw into soft aluminium or cast iron. For this reason it might be found that the top end of a cylinder head stud has a fine thread and the lower end a coarse thread to screw into the cylinder block. If the car has mainly UNF threads then it is likely that any coarse threads will be UNC, which are not the same as Whitworth. Small sizes have the same number of threads in Whitworth and UNC, but in the 1/4 inch size for example, there are twelve threads to the inch in the former and thirteen in the latter.

7. After a major overhaul, particularly if a great deal of work has been done on the braking, steering and suspension systems, it is advisable to approach the problem of testing with care. If the braking system has been overhauled, apply heavy pressure to the brake pedal and get a second operator to check every possible source of leakage. The brakes may work extremely well, but a leak could cause complete failure after a few miles.

   Do not fit the hub caps until every wheel nut has been checked for tightness, and make sure the tyre pressures are correct. Check the levels of coolant, lubricants and hydraulic fluids. Being satisfied that all is well, take the car on the road and test the brakes at once. Check the steering and the action of the handbrake. Do all this at moderate speeds on quiet roads, and make sure there is no other vehicle behind you when you try a rapid stop.

   Finally, remember that many parts settle down after a time, so check for tightness of all fixings after the car has been on the road for a hundred miles or so.

8. It is useless to tune an engine which has not reached its normal running temperature. In the same way, the tune of an engine which is stiff after a re-bore will be different when the engine is again running free. Remember too, that rocker clearances on push-rod operated valve gear will change when the cylinder head nuts are tightened after an initial period of running with a new head gasket. Trouble may not always be due to what seems the obvious cause. Ignition, carburation and mechanical condition are interdependent and spitting back through the carburetter, which might be attributed to a weak mixture, can be caused by a sticking inlet valve. For one final hint on tuning, never adjust more than one thing at a time or it will be impossible to tell which adjustment produced the desired result.

CHAPTER 1

THE ENGINE

1 :1 Description

These models are fitted with a straight six cylinder water cooled engine with pushrod operated overhead valves. Three different engines are available, having cubic capacities of 170, 200 or 250 cu inches. Adjustment and repair procedures for the three engines are basically the same except that the 170 cu inch engine has five main bearings while the two larger engines have seven. Other differences between the two engines will be mentioned as required. Details of bore, stroke and other engine dimensions are given in Technical Data at the end of this manual.

A sectional view of a 200 engine is shown in **FIG 1 :1**. Both the crankshaft and camshaft rotate in renewable bearing inserts in the crankcase. Split inserts are used on the main bearings and they are held in place by bearing caps attaching directly to the crankcase. The camshaft is driven, at half engine speed, by two sprockets and a timing chain from the front of the crankshaft. On the very front of the crankshaft is installed a damper (to damp out torsional vibrations in the crankshaft) and the pulleys for driving the engine accessories. Hydraulic valve lifters operate in bores in the crankcase and they are raised by the cams on the camshaft. The valve lifters automatically take up the lash in the valve train and the motion is transmitted to the rocker arms by pushrods. The rocker arms pivot around a rocker arm shaft, running the length of the engine, and their opposite ends act directly onto the tops of the valve stems to open the valves. The valves are closed by the combined pressures of the gases in the cylinder and the valve return springs. Gears on the camshaft also drive the distributor and oil pump. The connecting rods are an interference fit with the piston pins at the top end but the lower end is split, using a cap, and renewable bearing inserts are used for rotation about the crankshaft.

Lubrication:

A typical lubrication system is shown in **FIG 1 :2**. The oil runs down inside the engine to the sump of the oil pan, where it is cooled by the airflow over the engine. From the sump the oil is drawn into the oil pump through a gauze strainer. The oil pump is fitted with a relief valve which ensures that the maximum oil pressure is limited and the surplus oil is returned directly to the to the inlet of the pump. From the pump the oil flows through a full-flow oil filter. The filter contains a bypass valve which opens under the differential pressure if the filter element

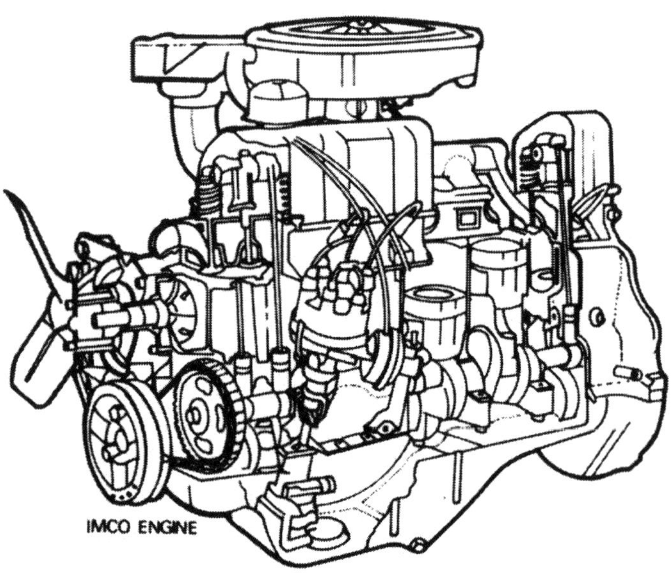

FIG 1:1 Sectional view of the 200 engine

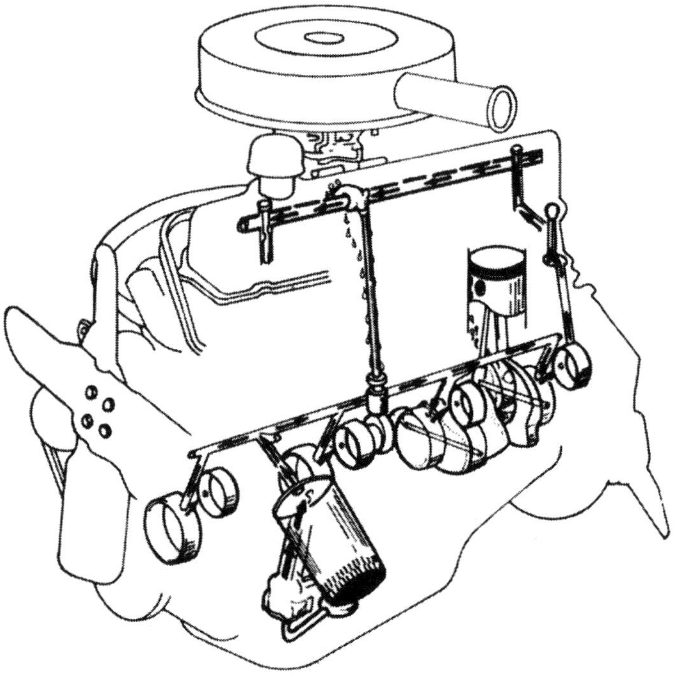

FIG 1:2 Typical lubrication system

is clogged. This is a safety precaution to ensure that the engine will not be starved of oil but if the valve does operate, the oil to the engine will be unfiltered. The filter also contains an anti-drain-back diaphragm which prevents oil from draining down out of the system when the engine is stationary. From the filter the oil passes to the engine main oil gallery. Internal passages feed oil directly from the main gallery to the crankshaft main bearings and hydraulic valve lifters. Other internal passages then lead the oil from the main bearings to the camshaft bearings. The rear camshaft journal has a groove machined in it and this groove allows a metered quantity of oil to pass from the rear camshaft bearing up through passages to the rocker shaft running along the cylinder head. From the rocker shaft the oil flows out through drilled holes to lubricate the rocker arms.

The crankshaft has internal oilways drilled in it and these lead oil from the main bearings to the connecting rod bearings for their lubrication. The connecting rods each have a squirt hole machined in them and when this hole indexes with the oil hole in the crankshaft a jet of oil lubricates and cools the cylinder walls. Oil flingers and oil seals are fitted at either end of the crankshaft to prevent the oil from leaking out.

The oil to the valve lifters lubricates their bores and also supplies the oil for their operation. A reservoir is fitted beside each valve lifter to ensure that each lifter will have an adequate supply of oil immediately on start-up.

The remainder of the parts are lubricated by the oil mist in the engine and splash thrown from the bearings.

Crankcase ventilation:

A typical ventilation system is shown in **FIG 1:3**. Air is drawn in from under the hood through the inlet duct to the carburetor air cleaner and it then passes through a hose connected to the oil filler cap on the rocker cover. The air for crankcase ventilation does not pass through the carburetor air cleaner element but a filter is fitted into the filler cap to ensure that the air entering the engine crankcase is clean. The rotation of the crankshaft sweeps the air around the engine from front to rear and it then passes up through the pushrod tubes at the rear and passes out of the rocker cover through a regulator valve. The regulator valve is shown in **FIG 1:4**. During normal running, the return spring holds the plunger in the open position providing an unrestricted flow for the air. When the throttle is closed the vacuum in the manifold increases and if fullflow were allowed to take place the extra air would be sufficient to alter the slow-running. At high vacuum the plunger in the valve is pulled across against the pressure of the spring, making the flow through the valve restricted. The

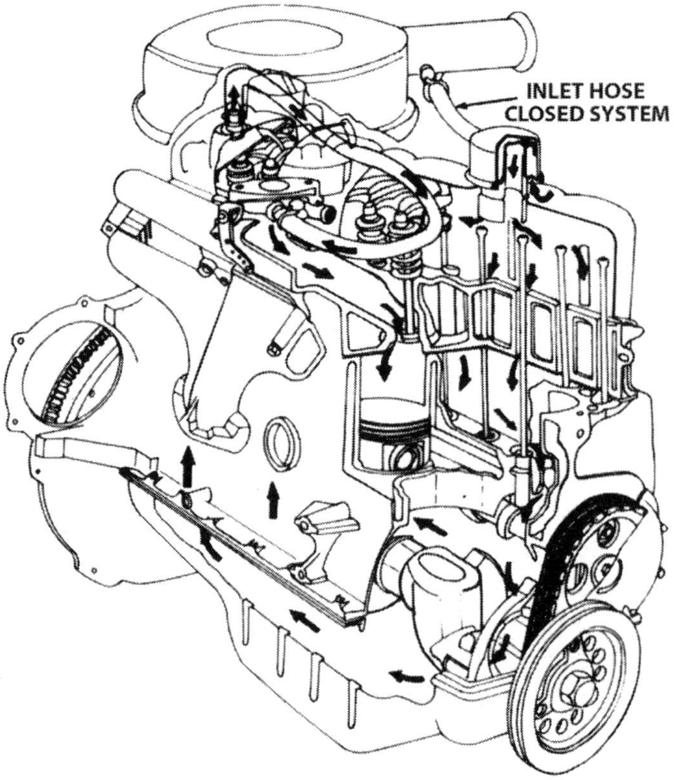

FIG 1:3 Typical crankcase ventilation system

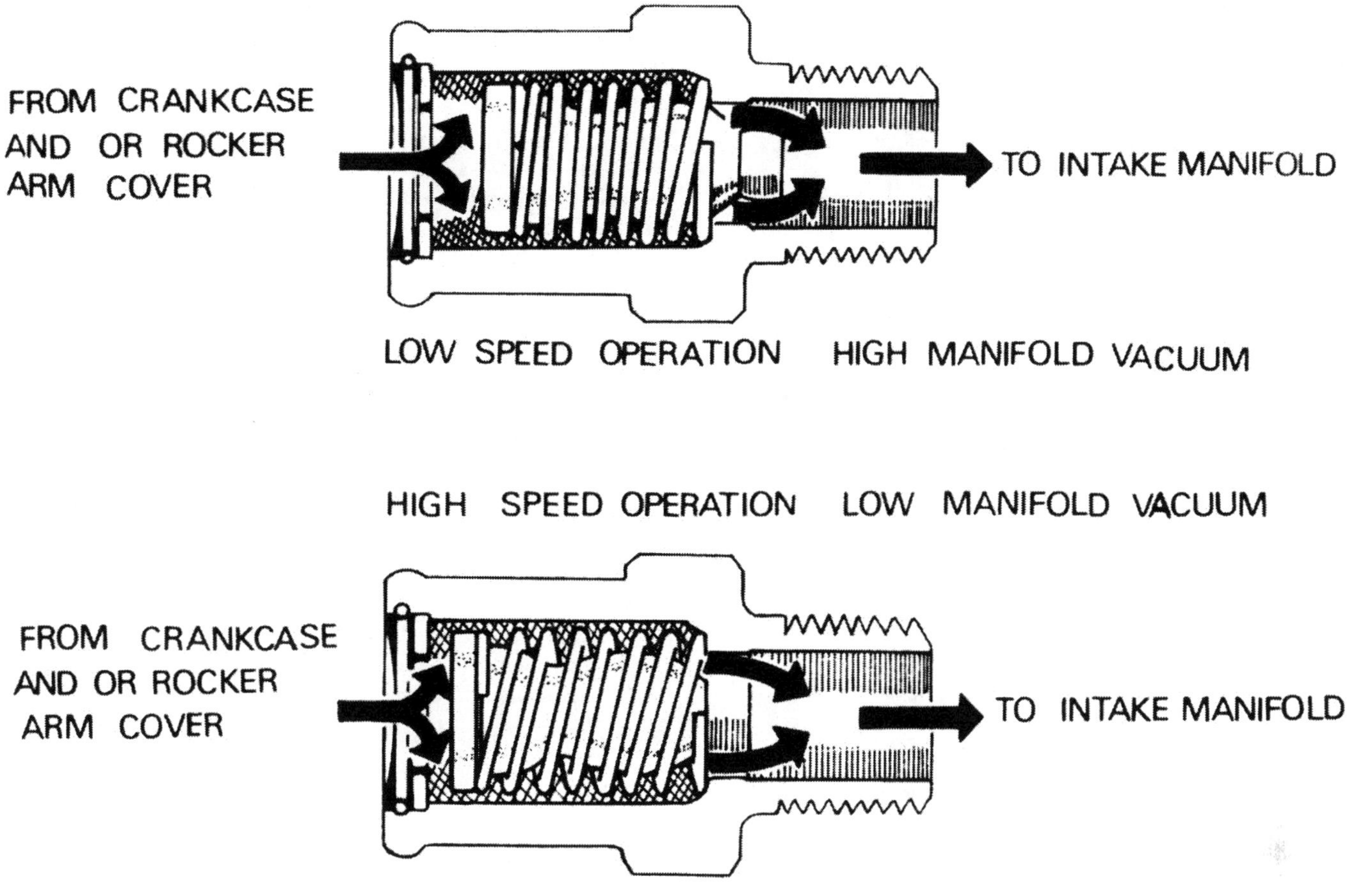

FIG 1:4 Crankcase ventilation regulator valve

valve can neither by dismantled or cleaned and when it is time-expired or defective a new valve must be fitted in its place. After having passed through the regulator valve the air and fumes are drawn into the intake manifold to mix with the ingoing fuel/air and be burnt in the combustion chambers.

Emission control:

Usually the Imco system is used on models fitted with automatic transmission and this system relies on specially made and adjusted carburetors to give efficient combustion and reduce emissions. On models fitted with a manual-shift transmission the Thermactor system is usually used. The Thermactor system uses an air pump driven by a belt from the engine, to inject excess cold air into the exhaust manifold where it mixes with the very hot exhaust gases to burn all surplus hydrocarbons. The Thermactor system is fully described in **Section 1:12.**

1:2 Engine removal and installation

The only operations for which it is essential to remove the engine from the vehicle are to remove the crankshaft or to renew the camshaft bearings. All other operations, including crankshaft main bearing renewal, can be carried out with the engine installed in the vehicle. However, if major repair work is to be carried out it will often be found more convenient to have the engine out and on the bench.

If the owner is not a skilled automobile mechanic, it is suggested that he reads the 'Hints on Maintenance and Overhaul' at the start of this manual as much useful information is contained there.

This section only refers to ancillaries and components connected to the engine without giving full details. For these the relevant chapter should be consulted. The cooling system and drive belts are fully dealt with in **Chapter 4,** and the drive shaft in **Chapter 8,** while the manual-shift transmission is in **Chapter 6** and the automatic transmission in **Chapter 7.** Ignition and distributor is dealt with in **Chapter 3.**

Removal:

1 Disconnect the battery. Drain the cooling system and crankcase. Remove the hood. Remove the radiator, cooling fan and hub after disconnecting the water hoses. On a vehicle fitted with automatic transmission it will be necessary to disconnect the oil cooler lines before removing the radiator. The oil cooler and lines should be blanked off to prevent the entry of dirt. Slacken all drive belts.

2 Disconnect the automatic choke air tubes and the crankcase ventilation system hoses from the air cleaner and remove the air cleaner assembly. Disconnect all vacuum hoses from the intake manifold and carburetor.

3 On models fitted with power steering free the pump from the engine and hang it safely out of the way so that fluid is not spilled from the reservoir. On models

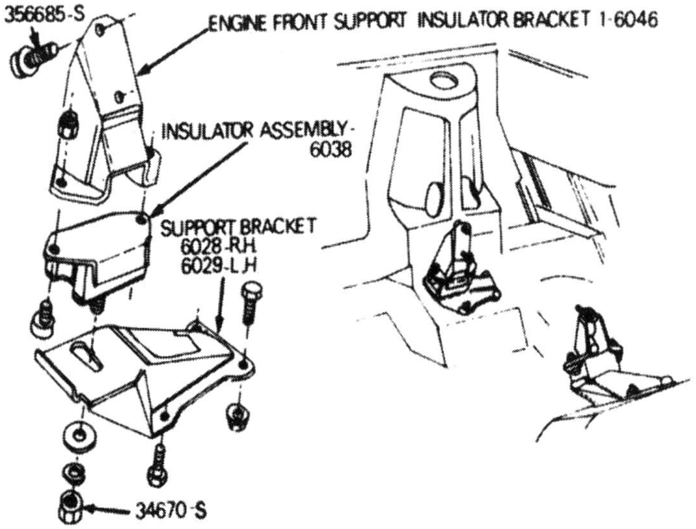

FIG 1:5 Typical front engine mountings

FIG 1:6 The engine lifting hook

fitted with Thermactor, remove the air cleaner if it is not mounted on the pump or engine. On models fitted with air-conditioning, remove the compressor from the block but leave the lines connected (see **Chapter 13**).

4 Disconnect the heater hoses. Disconnect all electrical wires and cables connecting the frame to the engine. These include the alternator wires, ignition coil, temperature and oil pressure sender units and the starter motor. If the colours have faded on the wires it is advisable to label them to ensure correct installation.

5 Disconnect the main fuel feed to the fuel pump and blank it off to prevent fuel from syphoning through. Disconnect the throttle rod and choke operating cable from the carburetter.

6 Raise the vehicle and place it securely on stands. Remove the starter motor and its dust shield. On models with manual-shift transmission, disconnect the clutch operating linkage-parts from the release lever and housing and then remove the bolts that secure the flywheel housing to the engine. On models fitted with automatic transmission, remove the inspection cover so that the converter can be disconnected from the engine drive plate, then remove the bolts that secure the converter housing to the engine. On all models disconnect the muffler inlet pipe from the exhaust manifold and free the clip that secures the pipe to the engine.

7 Lower the vehicle back to the ground and support the transmission with a jack and pad of wood underneath it. Disconnect the front engine mountings at the brackets on the underbody. Front engine mountings are shown in **FIG 1:5.** Check that there are no connections left holding the engine to the frame or body. Attach an engine lifting hook as shown in **FIG 1:6.** Use the hook in conjunction with a suitable hoist and raise the engine slightly so that it clears the front mountings. Pull the engine forward and then lift it out of the vehicle. On models fitted with manual-shift transmission, make sure that the transmission input shaft is well clear of the clutch before lifting out the engine. On models with automatic transmission make sure that the converter does not fall out of the transmission when the engine is free.

Installation:

The engine is installed in the reverse order of removal. Before installing the engine, check that the mounting insulators are in good condition and renew them if they are defective, distorted or perished.

The main point to watch when installing the engine is the mating to the transmission. **Do not attempt to force the engine back into place. Make sure that the engine and transmission are accurately aligned.** Rotate the crankshaft, using the damper and pulley until the splines on the input shaft slide through the clutch and into the spigot bearing in the crankshaft. **Do not allow the weight of the engine to hang on the input shaft of the transmission otherwise the clutch may be severely damaged.**

Make sure that all connections are tight enough to prevent leaks or faults. Fill the crankcase to the correct level with fresh oil and check the transmission fluid level. Fill and bleed the cooling system and adjust all drive belt tensions correctly. Check the clutch adjustment.

Check for leaks both before and after starting the engine. Allow the engine to run at a fast idle until it has reached its normal operating temperature and make final adjustments to carburetter and ignition.

1:3 Cylinder head removal

Normally the cylinder head should only require removal for decarbonizing or attention to the pistons and cylinder bores. It should be noted that it is possible to renew the valve springs and valve stem seals without removing the cylinder head from the engine. Air under pressure is injected through the spark plug hole, using a suitable adaptor, and this keeps the valves closed while the rocker assembly and valve spring parts are removed. The engine may turn until the piston is at BDC under the action of the air pressure and **if the air supply is cut off it is essential that the valves are held up using elastic bands or adhesive tape to prevent them from falling through into the engine.** To prevent distortion the cylinder head should not be removed while the engine is hot.

Removal:

1 Disconnect the battery and drain the cooling system. Remove the air cleaner after disconnecting all hoses to it. Remove the crankcase ventilation system and on models fitted with Thermactor, disconnect the air hose at the check valve and remove the intermediate exhaust air supply tube assembly.

2 Disconnect the vacuum line between carburetter and distributor at the carburetor and identify them as necessary. Disconnect the choke and throttle controls from the carburetor and on models fitted with automatic transmission also disconnect the kick-down rod. Disconnect the upper radiator hose at the engine. Disconnect the fuel line at the line filter. It is advisable to disconnect and remove the distributor vacuum lines and the carburetor fuel line as an assembly so that they will not become damaged or distorted.

3 Label the spark plug leads and disconnect them from the spark plugs by pulling and twisting on the moulded caps only. Disconnect the wire from the temperature send unit.

4 Remove the valve rocker cover. In order, slacken each rocker shaft holding pedestal bolt nut by two turns until all the bolts are slack and can be removed. Lift off the rocker parts as an assembly and withdraw the pushrods. The pushrods must be stored in the correct order as they must be installed back into their original positions.

5 Evenly slacken and remove the remaining cylinder head holding bolts, preferably in the sequence shown in **FIG 1:7**. Remove the cylinder head. If the head sticks, do not pry between the faces as this may damage them. Try tapping the heads up using a wooden drift and if this fails, crank the engine over on the starter motor with the plugs in place to break the seal.

Cleanliness:

Service the cylinder head as described in the next section. When all the parts have been checked and serviced, make sure that they are scrupulously clean before assembling them. Before installing the head, make sure that all the gasket surfaces are clean and have not picked up specks of dirt.

Distortion:

If the cylinder head gasket has failed, the cylinder head should be checked for distortion, as shown in **FIG 1:8**. Distortion must not exceed .003 inch in any 6 inches or .007 inch overall. If distortion is excessive the head may be machined down provided that not more than .010 inch of stock is removed.

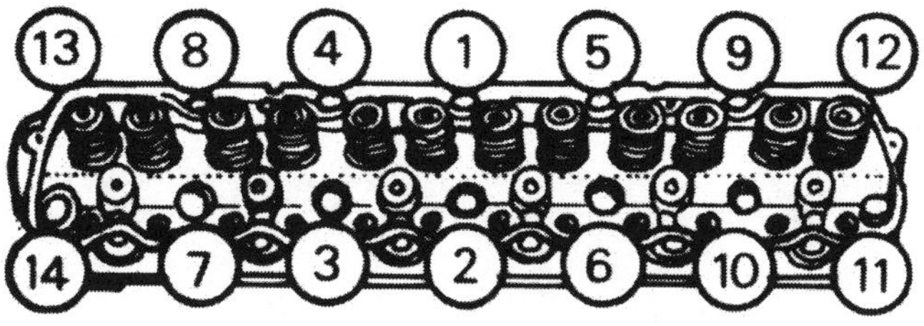

FIG 1:7 Cylinder head bolt torque sequence

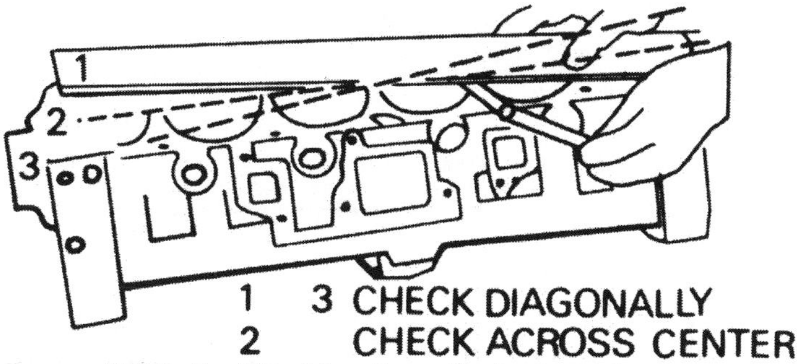

FIG 1:8 Checking the cylinder head for flames

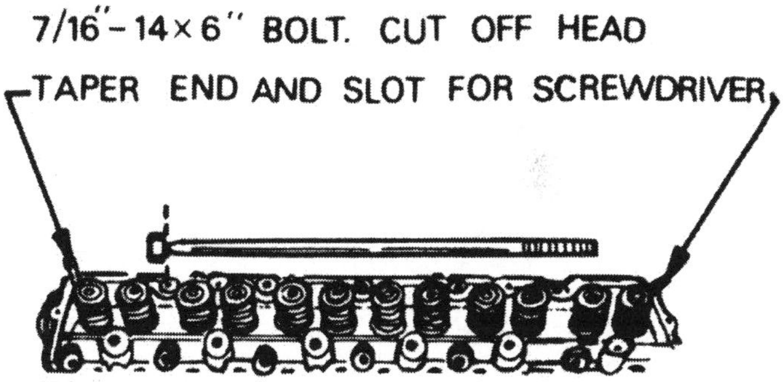

FIG 1:9 Guide studs for installing the cylinder head

Gaskets:

All gaskets and seals should be discarded and new ones fitted on installation. The rocker cover gasket should be cemented into the cover and the cover laid on a flat surface with weights on it until the cement has dried.

No gasket is required on the exhaust manifold but the face and threads should be smeared with graphite grease. Do not use any other form of grease as it will bake hard with the heat and lock the nuts in place.

On the 200 engine a composition gasket is generally used and this must **not** have any sealer applied to it. On the 170 engine which is fitted with an ordinary gasket, a light smear of sealer should be applied to both sides of the gasket. On both engines, guide studs should be screwed into the block and the gasket and head laid into place over these studs.

Installation:

Make up two guide studs, as shown in **FIG 1:9**, and fit them to the block.

1 Lay the gasket back into place, making sure that any markings are facing in the correct direction. **Do not use sealer on composition gaskets.** Lower the cylinder head into place down the studs and secure

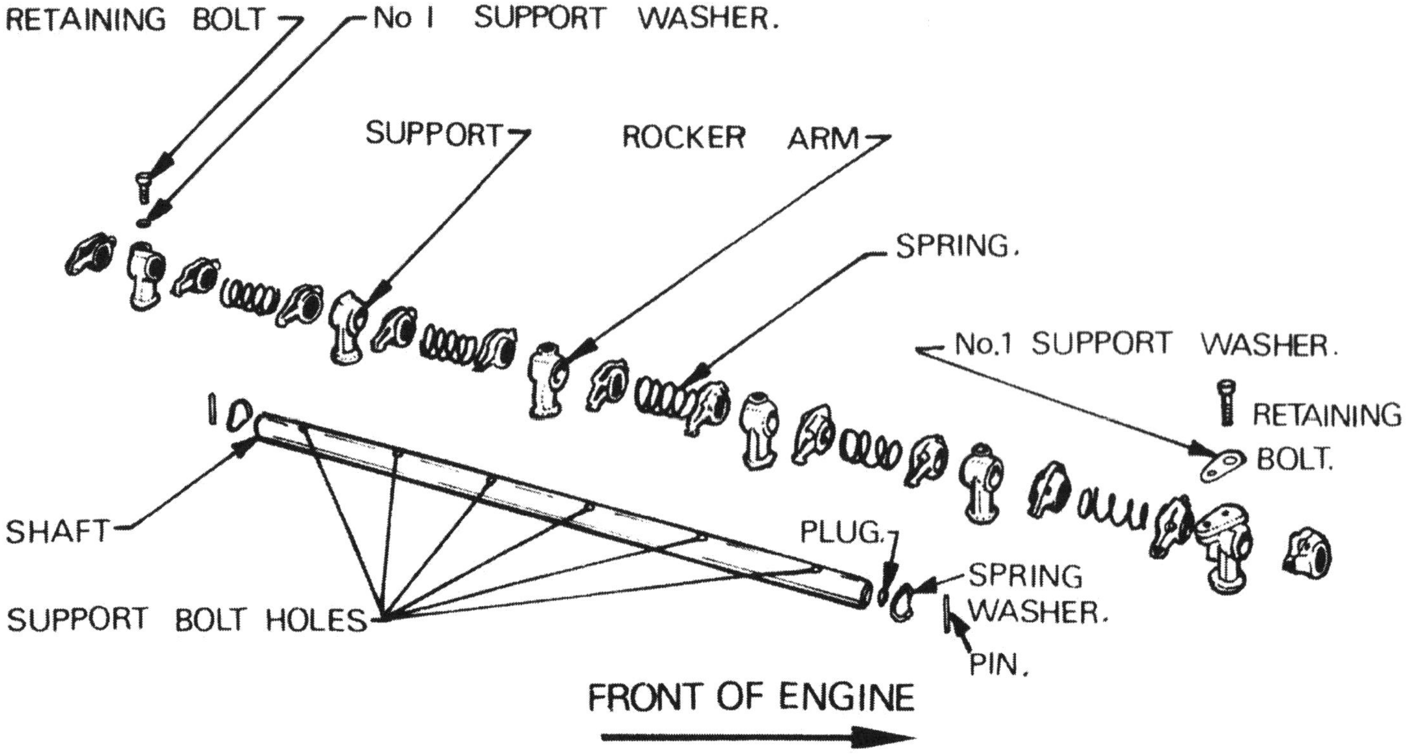

FIG 1:10 The valve rocker arm shaft components

it with two end bolts finger tight. Remove the guide studs and install the remainder of the cylinder head bolts finger tight.

2 Tighten the cylinder head bolts progressively in the sequence shown in **FIG 1:7** to a torque load of 55 lb ft and then tighten them again in sequence to a torque load of 65 lb ft. Finally tighten the bolts, again in sequence to a torque load of 70 to 75 lb ft. The bolts will not require torque-loading again but it may be carried out on the composition gaskets, if required.

3 Insert all the pushrods into their original positions after lubricating their ends with Lubriplate. Install the rocker assembly and make sure that all the pushrods are correctly located on the rocker arms and in the valve lifters. Tighten the rocker pedestal bolts in order and two turns at a time until they are all at 30 to 35 lb ft. The valve clearances are checked as described in **Section 1:5.**

4 The remainder of the parts are then installed in the reverse order of disassembly. Fill and bleed the cooling system. Check for leaks both before and after starting the engine. Set all adjustments after the engine has reached its normal operating temperature.

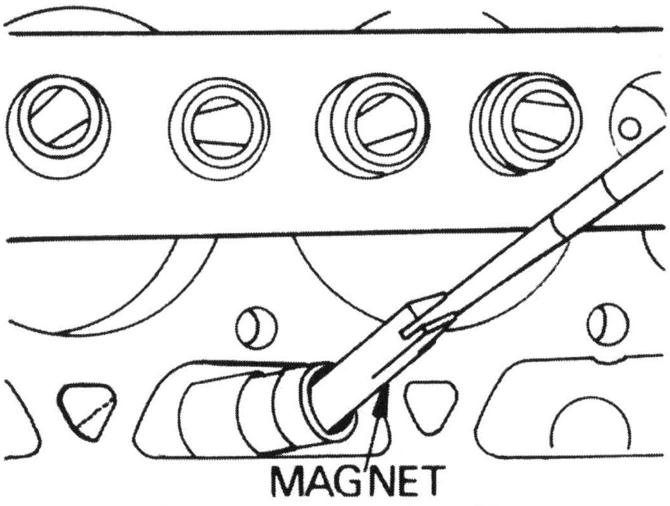

FIG 1:11 Removing a valve lifter

1:4 Servicing the cylinder head

Remove the cylinder head as described in the previous section.

Decarbonizing:

Normally decarbonizing will only be required after the vehicle has been used for at least 30,000 miles. When the cylinder head has been removed, a check should be made on the wear in the bores. Scrape away the carbon from the combustion chambers using a sharpened piece of perspex or hardwood and then polish the chambers using old emerycloth dipped in kerosene. On cast iron components, a rotary wire brush will make the removal of the carbon much easier but such treatment must not be given to light alloy parts. Remove the deposits from the combustion chambers before removing the valves, as the valves will protect the seats in the cylinder head.

Remove the valves as described and take off the manifolds. Clean out the ports in the cylinder head and manifolds, taking great care not to score or damage the valve seats in the cylinder head. Blow away all loose dust and dirt with compressed air or a tire pump. Clean the valve guides in the cylinder head. Special wire rotary

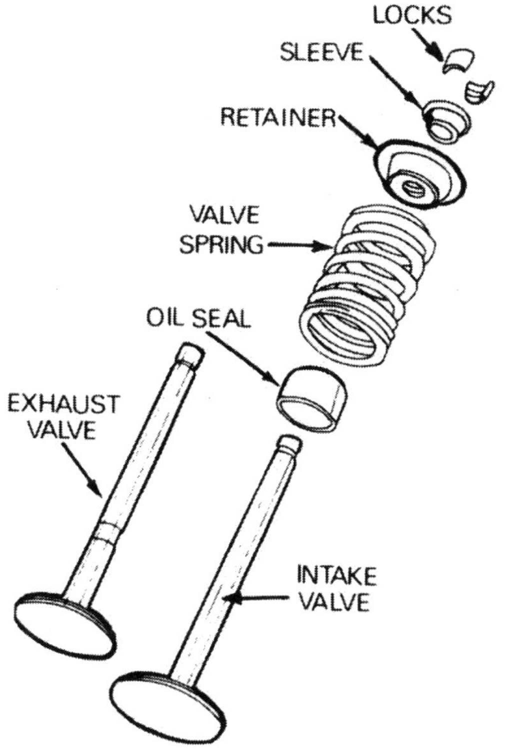

FIG 1:12 Valve and valve spring assembly

brushes are made for this purpose but if these are not available, pull strips of rag soaked in gasoline through them. Wash all the parts in a suitable solvent to remove all dirt, grease and oil. On models fitted with Thermactor, check the air nozzles and clean them with a suitable brush. Renew any nozzles that are eroded or damaged.

Plug all water and oilways in the top of the block with suitable pieces of rag. The pieces must be large enough not to be pushed through into the block or forgotten on reassembly. Smear a little grease around the top of each bore. Crank the engine until one pair of pistons is at TDC. Remove the carbon from the piston crowns with a sharpened piece of hardwood or perspex. Do not use any form of abrasive on the piston crowns as particles cannot be effectively removed. Blow away loose dust and dirt. Crank the engine until the next pair of pistons is at TDC and repeat the procedure. When all the piston crowns have been cleaned, crank the engine for a few rotations and wipe away the grease and dirt that the pistons have swept to the upper portion of the bores. Remove all the pieces of rag from the water and oilways.

The valve rocker assembly:

The components are shown in **FIG 1:10**. The assembly is removed from the cylinder head after evenly and progressively slackening the support bolts (two turns at a time).

To disassemble the parts, remove the cotterpins from either end and slide the parts off. The rear support is locked to the shaft by a bolt. The parts must be stored in the correct order as they are slid off so that they will be refitted into their original positions. Wash the parts in suitable solvent and make sure that all the oil holes are clear. If it is necessary to remove the plugs from the ends of the shaft, drill a hole in one and use a suitable hook to draw it out driving out the other plug with a suitable piece of rod. New plugs must be driven back into place if the old ones have been removed.

Check the shaft for wear or fret marks and check the fit of the rocker arms on the shaft. Renew parts as required if they are defective or worn. Check the pads on the ends of the rocker arms for wear. Light scoring or wear can be smoothed down with an oil-stone but if the wear is deep, new rocker arms must be fitted. Lubricate the parts liberally with fresh oil as they are reassembled and store them so that the assembly does not pick up dust or dirt.

Pushrods:

Remove the pushrods after the rocker assembly has been taken off. Store the pushrods in the correct order for installation.

Wash the pushrods with a suitable solvent and dry them with compressed air. Check the ends for wear, damage, or scoring. Check the pushrods for bends. The best method is to lay each rod onto two V-blocks, one at either end, and rotate the pushrod while measuring the runout with a dial gauge at the centre. Renew any pushrods which have damaged ends or where the runout exceeds .025 inch. The pushrods can also be checked by rolling them along a flat surface.

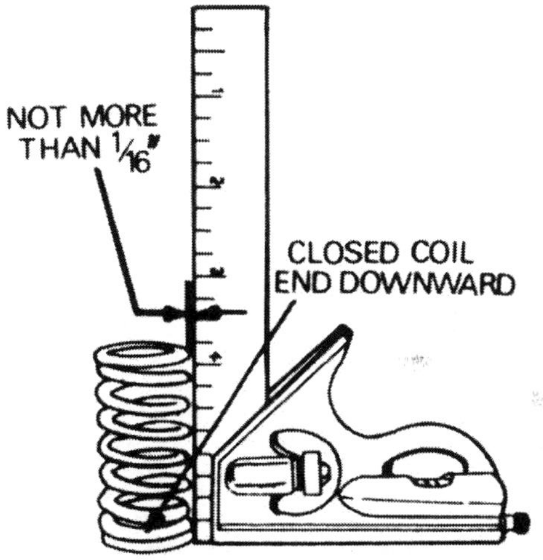

FIG 1:13 Checking a valve spring for squareness

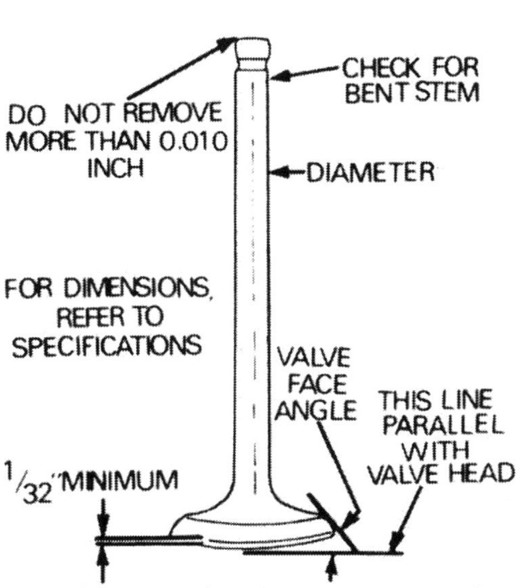

FIG 1:14 Critical valve dimensions and tolerances

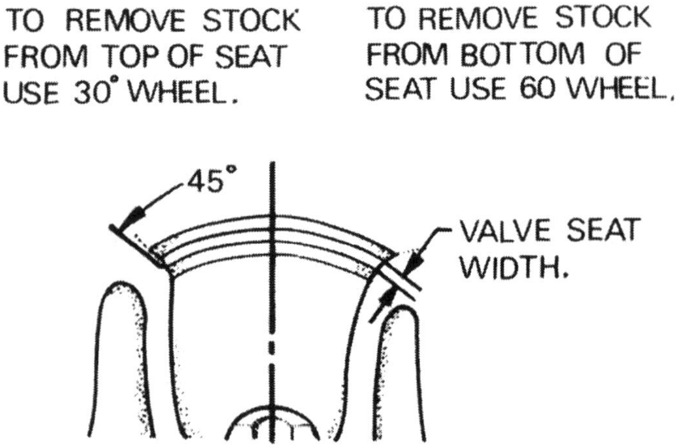

FIG 1:15 **The valve seat width**

FIG 1:16 **Checking a valve spring installed height**

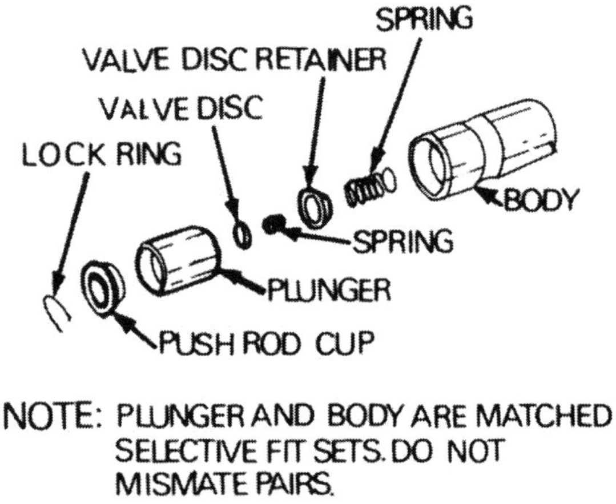

FIG 1:17 **Typical valve lifter components**

Valve lifters:

These will be more fully dealt with in the next section. The method of removing the valve lifters, after the cylinder head has been taken off, is shown in **FIG 1:11.** Note that the lifters must slide freely in their bores and that they must be stored in the correct order.

Valve removal:

The components of the valve and its spring are shown in **FIG 1:12.** A spring compressor is required to remove or install the valves. Remove the cylinder head and use the tool to compress the valve spring, by squeezing down the retainer and holding the head of the valve up in the combustion chamber. It may be necessary to drive down the sleeve with a suitable piece of tube and hammer. Remove the locks and release the spring compressor slowly. The sleeve, retainer and spring can then be removed. Take off the valve stem oil seal and discard it. Slide the valve out through the combustion chamber. **The valves must be stored in the correct order for installation,** so use a suitable rack or even a cardboard box with twelve holes punched in it for the valve stems.

Valve springs:

Special rigs are made for testing the valve springs. The data is given in **Technical Data** and the complete set should be renewed if any springs fall below tolerance.

The springs can also be roughly tested, provided that one new spring is available. Mount the old spring end onto the new one and compress the pair between the jaws of a vice. The spring is defective if it is 10 per cent shorter than the new one, as both have the same pressure applied by the vice.

Before installing a spring, stand it with the closed coils downwards on a flat surface and check spring with a square. Rotate the spring and if the gap, shown in **FIG 1:13** exceeds $\frac{5}{64}$ inch then the spring should be rejected.

Valves:

The critical dimensions to be checked on each valve are shown in **FIG 1:14.** Use a steel straightedge to check the stem for bends. Renew worn or burnt valves and any that have scored stems.

Remove carbon from the valve using a suitable scraper and worn emerycloth. Take great care not to damage the seat face or the ground portion of the valve stem. If the tip of the valve stem is damaged or bruised it may be ground down flat, provided that the minimum dimension is not reached.

Valve guides:

These are machined directly into the cylinder head. Wear can be checked using a special dial gauge but an estimate can also be made by partially inserting a new valve and noting the amount of rock on the head.

Valves with oversized diameter stems are available and these are used when the guides are worn. Before fitting new valves with oversized stems, ream out the guide to the correct diameter using the special piloted reamers made for the purpose. If the wear is such that the next oversize stem cannot be used, the guides must still be reamed out using the reamers in sequence until the

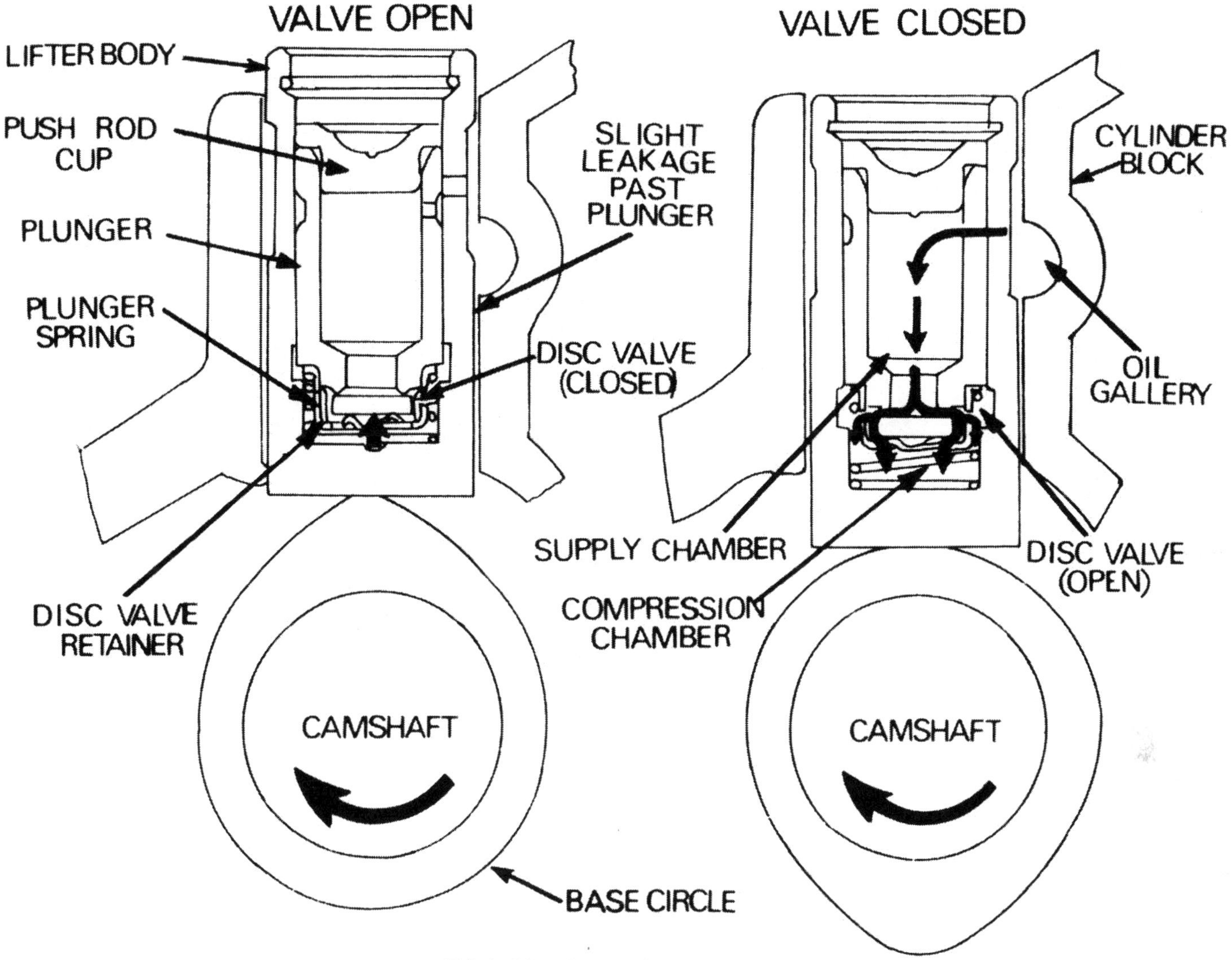

FIG 1:18 Valve lifter operation

correct diameter is reached. Do not ream out more than one size at a time. When the guides have been reamed, the valve seats must be recut to ensure that they are concentric with the guides.

Valve seats:

Check the efficiency of the seats, with the valve installed and held closed by its spring, by making a dam of plasticine around the seat area. Invert the head and fill the dam with gasoline and leaks will be immediately apparent.

Leaking seats or pitted seats should be recut using garage equipment. The seats on the valves are cut to a 44 deg. angle while those in the cylinder head are cut to 45 deg.

When the seats in the cylinder head have been recut, they should be checked for width as shown in **FIG 1:15.** On inlet valves the seat width should be .040 to .060 inch and on exhaust valves it should be .070 to .090 inch. Reduce the width with suitable cutters if it is too wide. Use Prussian blue to ensure that the head face meets the valve face centrally, adjusting the position by the use of the cutters.

The use of grinding paste for lapping in the valves is not recommended.

Assembly and installation:

Clean all the parts in suitable solvent and blow away dust, dirt and chippings with compressed air. Take great care not to damage the gasket surface of the cylinder head.

Install the valves, into their correct positions, in the reverse order of disassembly and using new stem seals. Lubricate the stems liberally with engine oil before inserting them into the valve guides.

When the valves are in place, tap the ends of the stems lightly to ensure that the locks are fully seated. Use dividers to measure the installed height of the valve spring, as shown in **FIG 1:16,** and transfer the measurement to a steel ruler for checking the actual dimension. If the height is not within limits, install .030 inch thick spacers between the cylinder head and spring to bring the height within limits. Spacers must not be installed unnecessarily otherwise the valve spring will be overstressed, leading to failure or excessive wear in the valve train. Lubricate the valve lifters with clean engine oil and install them back into their correct positions. Install the cylinder head followed by the pushrods. Note that both ends of the pushrods should be smeared with Lubriplate and they must seat fully into the valve lifters. Smear the tips of the valve stems with Lubriplate and install the rocker gear assembly. Check the valve lash as described

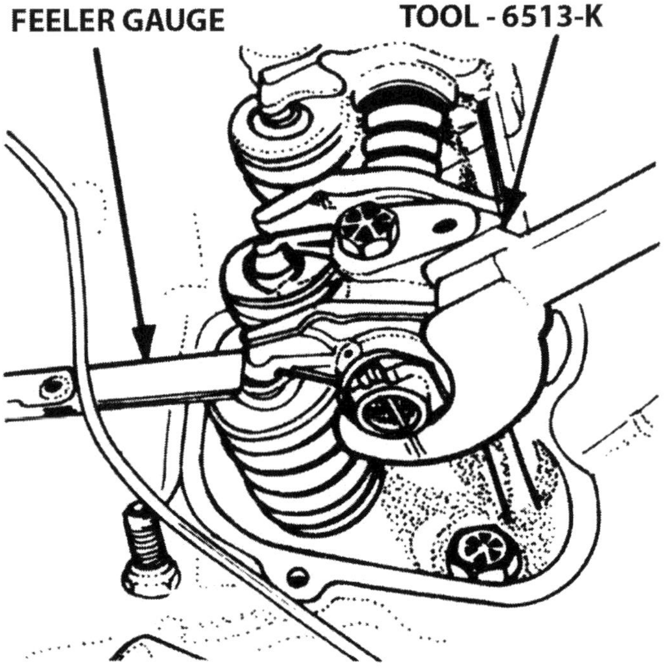

FIG 1 :19 Checking the valve clearance

in the next section. When installing the rocker cover using a new gasket, tighten the attachment bolts to the correct torque, and then wait two minutes before tightening them again to the correct torque.

1 :5 The hydraulic valve lifters

These can be removed from the engine after removing the cylinder head as shown in **FIG 1 :11**. Store the lifters in a rack in the correct order for installation. The components of a valve lifter are shown in **FIG 1 :17** and the operation in **FIG 1 :18**. When the valve is closed the lifter is on the base circle of the cam and the plunger spring moves the plunger upwards to take up all lash in the train. When the plunger moves upwards there is a reduced pressure in the compression chamber and the check valve opens, allowing oil from the gallery to fill the compression chamber. As the cam rotates and raises the lifter, the pressure in the compression chamber rises and the check valve closes. The oil being non-compressible makes the lifter act as a hydraulic ram transferring the lift directly to the pushrod without any lash. A slight amount of leakage is allowed as this ensures that the lifter will not overcompensate and cause the valve to be held open. The leakage is replenished on the next stroke.

Noisy operation:

When the engine has stood stationary for some time, such as overnight, the oil will drain down from the lifters—particularly those which are holding valves open. On start-up there will be excessive lash and the lifters will sound noisy. As the compression chambers fill, the noise will die away. This noisy operation on start-up is perfectly normal and no action need be taken, unless the noise does not die away after a few minutes of running.

Intermittent clickings are mainly caused by incorrect initial adjustment or a piece of dirt momentarily holding the check valve open.

General noise through the valve train is caused by incorrect initial settings or a lack of oil to the valve lifters. If the crankcase is overfilled, the connecting rods thrashing in the oil will make foam and this foam instead of oil may be drawn into the lifters. As air is compressible, all the lash will not be taken up and the valve train will sound noisy.

Moderate rapping is caused by a worn valve lifter or incorrect initial adjustment. Heavy rapping noises are caused by the plunger sticking in the body of the lifter.

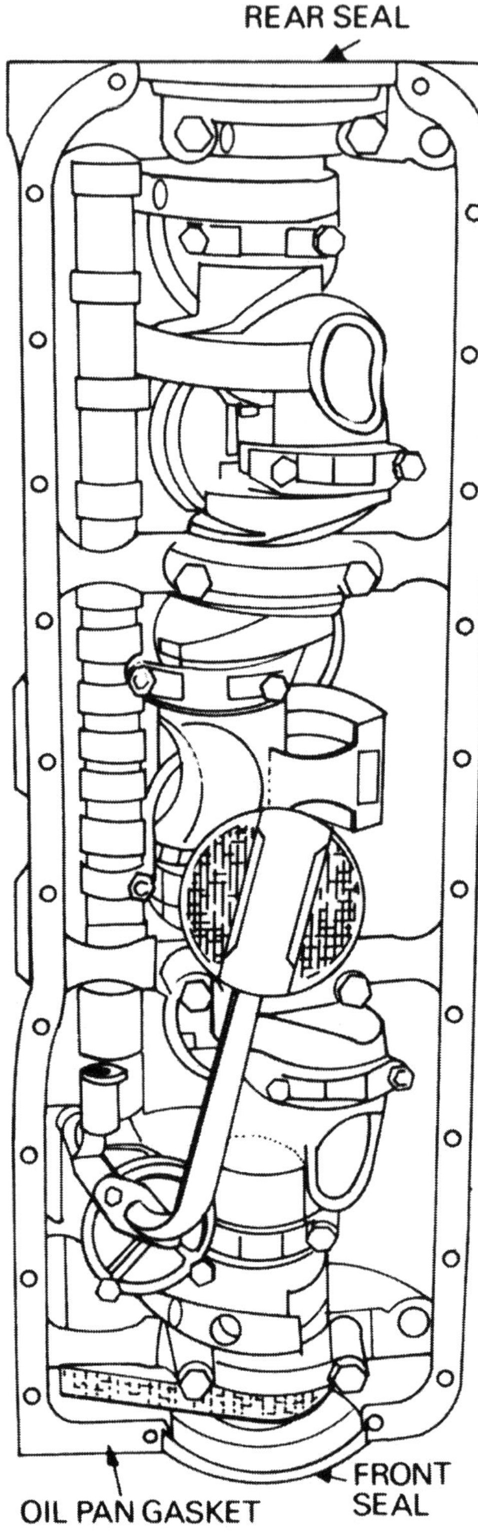

FIG 1 :20 170 engine with oil pan seals and gaskets installed

If the operation is noisy and continues to be noisy, first check the oil level in the crankcase and top up (or drain) until it is at the correct level. If the level was correct on filling but it is found to rise with use, and the coolant level drops, there is an internal leak allowing coolant into the crankcase. If the oil level is correct, check the initial adjustment of the valve train. If the adjustments are satisfactory it will be necessary to remove the cylinder head and withdraw the lifters for cleaning and inspection.

Disassembly:

Refer to **FIG 1:17**. If any part of a valve lifter is defective the complete assembly must be renewed as the parts are matched on manufacture. To ensure that parts are not accidentally interchanged, work on one lifter at a time and complete it before starting the next. Discard any lifters that have scored or chipped bodies. Remove the

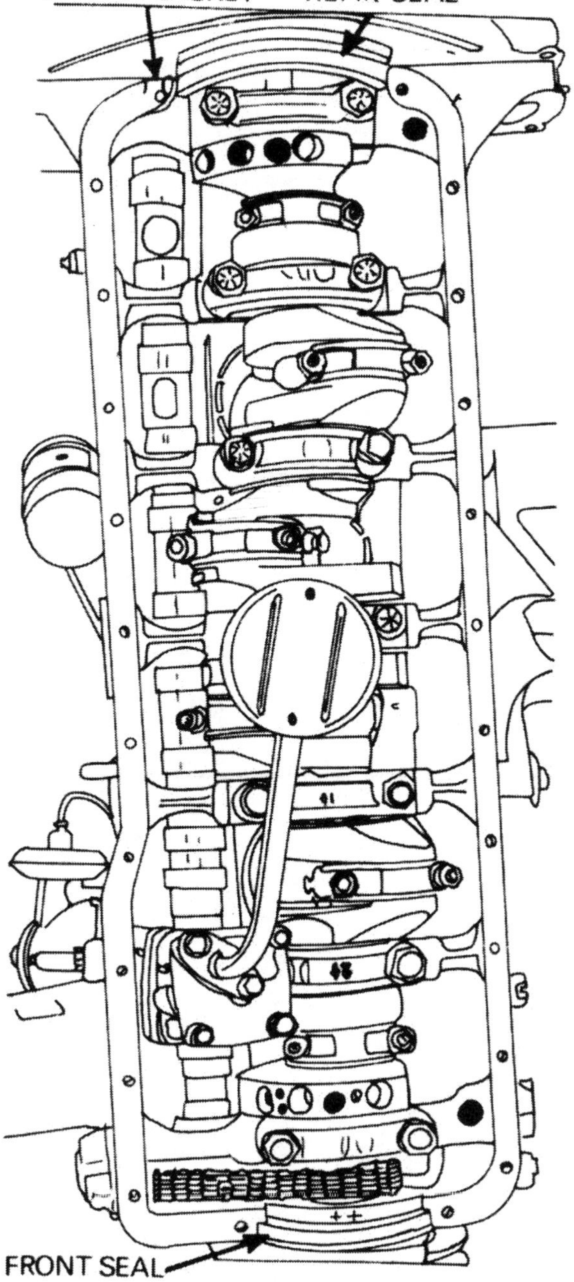

FIG 1:21 200 engine with oil pan seals and gaskets installed

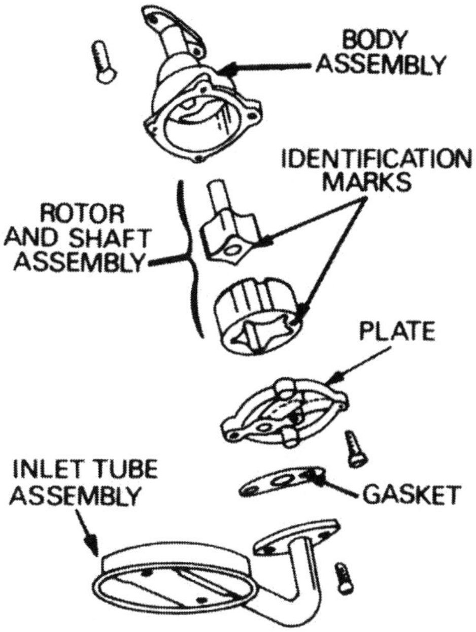

FIG 1:22 Oil pump components

lock ring using a pair of needle-nosed pliers. It may be necessary to partially depress the plunger when removing the lock ring. Remove the pushrod cup, plunger and spring. Invert the plunger and carefully remove the check valve retainer by prying it up with a screwdriver. Shake out the check valve disc and its spring.

Wash all the parts in a suitable harsh solvent, such as trichlorethylene, to remove oil and gum. Check the parts for wear, scoring, or distortion and make sure that the plunger slides freely in the body. Renew the lifter complete if defects are found.

Brush down the bench and lay clean paper on it to prevent the parts from picking up dirt. Invert the plunger and lay clean paper on it to prevent the parts from picking up dirt. Invert the plunger on the bench. Place the valve disc in position over the oil hole in the plunger and position the spring over it. Insert the check valve retainer into the plunger and press it down into position. Place the plunger spring followed by the plunger assembly into the body and install the pushrod cup. Secure the parts in place using the lock ring.

If the lifter is to be installed directly into the engine, immerse it in a bath of engine oil and pump the plunger up and down to fill the unit with oil. When the compression chamber is full, extremely heavy pressure will be required to move the plunger.

If the lifter is suspect and requires testing, it should be taken to a service garage without filling with oil. The leak down rate is checked on a special test machine after the unit has been filled with special test fluid. Any engine oil in the lifter would negate the test. On test, the machine applies a constant load to the plunger and the time is noted for the plunger to move through its full range of movement.

Adjustment:

Correctly install the valve lifters, cylinder head, pushrods and valve rocker gear. Crank the engine until it is at TDC after the compression stroke on No. 1 (front cylinder). The timing marks on the cylinder front cover

FIG 1:23 Checking the rotor end play on the oil pump

and crankshaft damper will give the TDC position for Nos. 1 and 6 cylinders. The compression stroke can be found by removing the spark plug and blocking the hole with a thumb as the engine is cranked. A rise in pressure in the cylinder will indicate that the piston is rising on the compression stroke. The correct position can also be found by removing the distributor cap and cranking the engine until the timing marks are aligned. The brass end of the rotor arm will be pointing to the position normally occupied by the electrode in the cap connected to the correct cylinder spark plug. With the engine set at TDC after the compression stroke on No. 1 (front cylinder) check the following valves:—

No. 1 Intake
No. 1 Exhaust
No. 2 Intake
No. 3 Exhaust
No. 4 Intake
No. 5 Exhaust

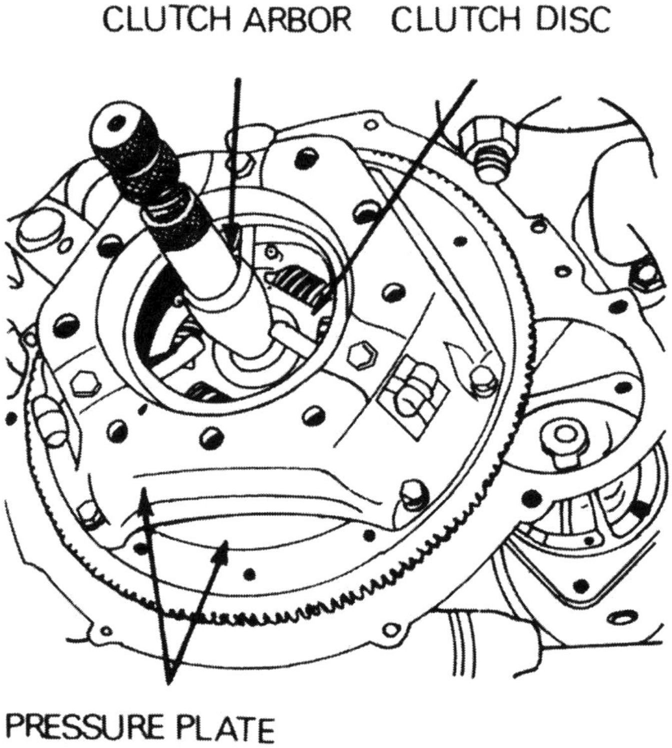

FIG 1:24 Installing the clutch assembly

When these valves have been checked, crank the engine through one complete turn until it is at TDC after the compression stroke on No. 6 (rear) cylinder, and check the following valves:

No. 2 Exhaust
No. 3 Intake
No. 4 Exhaust
No. 5 Intake
No. 6 Intake
No. 6 Exhaust

The method of checking the lash is shown in **FIG 1:19.** Use the tool shown to fully collapse the valve lifter and measure the gap between the rocker arm and tip of the valve stem with feeler gauges.

See **Technical Data** for the correct clearances. If the clearance is incorrect the pushrod will have to be changed. Pushrods .060 inch longer than standard are available to reduce excessive clearance, and pushrods .060 inch shorter than standard are available to increase the clearance if it is too small. The rocker assembly need not be removed in order to change the pushrod. Turn the engine past TDC so that there is no danger of the valve hitting the piston crown, and use the special tool to compress the valve spring so that the old pushrod can be removed and a new one of the correct length fitted. Make sure that both ends of the pushrod are correctly seated and lubricated with Lubriplate.

1:6 The oil filter

The engine oil and filter element should be renewed at intervals not exceeding 6000 miles. If the vehicle is used in dusty or adverse conditions, used for short-mileage journeys only, or kept at curb idle for long periods, the interval between oil changes should be reduced.

The oil does not lose its lubricating properties but it does collect sludge from the engine and combustion products which pass by the piston rings. The filter

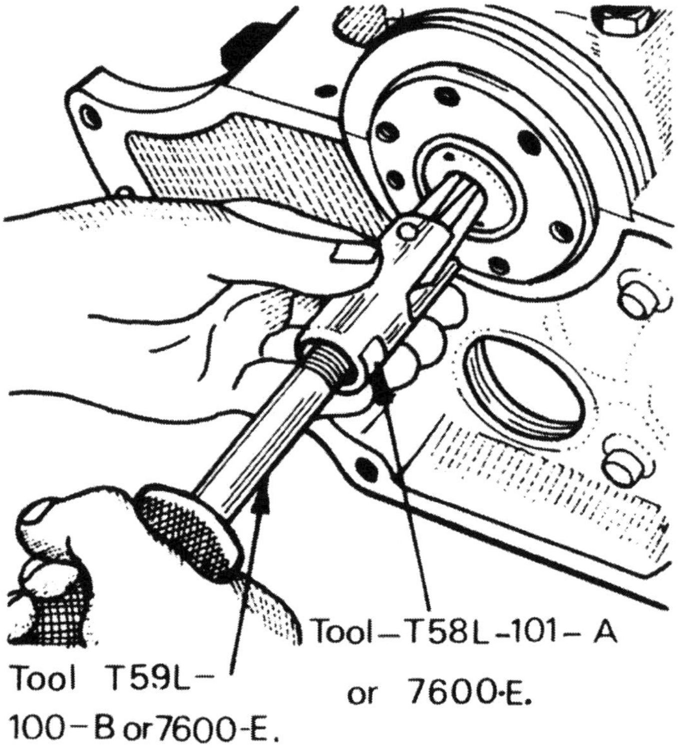

FIG 1:25 Removing the clutch pilot bushing

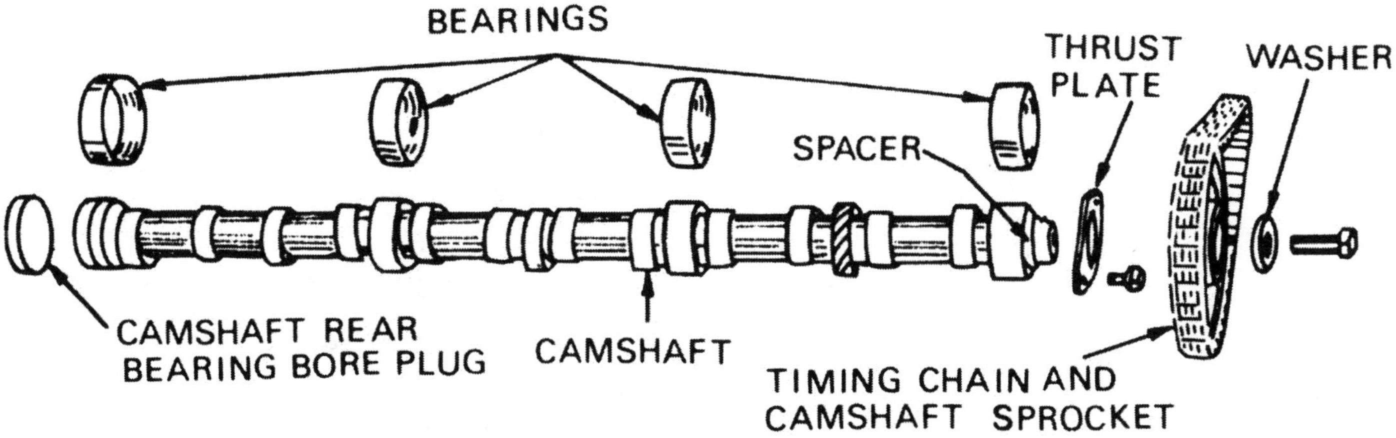

FIG 1 : 26 The camshaft and its related parts

element will remove the larger particles from the oil and therefore help to keep it clean. If the engine never reaches its full operating temperature, the combustion products will rapidly build up in the oil and cause wear and erosion of the metal parts. When the oil is at its full normal operating temperature for long periods, many of the combustion products are boiled out of the oil and taken away by the air ventilating the crankcase. It is for this reason that cars used under adverse conditions must have their oil changed at shorter intervals.

Before draining the oil from the crankcase take the vehicle for a short run to warm the oil and stir dirt up into suspension. If the oil is drained after a long run, take great care as the oil will be hot enough to cause scalding. Stop the engine and slide a container under the oil pan. Remove the drain plug from under the oil pan and allow the old oil to drain out into the container. Leave the oil to drain while changing the oil filter, but do not forget to fit the drain plug before filling the engine with fresh oil. Slide the drip tray or container under the oil filter, on the lefthand side of the crankcase, so that any spillage will be caught. Unscrew the complete cartridge from the engine and discard it.

Clean the area where the cartridge fits. Check that the gasket on the new cartridge is in good condition and smear it lightly with a little engine oil. Screw the new cartridge back into place until its gasket just touches the crankcase. Hand-tighten the cartridge by a further $\frac{1}{2}$ turn.

Install the drain plug in the oil pan. Fill the engine with fresh oil through the filler cap on the rocker cover. Approximately $4\frac{1}{2}$ quarts of oil will be required to fill the crankcase and oil filter, allowing for the quart held by the oil filter. The level may be allowed to go over the top mark on the dipstick in this case. Stand the car on level ground and start the engine. Check for oil leaks and then switch off the engine. Leave the vehicle standing for at least five minutes, to allow the oil to run down inside the engine, and again check the level on the dipstick. Top up as required but do not overfill the engine.

It should be noted that an oil strainer is fitted to the inlet to the oil pump. If regular oil changes are carried out the strainer should not block. Whenever the oil pan is removed, for whatever reason, the oil pump strainer should be removed, cleaned and inspected before installing it using a new gasket.

1 : 7 The oil pan and oil pump

Both the oil pan and oil pump can be removed with the engine installed in the vehicle. Before removing the oil pan drain the oil from the crankcase by removing the drain plug, using a $\frac{7}{8}$ inch socket wrench.

Oil pan removal:

With the oil drained from the crankcase, remove the flywheel housing inspection cover. Take out the screws securing the oil pan and lower it out of the vehicle. Remove the old gaskets and seals. Remove the oil pump inlet strainer.

Oil pan installation:

Clean the oil pump inlet strainer and install it using a new gasket. The seals and gaskets on a 170 engine are shown in **FIG 1 : 20** and those on a 200 engine in **FIG 1 : 21**. Remove all traces of old seal, gasket or sealing compound from the mounting faces, making sure that the grooves for the oil seals are also cleaned. Smear the gasket surfaces of the block and pan with oil-resistant sealer. Install the two-piece gasket set into place on the cylinder block. Install the front oil seal to the cylinder

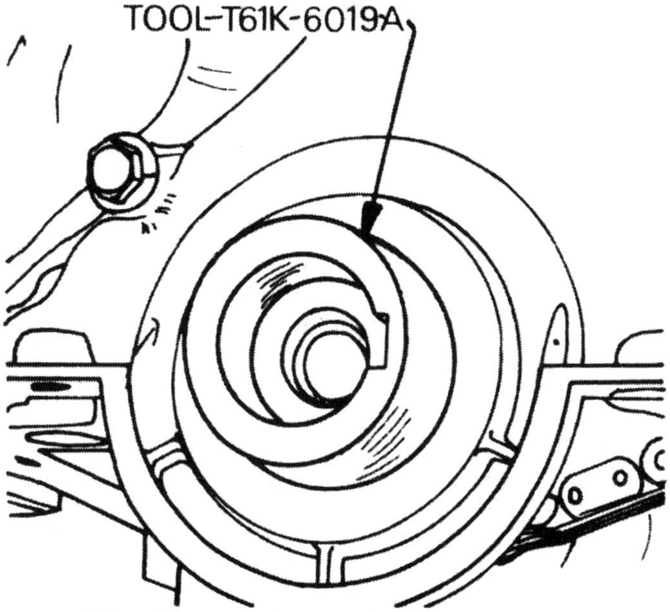

FIG 1 : 27 Installing the front cylinder cover

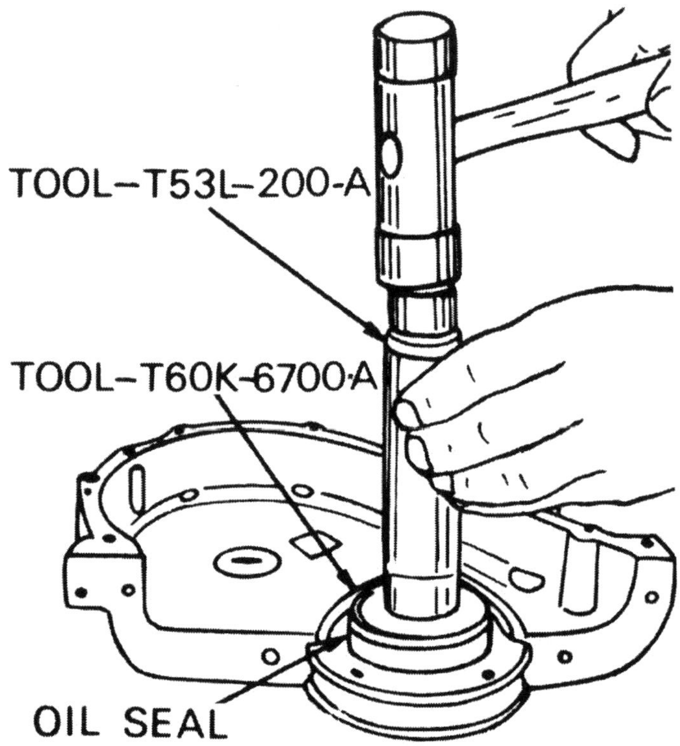

FIG 1 : 28 Installing the front crankshaft oil seal

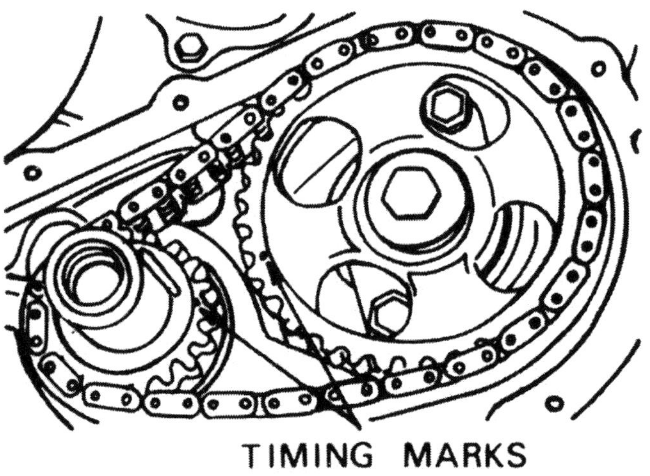

FIG 1 : 29 The valve timing marks

front cover and the rear seal to the rear main bearing cap, making sure that the tabs on the seal are over the oil pan gaskets. Raise the oil pan into position and secure it with one screw on either side. Fit the remainder of the screws without disturbing the positions of the gaskets and then tighten all the screws to a torque of 7 to 9 lb ft, working from the centre outwards. Install the drain plug and tighten it to a torque of 15 to 20 lb ft. Fill the engine with fresh oil through the filter cap. **Do not overfill the engine as the oil filter will be full and will not take any surplus oil.** Start the engine and check for leaks. Top up to the correct level after the engine has stood stationary for at least five minutes.

Oil pump removal:

Remove the oil pan as described earlier. The components of a typical oil pump are shown in **FIG 1 : 22.** Take out the pump attachment bolts and withdraw the pump complete with the intermediate drive shaft.

Oil pump servicing:

Remove the inlet tube assembly and discard the gasket. Take out the coverplate securing screws and remove the plate. Slide out the inner rotor and shaft assembly followed by the outer rotor.

Insert a self-threading screw into the cap for the oil relief valve and use a pair of grips on the screw to draw out the cap. The spring and plunger can then be removed from the body of the pump.

Wash all the parts in suitable solvent and examine them for obvious wear or scoring. If the rotors are worn or scored they must be renewed as a set as they are mated on manufacture.

Install the rotors back into the body assembly, making sure that the identification marks are outwards. Insert feeler gauges between the outer rotor side and the body assembly to measure the clearance. If the clearance is outside the limits of .006 to .013 inch, new rotors must be fitted and if the clearance is still not within the limits the complete pump must be renewed.

Lay a straightedge across the pump, as shown in **FIG 1 : 23,** and use feeler gauges to measure the end play of the rotors. If the end play is outside the limits of .0011 to .0041 inch, try fitting new rotors and if this does not cure the end play then a new pump must be fitted.

Check the end plate for scores and install a new plate if the old one is worn or scored.

Examine the relief valve plunger and bore for fretting or wear. Check the spring and renew it if it is weak. If the spring is satisfactory it will take a load between 9.0 to 10.1 lb to compress it to a length of 1.078 inches.

Assemble the pump in the reverse order of disassembly. Inject engine oil through the inlet or outlet port and rotate the pump to spread the oil evenly. Check that the pump rotates freely.

Oil pump installation:

Position the intermediate drive shaft in the distributor socket and place a new gasket onto the oil pump body. Fit the pump back into place, without trying to force it. If the pump sticks, turn the intermediate shaft to a new position to allow the hex to align. Secure the pump in place with the attachment bolts and fit the inlet tube assembly to the pump, using a new gasket. Install the oil pan and fill the crankcase to the correct level with fresh oil. If possible, crank the engine over on the starter motor without allowing it to start for a short while to make sure that the oil pump and oil system is fully primed. The engine can be prevented from starting by disconnecting the ignition low-tension circuit or by pumping the accelerator pedal 8 to 10 times to flood the engine.

1 : 8 The clutch and flywheel

These are only accessible after the transmission has been removed from the vehicle (see **Chapter 6**) or the engine has been taken out. This section deals only with the removal and installation of the clutch. All details of clutch maintenance or servicing are dealt with in **Chapter 5.**

Clutch removal:

Remove the transmission and flywheel housing as described in **Chapter 6.** The attachments of the clutch are shown in **FIG 1 : 24.** The clutch arbor shown in the figure is not an integral part of the clutch and is used for aligning the clutch disc when installing the clutch.

If the clutch is to be reused, make aligning marks on the cover and flywheel to ensure that the cover assembly will be installed back into its original position, and the balance not lost. **When handling the clutch components, particularly the disc, avoid using dirty or greasy hands and prevent oil or grease from coming into contact with the disc friction linings.**

Evenly and in a diagonal sequence slacken the six bolts that secure the cover and when they are all free remove them, while holding the cover in place. Remove the cover and collect the clutch disc without allowing it to drop.

Install the clutch in the reverse order of removal. Fit the arbor through the disc and into the pilot service bearing in the crankshaft so that the disc is held centrally in place. If the disc is not centralized it will be impossible to fit the transmission input shaft back through the clutch. Install the cover so that the previously made marks are again in alignment. Secure the cover by progressively tightening the attachment bolts to a torque of 12 to 20 lb ft. The bolts must be progressively tightened in a diagonal sequence to prevent the pressure of the clutch springs from distorting the cover.

Clutch pilot bushing:

Whenever the clutch is removed, the bushing should be examined and measured for wear or fretting. If the bushing is worn, extract it using the tools shown in **FIG 1:25.** If the special tools are not available, the bushing can still be removed provided that a snug fitting piece of rod can be found or turned. Pack the recess in the crankshaft with grease so that there are no air pockets. Insert the rod into the bushing and drive it in with a hammer. The hydraulic action of the grease will force the old bushing out. Drive the new bushing back into place, preferably with the special drift No. T65L-7600-A.

The flywheel:

This can be removed, after the clutch has been taken off, by undoing the bolts that secure it to the crankshaft and then drawing it back off the crankshaft, taking care not to drop it as it comes free.

Before installing the flywheel, clean the mating surfaces of the flywheel and crankshaft and make sure that there are no specks of dirt which would cause excessive run-out. Install the flywheel and progressively tighten the attachment bolts to a torque of 75 to 85 lb ft.

Use a dial gauge mounted on the crankcase to check the runout of the flywheel friction surface and the periphery of the flywheel.

If the flywheel friction face is scored or worn, up to .045 inch of stock may be removed from the friction surface in order to remove the wear, damage or to cure excessive runout. If the surface cannot be restored by machining to the maximum depth a new flywheel must be fitted.

Flywheel ring gear:

If the teeth on the ring are worn or damaged a new ring should be fitted. To remove the old ring, partially drill through at the root of a tooth and then split the ring

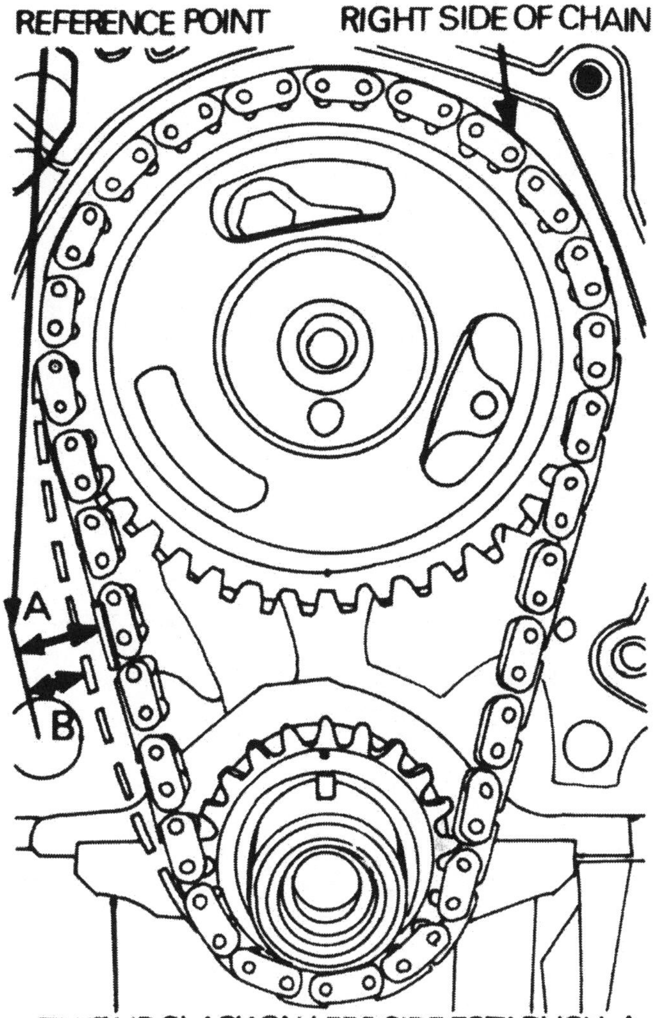

FIG 1:30 Checking the timing chain deflection

at the weak spot using a cold chisel. **Take great care not to damage or mark the flywheel itself.**

Thoroughly clean the periphery of the flywheel, using a wire brush to remove dirt and scale and using emery-cloth to remove burrs and high spots. Heat the new ring evenly to a maximum temperature of 500°F. If this temperature is exceeded the temper of the metal will be destroyed. When the ring is hot enough and has expanded, use tongs to lay into place on the flywheel. Tap the ring fully home using drifts made of soft-metal. Leave the flywheel and ring to cool naturally to room temperature before moving the flywheel.

1:9 The valve timing gear and camshaft

All the parts, except the camshaft bearings, can be removed and installed with the engine still fitted to the vehicle.

A typical camshaft and its associated parts are shown in **FIG 1:26.**

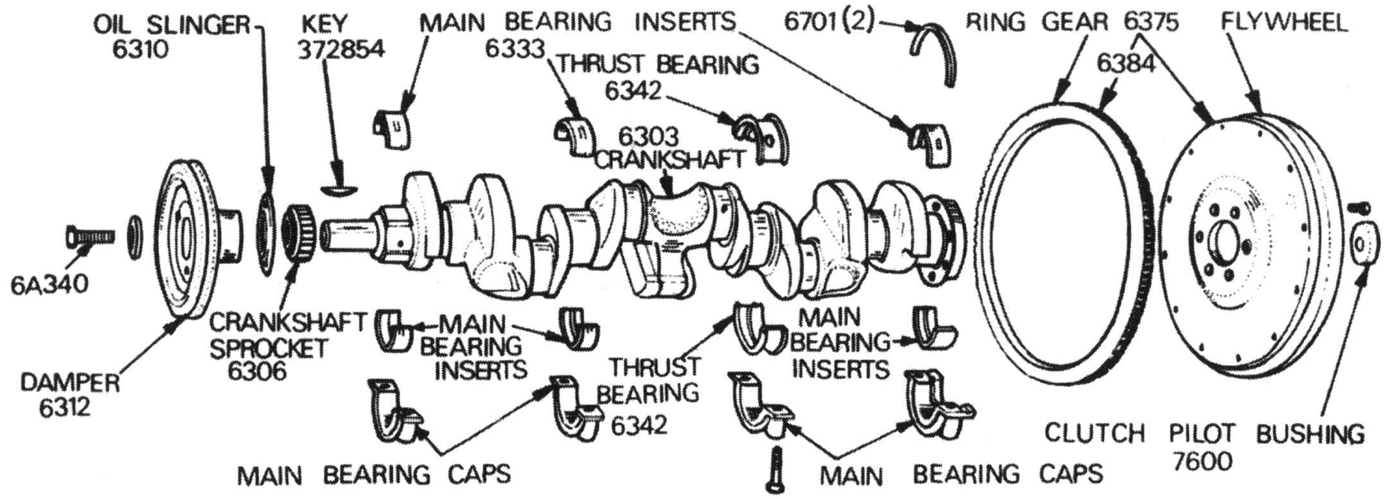

FIG 1:31 170 engine crankshaft and parts

Cylinder front cover:

1 Drain the crankcase and cooling system. Disconnect the battery. Remove the radiator (see **Chapter 4**) noting that on models fitted with automatic transmission it will be necessary to disconnect the oil cooler lines. Remove all drive belts, the cooling fan and water pump pulley.

2 On models fitted with air-conditioning, free the compressor from the engine but do not disconnect the refrigerant lines. On models fitted with power-steering, free the pump from the engine and hang it safely out of the way so that fluid does not spill from the reservoir.

3 Remove the bolts that secure the front cover, not forgetting the bolts that secure the oil pan to the front cover. Remove the crankshaft damper and pulley, preferably using the special tool T58P-6313-A for the 170 engine and -B for the 200 engine.

4 Pry the front of the cover slightly away from the top of the block and use a thin sharp knife to cut the oil pan gasket flush with the front face of the cylinder block. Remove the front cover.

After carrying out servicing as required, the front cover is installed in the reverse order of removal. It is most advisable to renew the oil seal in the cover. Clean all the parts and remove old gasket or sealer. Coat the surface on the block with sealer. Cut a new gasket to the required shape and attach it to the block using the sealer. Install pan seal as required. Smear the gasket surfaces of the cover with sealer and install it back into position, taking care not to disturb the gasket. Install a pilot tool on the crankshaft, as shown in **FIG 1:27**, to centralize the cover about the crankshaft. Downward pressure may

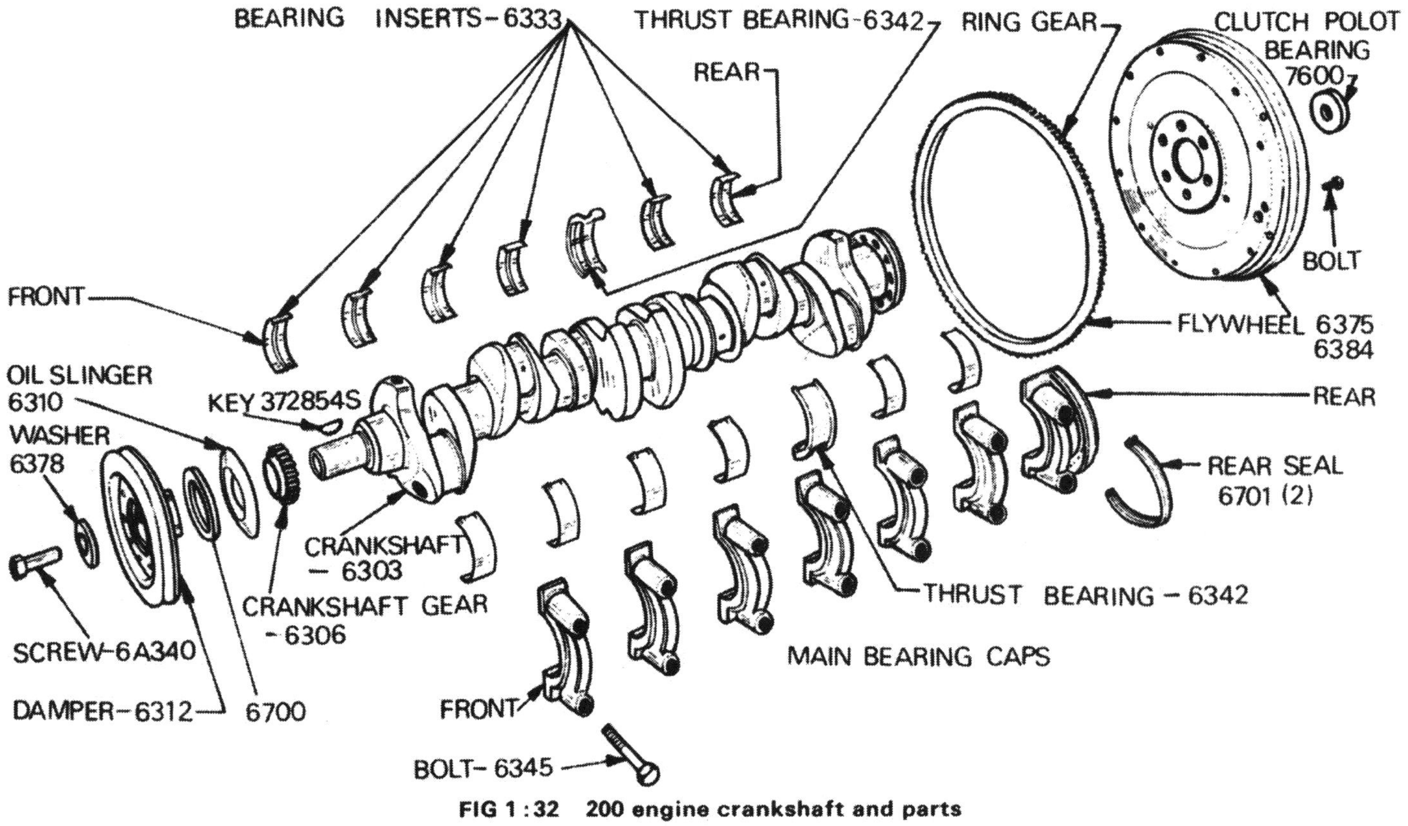

FIG 1:32 200 engine crankshaft and parts

have to be applied to the cover in order to align it correctly. Secure the cover in place by tightening the attachment bolts to a torque of 7 to 9 lb ft. Remove the pilot and install the damper, preferably using the tool T52L-6306-AEE, after lubricating the hub with Lubriplate to prevent scuffing.

The remainder of the parts are then installed in the reverse order of disassembly. Set all drive belt tensions correctly. Fill the cooling system. If coolant has run down into the engine when removing the front cover, it will be necessary to drain the crankcase and refill it with fresh oil ($3\frac{1}{2}$ quarts only as the oil filter need not be changed).

Front oil seal:

When the front cylinder cover has been removed, drive out the old oil seal with a suitable pin punch, taking care not to damage the cover itself. Coat the new seal with grease and drive it in, as shown in **FIG 1 : 28**, until it is fully in its recess. Check that the spring is correctly located and that the lips face into the engine.

Timing chain and sprockets:

Remove the cylinder front cover as described earlier. Crank the engine until the timing marks, shown in **FIG 1 : 29**, are in alignment. Turn the engine until the lefthand side (righthand side when viewed from in front of the car) is taut and without slack. The method of checking the chain total deflection is shown in **FIG 1 : 30**. Note that this figure shows the parts on a V-8 engine but the method is still the same. If the maximum deflection of the chain (A-B) exceeds .5 inch the timing chain and sprockets are worn and must be renewed as a set.

Remove the bolt that secures the camshaft sprocket to the camshaft. Grip both sprockets, one in either hand, and slide them off together so that the timing chain is kept in rough alignment. Once both sprockets are free the timing chain can be removed from around them.

Install the parts in the reverse order of disassembly and removal. Make sure that the timing marks on the sprockets are again in alignment after the parts are installed.

It is essential that neither the camshaft or the crankshaft is rotated while the sprockets and timing chain are off. If one is rotated there is a danger that the valves will hit the crowns of the piston, causing internal damage.

The camshaft:

Remove the front cylinder cover and sprockets with timing chain as described previously. Remove the cylinder head as described in **Section 1 : 3** and then lift

FIG 1 : 33 **Using Plastigage to check bearing clearance**

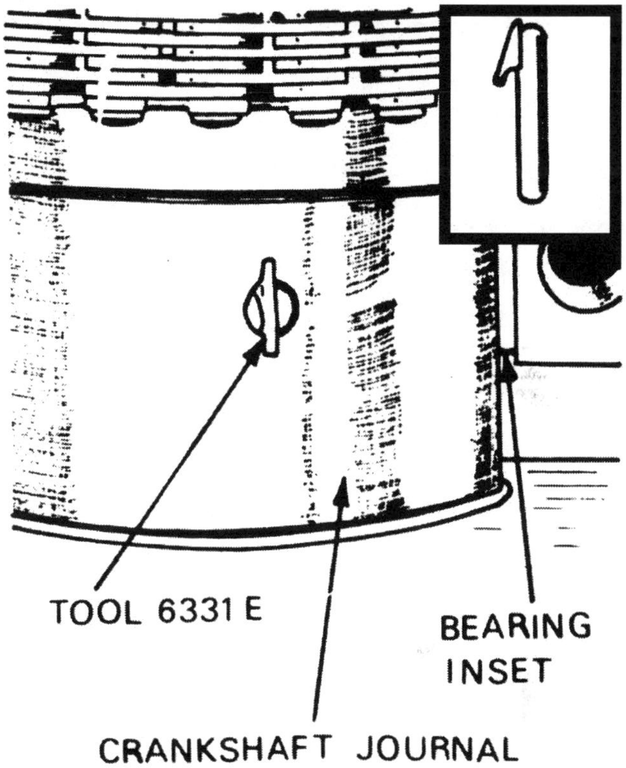

FIG 1 : 34 **Removing or installing an upper main bearing insert**

out the valve lifters as shown in **FIG 1 : 11**. Remove the oil pan and oil pump as described in **Section 1 : 7**. Remove the distributor and the fuel pump.

The camshaft is now ready for removal. Before removing the camshaft, install its sprocket correctly, omitting the timing chain and crankshaft sprocket. Mount a dial gauge onto the cylinder block so that its

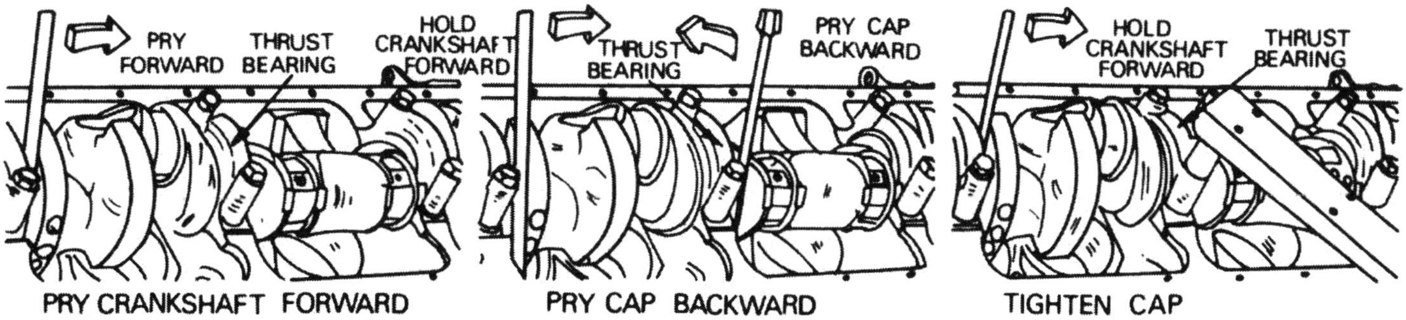

FIG 1 : 35 **Aligning the thrust bearing correctly**

stylus rests vertically on the end of the sprocket cap bolt. Lever the camshaft firmly in and out so that the end play of the camshaft can be measured. The end play should lie between .001 and .007 inch. If the end play is excessive, a new thrust plate must be fitted. Remove the air intake grille from the front of the vehicle to allow the camshaft to be drawn out. Take off the camshaft sprocket and remove the bolts that secure the camshaft thrust plate. Withdraw the camshaft, taking great care not to allow the journals or cams to damage the four bearings.

Install the camshaft in the reverse order of removal. The cam lobes should be lubricated with Lubriplate. It should be noted that slight pitting on the tips of the lobes is normal wear and no action need be taken unless the loss of lift exceeds .005 inch. The remainder of the parts are also installed in the reverse order of disassembly and removal.

Camshaft bearings:

The engine must be removed from the vehicle in order to renew the bearings. Before fully dismantling the engine, remove the rocker cover and rocker assembly but leave the pushrods in place. Mount a dial gauge on the cylinder head so that it can be used to measure the movement of each pushrod. Slowly crank the engine and note the pushrod lift. If the loss of lift from standard exceeds .005 inch, parts will have to be renewed to cure the fault. The standard lift on all valves should be .2320 inch.

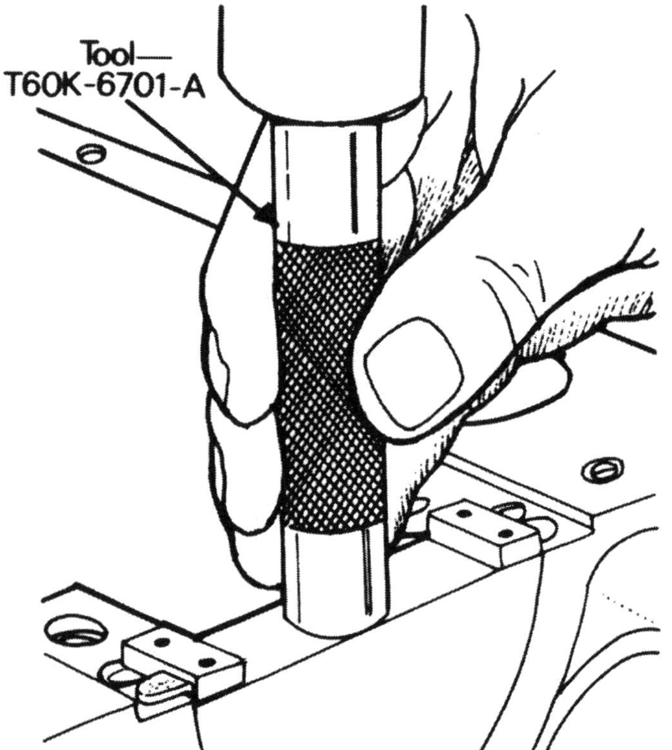

FIG 1:36 Installing the rear crankshaft oil seal

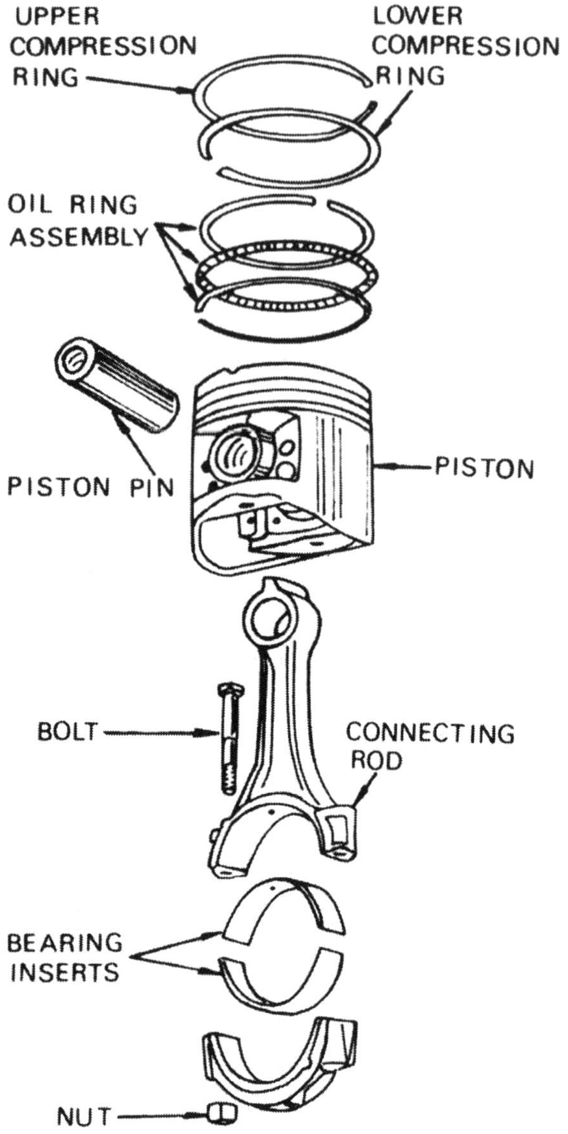

FIG 1:37 Piston, connecting rod and related components

The camshaft must be removed and the clutch and flywheel taken off (to gain access to the rear bearing bore plug) before the bearings can be renewed. The work should be left to a service garage as if the special tools are not used there is every likelihood of damaging the new bearings when drawing them back into place.

1:10 The crankshaft and bearings

All the crankshaft bearings, including the main bearings, can be renewed without removing the engine from the vehicle. If, on removing the bearings, it is found that the journal or crankpin is excessively worn or scored, the engine must be removed from the vehicle so that the crankshaft can be taken out.

The crankshaft for a 200 engine is shown in **FIG 1:31**. The crankshaft for a 170 engine is similarly shown in **FIG 1:32**.

Bearing inserts:

These are renewable and steel-backed with bearing material lining them. They are fitted exactly as received, apart from cleaning off protective, and do not require boring or scraping to make them fit.

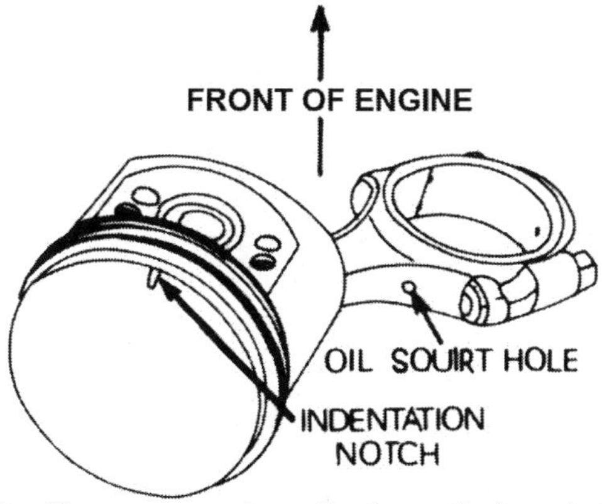

FIG 1 : 38 The correct installation of the piston and connecting rod assembly

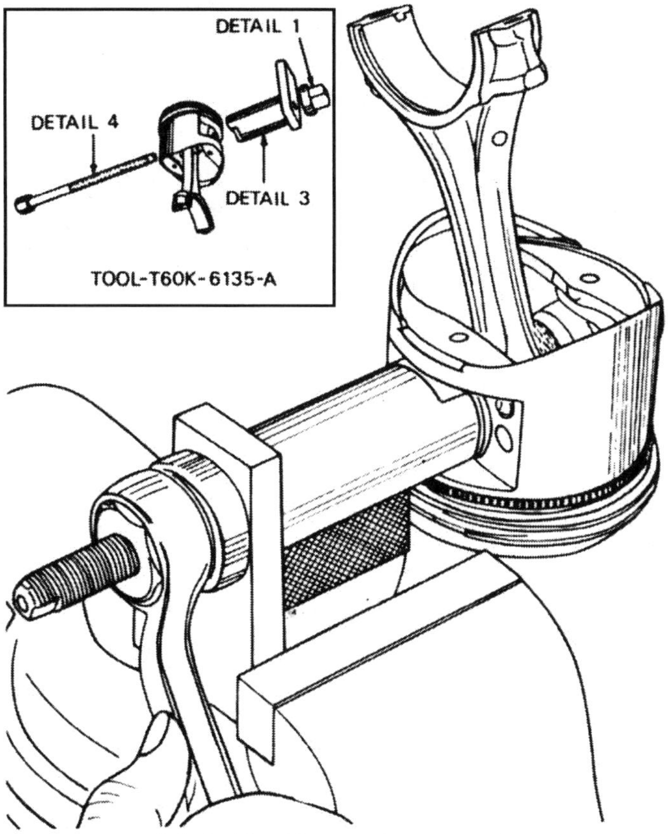

FIG 1 : 39 Piston pin removal

If a crankpin, journal, or insert is worn, under no circumstances file the connecting rod or bearing cap in an attempt to reduce the clearance. Similarly never fit shims under the cap to increase the clearance.

Undersize bearing inserts are obtainable to fit a crankshaft which has had its journals or crankpins ground down, and these are supplied as a standard fitting for each grade of undersize so that no selective fitting is possible.

Bearings for standard diameter crankpins or journals are supplied in three sizes; standard, —.001 inch and —.002 inch. This allows the inserts to be selectively fitted to give the correct bearing diametrical clearance. The tolerance for clearance on bearings is given in **Technical Data.** If .001 or .002 inch undersize bearings have to be used to bring the clearance within tolerances, make sure that all the undersize bearing inserts are fitted on the cylinder block side.

The clearance of a bearing is measured using Plastigage. Remove the cap and make sure that the cap, bearing, and pin are all scrupulously clean. Lay a strip of Plastigage into the cap, as shown in **FIG 1 : 33.** Install the cap and tighten its attachments to the correct torque load. Do not turn the crankshaft while the Plastigage is in place. Remove the cap and use the gauge to measure the bearing clearance. Measure at the widest and narrowest points so that any taper can be checked.

The oil pan must be removed to gain access to the bearings. Work will be easiest if the vehicle is positioned over a pit. If the car has to be raised to gain access room, make sure that all supports are firmly based and of sufficient strength.

Connecting rod bearings:

Remove the oil pan. Crank the engine to a convenient position for working on the connecting rod bearings. Undo the nuts that secure the bearing caps to the connecting rods. Pull off the bearing caps and remove the inserts from them. Store the bearing caps and inserts in the correct order for installation. The caps and connecting rods are normally marked with numerals to indicate position and order. If these numerals are not present, mark the caps and connecting rods with a code of light dot marks, carefully noting which side the marks face.

Press the connecting rods up the bores so that they clear the crankpins and remove the bearing inserts from the connecting rods. If the bearing inserts are to be

FIG 1 : 40 Checking the piston ring fitted gap

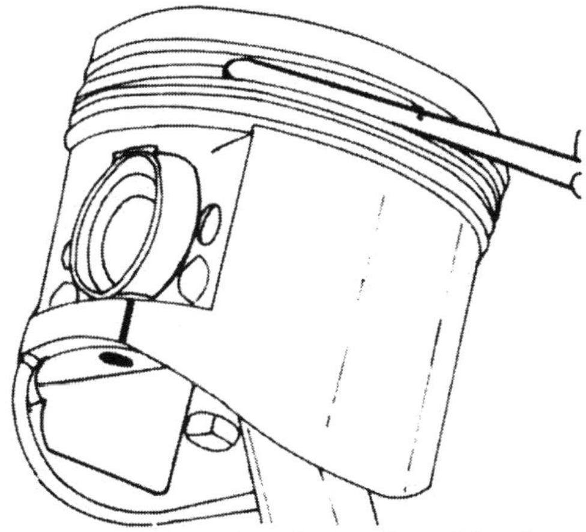
FIG 1 : 41 Checking the piston ring side clearance

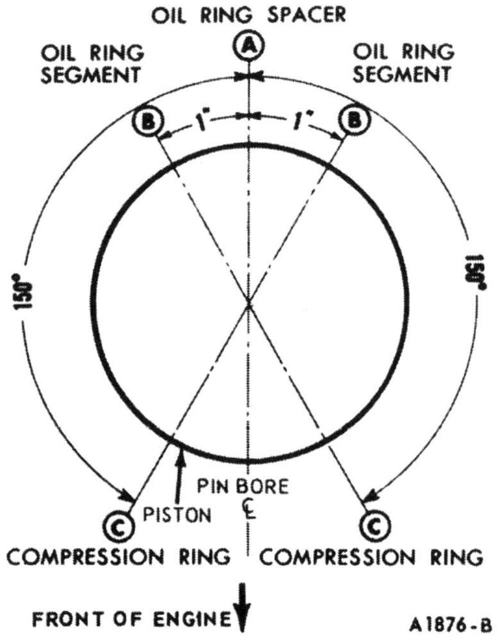

FIG 1 : 42 The piston ring gap spacing

used again, they must be stored in order so that there is no danger of interchanging them on assembly. If the crankshaft is rotated, take care not to trap the connecting rods between the throws and side of the crankcase.

Examine the bearing inserts and install a new set if any show signs of cracking, pitting, scoring, or wear. Measure each crankpin at several points using a micrometer gauge. If the pin is scored, the maximum out-of-round exceeds .0004 inch, or the taper exceeds .003 inch per inch, the crankshaft must be removed, the crankpins ground down to the next suitable undersize and the crankshaft installed using new undersize bearing inserts.

Thoroughly clean the crankpin and bores for the inserts in the connecting rod and bearing cap. Wipe the inserts clean with a piece of leather or lint-free cloth. Fit the insert back into position, making sure that their tags locate correctly. Lubricate the crankpins liberally with clean engine oil. Pull the connecting rods down onto the crankpins and fit the bearing caps, making sure that the identification marks are in the correct order and facing the correct side of the engine. Tighten the nuts to a torque of 19 to 24 lb ft and check that the crankshaft rotates freely.

Install the oil pump followed by the oil pan. Fill the crankcase to the correct level with fresh engine oil.

Main bearings:

If the engine is fitted to the vehicle, renew one bearing at a time, leaving the crankshaft supported by the remaining bearings. On the 200 engine No. 4 (counting from the front) bearing is fitted with special inserts which control the end play of the crankshaft, while on the 170 engine No. 3 main bearing is fitted with these special inserts.

Free and remove the bolts securing the bearing cap noting that the identification number will give the direction which the cap faces and pull off the cap. Remove the insert from the cap. Insert the special tool into the oil hole in the crankshaft (see **FIG 1 : 34**). Turn the crankshaft so that the tool abuts onto the end of the insert and drives it out from above the crankshaft. Make sure that the tag end of the insert comes out first.

Use a syringe filled with gasoline to clean the area behind the crankshaft. Lubricate the new insert with clean engine oil and drive it back into position using the special tool in the crankshaft oil hole, making sure that its tag locates correctly. Fit the lower insert back into the cap and secure the cap correctly, tightening the bolts to a torque of 60 to 70 lb ft. Check that the crankshaft rotates freely before servicing the next bearing.

When servicing the rear bearing, fit a new oil seal to the bearing cap.

When servicing the bearing which takes the crankshaft thrust, install the cap so that the thrust faces are in line. The method of ensuring this is shown in **FIG 1 : 35**. If the crankshaft journals are found to be worn or scored, the crankshaft must be removed and the journals ground down to the next suitable undersize. Install the crankshaft using new undersize bearing inserts.

Crankshaft end play:

Mount a dial indicator onto the crankcase so that the stylus rests vertically on the end of the crankshaft. Lever the crankshaft backwards and forwards so that the end play can be measured on the indicator. The limits of play are .004 to .008 inch but no action need be taken until the wear limit of .012 inch is exceeded. If the play is excessive, new thrust bearing inserts must be fitted to reduce the play. If the play is too small then it is likely that the procedure shown in **FIG 1 : 35** has not been correctly followed, though dirt can also be a cause.

Rear oil seal:

The portion in the bearing cap should always be renewed when the cap is removed. The portion in the crankcase can only be renewed after the crankshaft has been removed.

Peel out the old seal and clean the recess. Roughly shape the seal in the fingers and press it into the recess using the thumbs. Start at the centre and work outwards so that the seal protrudes evenly on both sides. Press the seal fully into position with the seal forming tool, as shown in **FIG 1 : 36**. Trim the ends of the seal flush with a sharp knife. Apply a little sealer in the area shown on the cap, without allowing sealer onto the seal itself.

Crankshaft removal:

Remove the engine from the vehicle (see **Section 1 : 2**). Invert the engine, using old tires to stand it on after removing vulnerable components. Remove the clutch and flywheel (see **Section 1 : 8**) as well as the oil pan and oil pump (see **Section 1 : 7**). Remove the cylinder front cover and the sprockets with timing chain (see **Section 1 : 9**). Disconnect all the connecting rod bearings and loosely fit the bearing caps back into place after the connecting rods have been pushed into the cylinder bores. Remove all the main bearing caps and carefully lift out the crankshaft.

Thoroughly clean all the parts. Blow through the oilways in the crankshaft with kerosene followed by compressed air to remove any swarf or dirt.

Put the inserts correctly into position so that the tags are located. Lubricate the journals liberally with fresh oil and lay the crankshaft back into place.

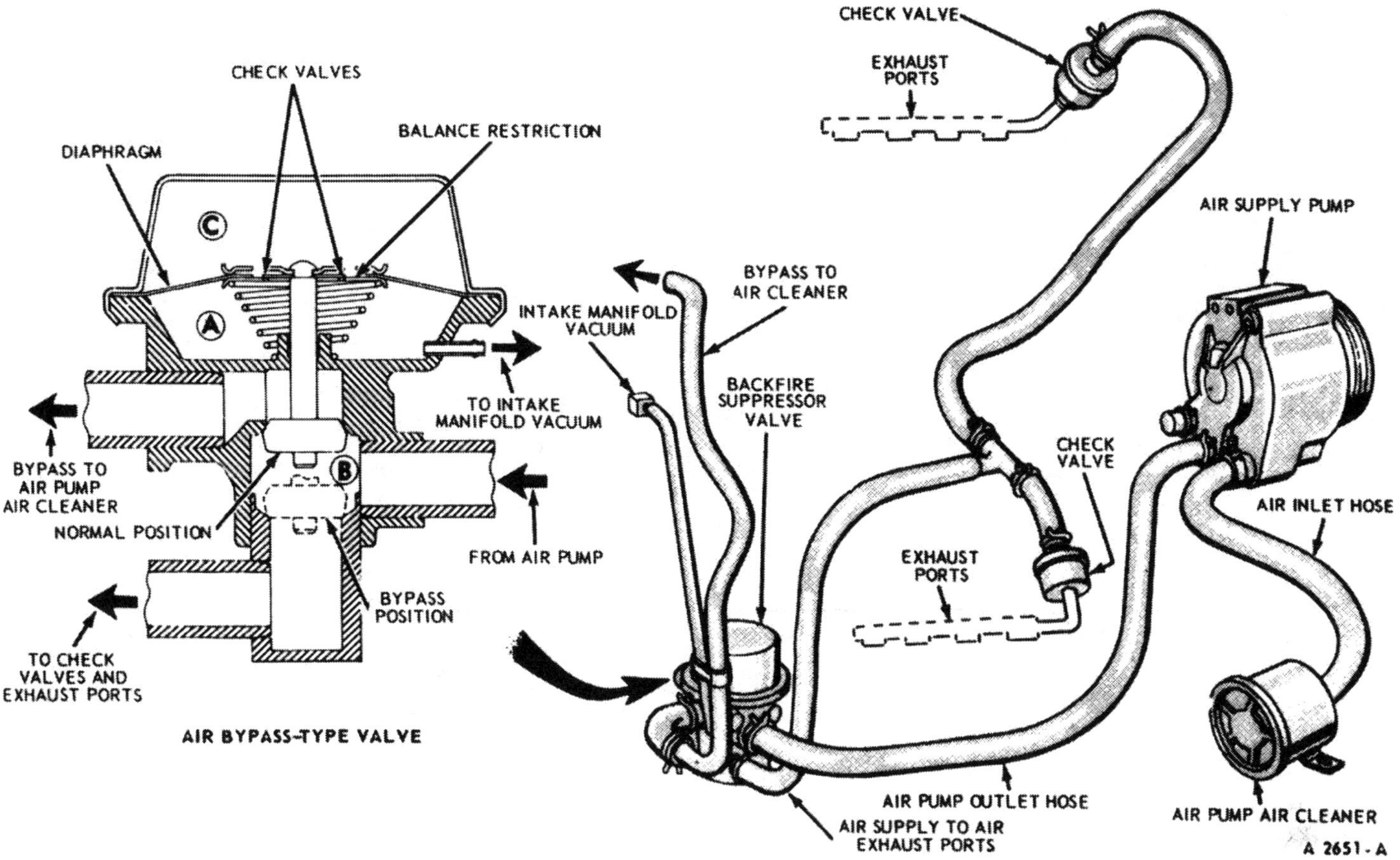

FIG 1:43 Schematic of the Thermactor emission control system

Install the bearing caps into their correct positions, making sure that they face in their original directions. Secure the caps by torque-loading the bolts. Check the end play and see that the crankshaft rotates freely before fitting the connecting rods to the crankpins.

The remainder of the parts are installed in the reverse order of removal and the engine is then fitted back into the vehicle. Leave accessories and vulnerable components until last. Check coolant and lubricant levels before and after starting the engine. Check for leaks with the engine running at a fast idle.

1:11 The pistons and connecting rods

The parts can be removed from the engine, with the engine installed in the vehicle. A typical connecting rod and piston assembly is shown in **FIG 1:37.** When the parts have been installed, the notch on the piston crown must face forwards and the squirt hole in the connecting rod must then face as shown in **FIG 1:38**

To remove a piston and connecting rod, remove the cylinder head (see **Section 1:3**) and the oil pan and oil pump (see **Section 1:7**). Remove the unworn ridge and ring of carbon from around the top of the bore with garage equipment cutters. If the ridge and carbon are left in place they may damage the piston rings as the piston is removed. Lay rags into the bottom of the bore to catch the swarf and dust as it is cut away. Disconnect the appropriate connecting rods from the crankshaft, as described in the previous section, and push the connecting rod right up the bore so that the piston can be removed from the top.

Cylinder bores:

These should be examined whenever the cylinder head is removed. A good estimate of the wear can be made by judging the depth of the unworn ridge around the top of the bore. For accurate measurements a special dial gauge should be used and the diameter measured at several points up and down the bore, both in line with, and at right angles to, the piston pin. If the maximum ovality exceeds .005 inch or the taper exceeds .010 inch the bore must be bored out and an oversize piston fitted. Reboring must also be carried out if the bore is scored or damaged.

All pistons are of the same weight so if only one bore is damaged it may be rebored, leaving the remainder of the bores untouched. If the wear is not sufficient to warrant reboring but the high oil consumption requires new rings to cure it the unworn ridge around the top of the bore must be honed out. The old top rings will have worn with the cylinder bore, but if a new ring is fitted it will hit the ridge at every stroke and rapidly break up.

Piston pins:

Thses are an intereference fit in the connecting rod and a floating fit in the piston. The pins should not be removed unnecessarily. The tools required for removing or installing the pin are shown in **FIG 1:39.**

If the pistons are removed, and the original pistons used again, it is essential that they are fitted back into their original bores and connecting rods.

All cleaning and servicing operations can be carried out with the pistons fitted to the connecting rods.

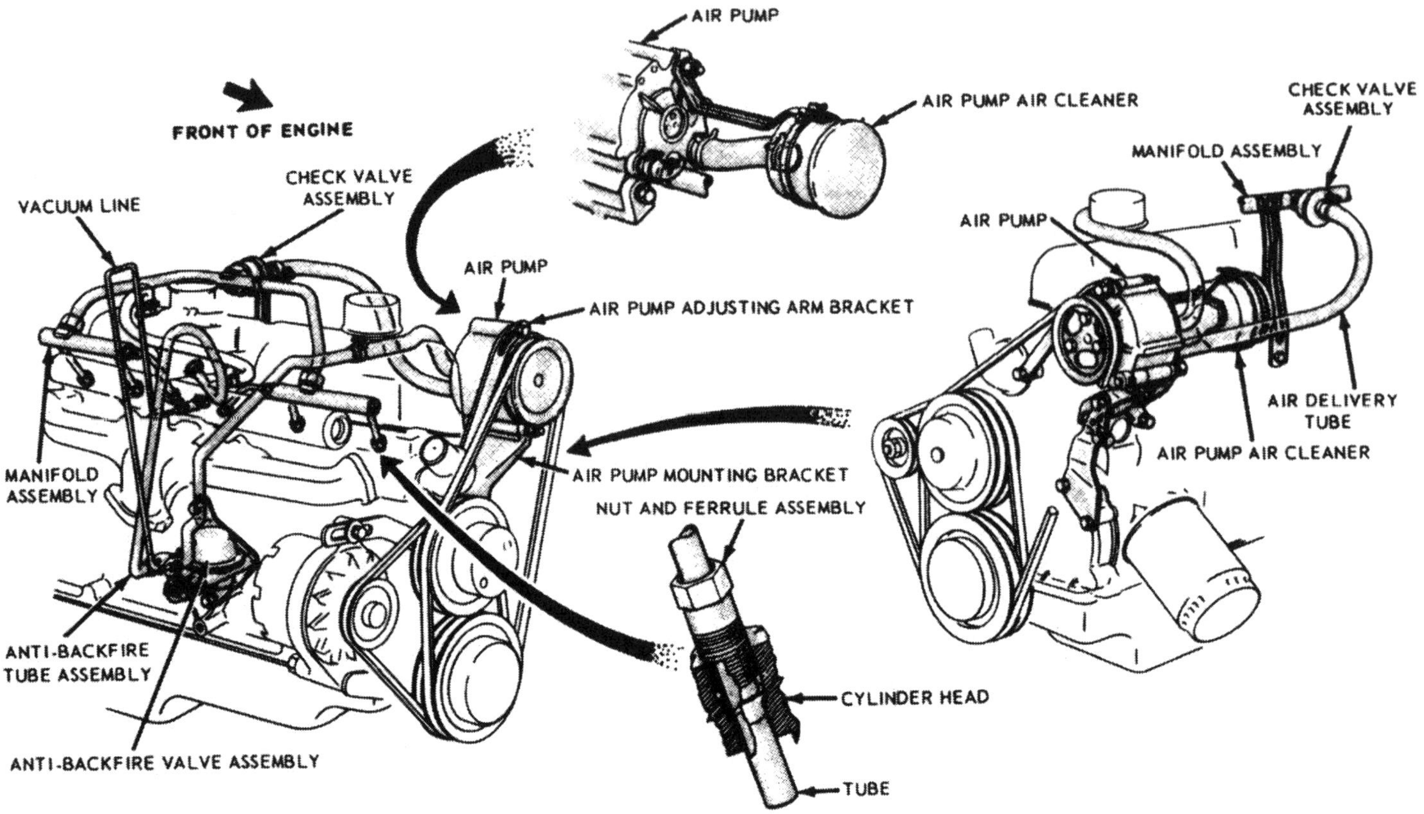

FIG 1:44 The Thermactor emission control system installed on the engine

Piston rings:

Special ring expanding tools are made for removing or installing the piston rings. The tool grips the ends of the ring and then gently parts them so that the piston can be lifted up and off the piston. If a special tool is not available three steel shims—discarded feeler gauges for instance—can be used instead. Slide a shim under one end of the ring and then work the shim round under the ring, while pressing the ring as it comes free onto the piston land above it. When all the ring is on the land, slide the remaining two shims under the ring so that all three are evenly spaced. Part the ends of the ring gently with the thumb nails and slide the ring up and off the piston, using the shims to protect the piston. Take care not to allow the back of the ring to scrape the piston.

Use a piece of broken ring, or a special tool, to clean the ring grooves in the piston. Remove all the carbon but take care not to remove metal from the piston or the oil consumption will be increased. Clean out the return holes behind the oil control ring with a piece of blunt-ended wire. If the grooves have a high step on the lower lands it is likely that the clearance will be excessive and a new piston will have to be fitted to cure the fault.

Spring each new ring in turn into the top of a bore and use an inverted piston to press it squarely down the bore to the point of maximum wear. Use feeler gauges, as shown in **FIG 1:40**, to measure the gap between the ends. Try another set of rings but if the gap is too small, file the ends carefully with a smooth file to set the correct gap. The gap for the compression rings is given in **Technical Data.**

Install the rings on the piston in the reverse order of removal, taking great care not to force or snap them. Use feeler gauges to check the side clearance as shown in **FIG 1:41**. The oil control ring should be a snug fit in all cases. The side clearance on compression rings should be .002 inch when parts are new but it may be allowed to reach the wear limit of .006 inch before taking action. Check the side clearance using a new piston ring, and if the clearance is still outside limits a new piston must be fitted otherwise oil consumption will be excessive.

Pistons:

The pistons are graded to ensure that the clearance between piston skirt and cylinder bore can be set with accuracy. Special measuring equipment is required for checking the clearance, though a skilled mechanic can check using a set of feeler gauges and spring balance gauge. The selection of pistons is best left to the specialist reboring the block or supplying new pistons. Make sure that the pistons are fitted to their correct bores and not interchanged.

When the pistons are out of the engine, the crowns may be lightly polished using worn emerycloth dipped in kerosene as all abrasive particles can be washed off before installation. **Do not use any form of abrasive on the piston skirt.** If there is lacquering on the skirt, indicative of gases blowing past the rings, wash it off with a suitable harsh solvent such as trichlorethylene.

The pistons must be installed so that the notch on the piston crown faces the front of the engine.

Installation:

Check that all the parts are clean and free from particles or dirt. If the piston has been removed from the connecting rod, assemble them as shown in **FIG 1:38**. Fit the rings into position and rotate them so that their gaps are positioned as shown in **FIG 1:42**. Compress the rings into their grooves with a piston ring clamp, after lightly oiling them and the piston skirt. Lower the connecting rod into its correct bore and insert the skirt of the piston. Gently press down on the piston, allowing the clamp to slide off the rings as they enter the bore. Do not force the rings into place and do not allow them to escape from the clamp before they enter the bore. The rings are very brittle and will easily snap if mishandled. If the correct ring clamp is not available, a large worm-driven hose clip can be used instead but more care must then be taken.

Install the connecting rod bearings. Refit the oil pump, oil pan and cylinder head in the reverse order of disassembly.

1:12 Thermactor emission control

This type of emission control is usually only used on models fitted with manual-shift transmission. An engine-driven air pump provides a constant flow of air under pressure. The pump is fitted with a relief valve to limit the maximum pressure of the air. Under normal running the air is injected through nozzles directly into the ports by the exhaust valves, so that the excess air mixes with the hot exhaust gases to burn any surplus hydrocarbons. A schematic layout of the system is shown in **FIG 1:43**. The check valve is fitted to prevent exhaust gases flowing back through the system if the exhaust pressure rises above the air pressure.

A dump air valve is fitted to prevent backfiring on deceleration, and a sectioned view of the valve is shown in **FIG 1:43.** Normally the pressure of the spring holds the valve open so that air flows directly to the ports. When the throttle is closed and intake manifold suction is high, the suction operates the valve so that the air no longer passes to the exhaust ports but is dumped through the air cleaner.

The installation of the parts on the engine is shown in **FIG 1:44.**

Air cleaner:

The earlier air cleaner components are shown in **FIG 1:45** and the renewal of the filter is self-explanatory.

On some later models a centrifugal fan filter is attached directly to the pump. To remove the centrifugal type filter, remove the drive pulley from the pump. Pry the outer disc loose and then use a pair of pliers to grip the fan by the vanes and pull it out. If the fan breaks, take care to prevent parts entering the air inlet hole of the air pump. Do not attempt to remove the metal drive hub. The new centrifugal type filter is pulled back into position using the pulley and the bolts to draw it back, making sure that it fits fully into place. The new filter may be tight, giving noisy operation at first, but noise should die away as the filter beds in.

Check valve test:

Start the engine and run it until it has reached its normal operating temperature. Disconnect the air supply hose from the check valve. At idle the valve should be closed with the plate away from the manifold side, though a little flutter may be apparent at idling because of the pulsations in the exhaust. Use a pencil to press the valve open and check that the valve returns immediately the pencil is removed. Slowly increase the engine speed to 1500 rev/min and check that the valve is tightly closed with no exhaust gas leakage.

If the valve does not return freely after being opened by hand, or if it leaks at the higher engine speed, the valve is defective and must be renewed.

Air leakage:

Air leaks in the system will make it impossible to set a satisfactory curb idle speed. Note that a fault in the crankcase ventilation system will also make it impossible to set curb idle.

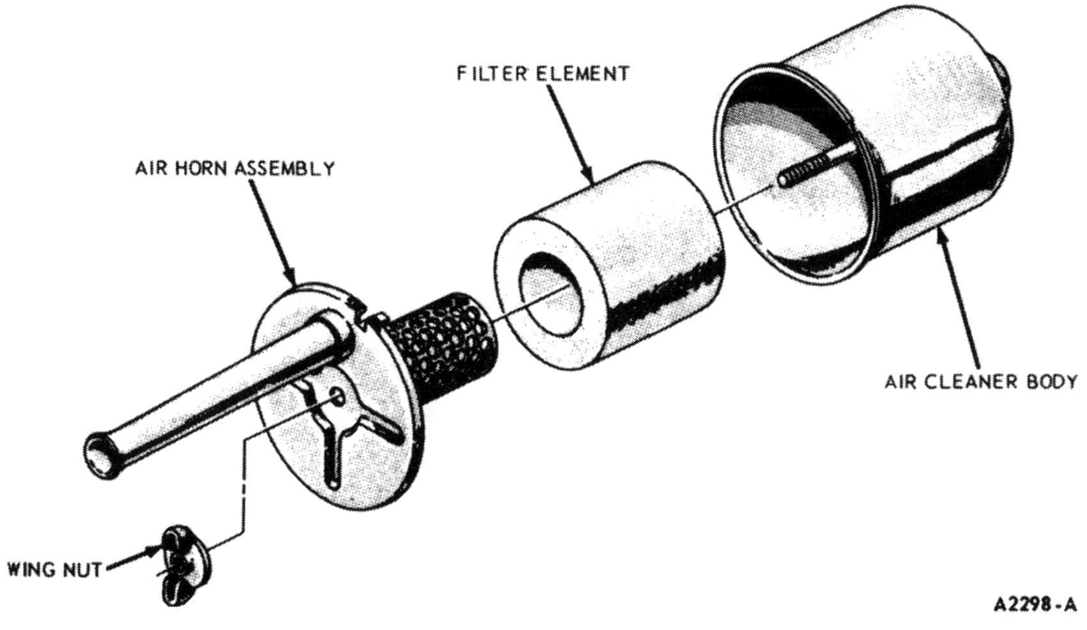

FIG 1:45 The early Thermactor air cleaner

Disconnect the vacuum line at the manifold and plug the connector on the manifold. If curb idle can now be satisfactorily set there is a leak and either the hose or the dump valve must be renewed, depending on which component is defective.

Dump valve functional test:

With the engine running, disconnect the hose from the pump at the valve. Snap the throttle open and shut. A sucking noise should be heard from the valve and there should be no backfiring on deceleration. If the engine backfires, the valve is defective and must be renewed.

Reconnect the hose from the pump to the valve and disconnect the dump air hose at the valve. Again snap the throttle open and shut. Compressed air should blow from the valve as the engine decelerates. If air does not blow from the valve it is defective and must be renewed.

1:13 Trouble shooting guide

(a) Engine will not start

1 Water or dirt on spark plug leads, distributor cap or ignition coil
2 Loose wires or defective insulation in ignition low-tension circuit
3 Defective ignition coil
4 Defective ignition contact points
5 Defective ignition condensor
6 Defective automatic choke
7 Faulty or jammed starter motor
8 Spark plug leads incorrectly connected
9 Blocked fuel filter or carburetor jets
10 Leaking valves
11 Sticking valves
12 Valve timing incorrect
13 Ignition timing incorrect

(b) Engine stalls

1 Check 1, 3, 4, 6, 9, 10 and 11 in (a)
2 Spark plugs defective
3 Retarded ignition
4 Mixture too weak
5 Water in fuel
6 Fuel tank vent system blocked
7 Incorrect initial adjustment of valve lifters

(c) Rough engine idle

1 Check 4, 6, 8, 10, 11, 12 and 13 in (a) and 2 and 7 in (b)
2 Incorrectly adjusted carburetor
3 Air leaks at carburetor flange
4 Leaking vacuum fittings
5 Defective crankcase ventilation system
6 Worn piston rings
7 Worn valve stems or guides
8 Weak exhaust valve springs
9 Defective dump valve (Thermactor only)

(d) Engine misfires

1 Check 1, 2, 3, 4, 8, 10, 11, 13 in (a); 2, 3, 4 and 7 in (b)
2 Weak or broken valve springs

(e) Engine overheats

1 Check Chapter 4

(f) Compression test pressure low*

1 Check 10, 11 and 12 in (a); 6 and 7 in (c) and 2 in (d)
2 Worn or scored cylinder bores
3 Worn piston rings and grooves
4 Choked air cleaner

*A high test pressure indicates excessive deposits in the combustion chamber.

(g) Engine lacks power

1 Check 4, 6, 10, 11, 12 and 13 in (a); 2, 3, 4 and 7 in (b); 6 and 7 in (c) and 2 in (d). Also check (e)
2 Leaking cylinder head gasket
3 Distributor automatic advance and retard not working correctly

(h) Burnt valves or seats

1 Check 10, 11 and 12 in (a) and 7 in (b). Also check (e)
2 Excessive carbon in combustion chamber

(j) Sticking valves

1 Weak valve springs
2 Bent valve stems
3 Scored stem or guide
4 Defective valve lifter
5 Incorrect initial valve lifter adjustment
6 Gum deposits on stem or guide

(k) Excessive cylinder wear

1 Check 6 in (a) and also check (e)
2 Lack of oil
3 Dirty oil
4 Piston rings gummed or broken
5 Badly installed piston rings
6 Connecting rod bent or twisted
7 Piston pin worked loose or incorrectly fitted

(l) Excessive oil consumption

1 Check 6 and 7 in (c); 2 and 3 in (f) and 4 and 5 in (k).
2 Oil return holes in piston blocked
3 Oil level too high in crankcase
4 External oil leaks

(m) Crankshaft bearing failure

1 Check 2, 3 and 6 in (k)
2 Restricted oilways
3 Worn journals or crankpins
4 Loose bearing caps
5 Dirt particles trapped during installation
6 Extremely low oil pressure

(n) Low oil pressure

1 Check 2 and 3 in (k) and also check 2, 3 and 4 in (m)
2 Choked oil filter or pump intake tube strainer
3 Relief valve defective
4 Very worn oil pump
5 Defective gauge or sender unit

(o) Internal water leakage (see **Chapter 4**)

(p) Poor water circulation (see **Chapter 4**)

(q) Corrosion (see **Chapter 4**)

(r) High fuel consumption (see **Chapter 2**)

Inches	Decimals	Millimetres	Inches to Millimetres: Inches	Inches to Millimetres: mm	Millimetres to Inches: mm	Millimetres to Inches: Inches
1/64	.015625	.3969	.001	.0254	.01	.00039
1/32	.03125	.7937	.002	.0508	.02	.00079
3/64	.046875	1.1906	.003	.0762	.03	.00118
1/16	.0625	1.5875	.004	.1016	.04	.00157
5/64	.078125	1.9844	.005	.1270	.05	.00197
3/32	.09375	2.3812	.006	.1524	.06	.00236
7/64	.109375	2.7781	.007	.1778	.07	.00276
1/8	.125	3.1750	.008	.2032	.08	.00315
9/64	.140625	3.5719	.009	.2286	.09	.00354
5/32	.15625	3.9687	.01	.254	.1	.00394
11/64	.171875	4.3656	.02	.508	.2	.00787
3/16	.1875	4.7625	.03	.762	.3	.01181
13/64	.203125	5·1594	.04	1.016	.4	.01575
7/32	.21875	5.5562	.05	1.270	.5	.01969
15/64	.234375	5.9531	.06	1.524	.6	.02362
1/4	.25	6.3500	.07	1.778	.7	.02756
17/64	.265625	6.7469	.08	2.032	.8	.03150
9/32	.28125	7.1437	.09	2.286	.9	.03543
19/64	.296875	7.5406	.1	2.54	1	.03937
5/16	.3125	7.9375	.2	5.08	2	.07874
21/64	.328125	8.3344	.3	7.62	3	.11811
11/32	.34375	8.7312	.4	10.16	4	.15748
23/64	.359375	9.1281	.5	12.70	5	.19685
3/8	.375	9.5250	.6	15.24	6	.23622
25/64	.390625	9.9219	.7	17.78	7	.27559
13/32	.40625	10.3187	.8	20.32	8	.31496
27/64	.421875	10.7156	.9	22.86	9	.35433
7/16	.4375	11.1125	1	25.4	10	.39370
29/64	.453125	11.5094	2	50.8	11	.43307
15/32	.46875	11.9062	3	76.2	12	.47244
31/64	.484375	12.3031	4	101.6	13	.51181
1/2	.5	12.7000	5	127.0	14	.55118
33/64	.515625	13.0969	6	152.4	15	.59055
17/32	.53125	13.4937	7	177.8	16	.62992
35/64	.546875	13.8906	8	203.2	17	.66929
9/16	.5625	14.2875	9	228.6	18	.70866
37/64	.578125	14.6844	10	254.0	19	.74803
19/32	.59375	15.0812	11	279.4	20	.78740
39/64	.609375	15.4781	12	304.8	21	.82677
5/8	.625	15.8750	13	330.2	22	.86614
41/64	.640625	16.2719	14	355.6	23	.90551
21/32	.65625	16.6687	15	381.0	24	.94488
43/64	.671875	17.0656	16	406.4	25	.98425
11/16	.6875	17.4625	17	431.8	26	1.02362
45/64	.703125	17.8594	18	457.2	27	1.06299
23/32	.71875	18.2562	19	482.6	28	1.10236
47/64	.734375	18.6531	20	508.0	29	1.14173
3/4	.75	19.0500	21	533.4	30	1.18110
49/64	.765625	19.4469	22	558.8	31	1.22047
25/32	.78125	19.8437	23	584.2	32	1.25984
51/64	.796875	20.2406	24	609.6	33	1.29921
13/16	.8125	20.6375	25	635.0	34	1.33858
53/64	.828125	21.0344	26	660.4	35	1.37795
27/32	.84375	21.4312	27	685.8	36	1.41732
55/64	.859375	21.8281	28	711.2	37	1.4567
7/8	.875	22.2250	29	736.6	38	1.4961
57/64	.890625	22.6219	30	762.0	39	1.5354
29/32	.90625	23.0187	31	787.4	40	1.5748
59/64	.921875	23.4156	32	812.8	41	1.6142
15/16	.9375	23.8125	33	838.2	42	1.6535
61/64	.953125	24.2094	34	863.6	43	1.6929
31/32	.96875	24.6062	35	889.0	44	1.7323
63/64	.984375	25.0031	36	914.4	45	1.7717

UNITS	Pints to Litres	Gallons to Litres	Litres to Pints	Litres to Gallons	Miles to Kilometres	Kilometres to Miles	Lbs. per sq. In. to Kg. per sq. Cm.	Kg. per sq. Cm. to Lbs. per sq. In.
1	.57	4.55	1.76	.22	1.61	.62	.07	14.22
2	1.14	9.09	3.52	.44	3.22	1.24	.14	28.50
3	1.70	13.64	5.28	.66	4.83	1.86	.21	42.67
4	2.27	18.18	7.04	.88	6.44	2.49	.28	56.89
5	2.84	22.73	8.80	1.10	8.05	3.11	.35	71.12
6	3.41	27.28	10.56	1.32	9.66	3.73	.42	85.34
7	3.98	31.82	12.32	1.54	11.27	4.35	.49	99.56
8	4.55	36.37	14.08	1.76	12.88	4.97	.56	113.79
9		40.91	15.84	1.98	14.48	5.59	.63	128.00
10		45.46	17.60	2.20	16.09	6.21	.70	142.23
20				4.40	32.19	12.43	1.41	284.47
30				6.60	48.28	18.64	2.11	426.70
40				8.80	64.37	24.85		
50					80.47	31.07		
60					96.56	37.28		
70					112.65	43.50		
80					128.75	49.71		
90					144.84	55.92		
100					160.93	62.14		

UNITS	Lb ft to kgm	Kgm to lb ft	UNITS	Lb ft to kgm	Kgm to lb ft
1	.138	7.233	7	.967	50.631
2	.276	14.466	8	1.106	57.864
3	.414	21.699	9	1.244	65.097
4	.553	28.932	10	1.382	72.330
5	.691	36.165	20	2.765	144.660
6	.829	43.398	30	4.147	216.990

CHAPTER 2

THE FUEL SYSTEM

2:1 Description

All models covered by this manual are fitted with a single barrel carburetor. Gasoline is stored in a tank at the rear of the vehicle. The fuel is drawn from the tank, through a system of pipes and hoses, by a mechanical engine-driven fuel pump.

On later models the fuel system is sealed in that a sealed filler cap is fitted to the tank and the tank is vented through a canister filled with carbon granules. The carbon granules absorb the gasoline vapor from the tank when the vehicle is stationary, and air passing into the air intake filter purges the gasoline vapor from the carbon granules when the engine is running.

The intake pipe from the fuel tank is combined with the fuel level transmitter as one unit and a strainer is fitted at the intake to ensure that only clean fuel passes to the fuel pump. A further filter is fitted to ensure that the fuel to the carburetor is perfectly clean and as a further precaution against dirt a filter screen is fitted at the inlet to the carburetor. On the earlier models the main fuel filter takes the form of a canister which is screwed onto the fuel pump but on later models a line filter is fitted between pump and carburetor.

An air cleaner is fitted to the intake of the carburetor. The air cleaner contains a renewable element which filters the air entering the engine. The unit also acts as a silencer in that it cuts out the roar from the carburetor intake.

Later models are fitted with a thermostatically controlled intake duct which controls the temperature of the air to the carburetor.

Later models of carburetor may be fitted with a limiter device to limit the range of adjustment of the idle mixture screw. This limiter must not be removed or damaged.

A typical late model fuel system is shown in **FIG 2:1**.

2:2 Maintenance

This is confined to renewing the filter elements at the correct intervals and checking that the throttle linkage operates fully and freely.

To ensure that the engine is not making excessive emission, beyond legal limits, it is most advisable to have the settings checked by a service agent using electronic test equipment. This checking should be carried out at regular intervals. Not only will this check

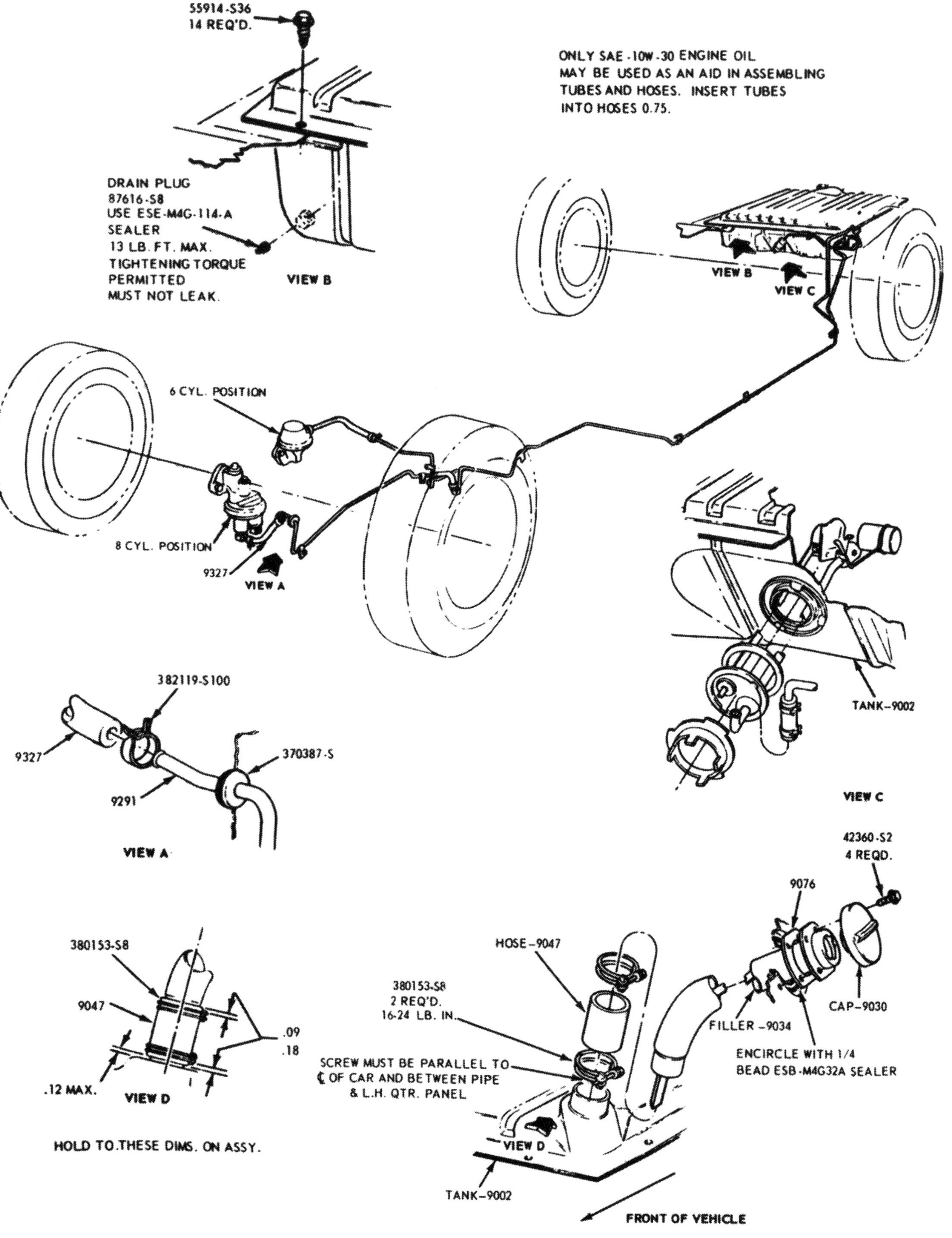

FIG 2:1 Late model fuel lines

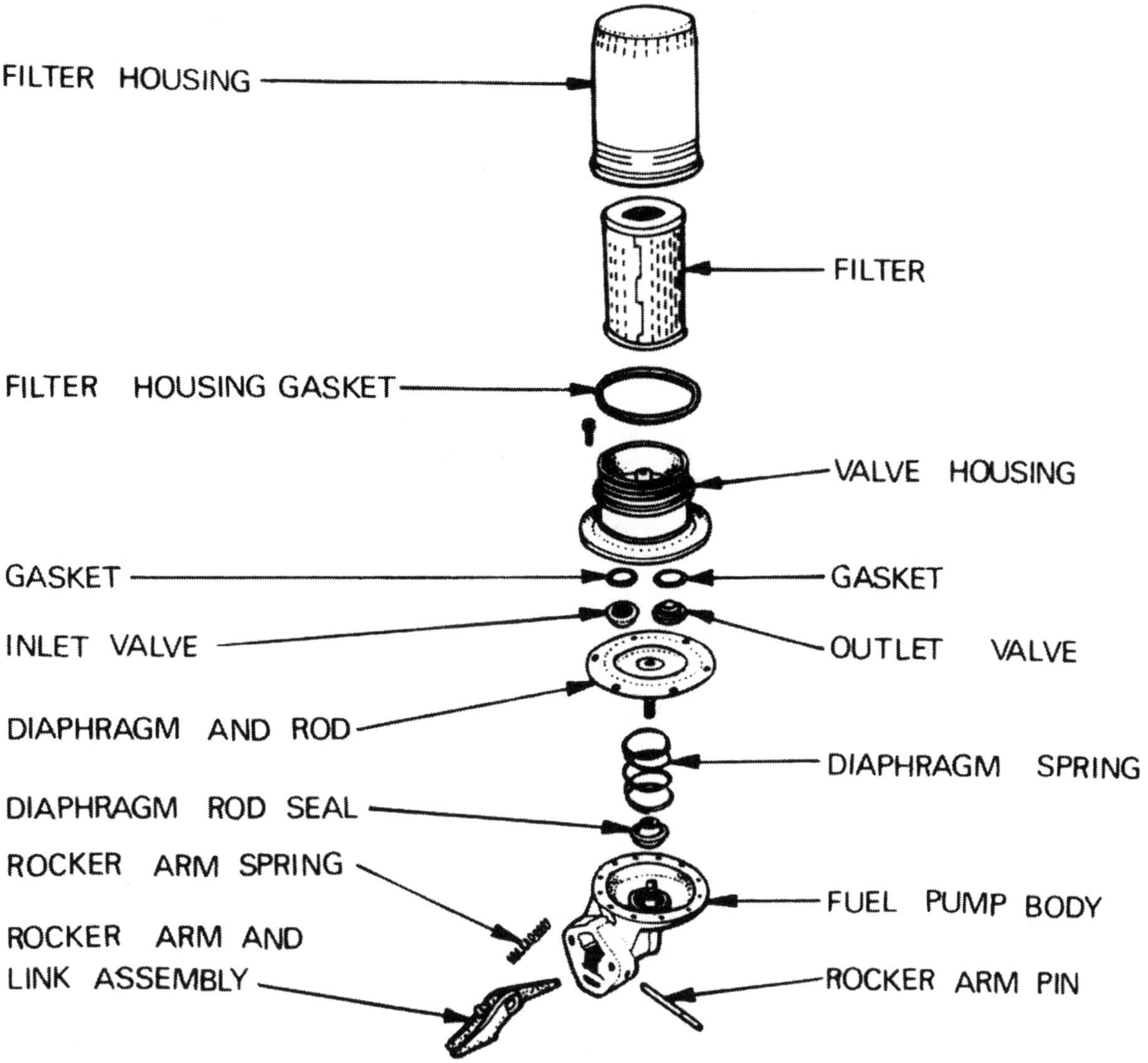

FIG 2:2 Early fuel pump and filter assembly

ensure that the engine is within legal limits for emission, but it will also ensure optimum economy and performance.

Fuel filter:

An early model pump is shown in **FIG 2:2.** With the pump installed on the engine, unscrew the filter housing from the pump—noting that some gasoline will spill out. Remove and discard the old filter element as well as the housing gasket. Be sure to obtain the correct replacement filter as those on V8 models are different. Lightly smear the new filter housing gasket with engine oil and position the gasket on the housing. Install the new element and tighten down the housing by hand until the gasket just contacts the pump. Tighten the housing by a further $\frac{1}{8}$ turn. Crank the engine until it starts and check for fuel leaks around the filter.

On later models an in-line filter is fitted and a typical installation is shown in **FIG 2:3.** The filter cannot be cleaned or dismantled and it must be renewed. Remove the air cleaner to gain access. Loosen the retaining clamp which secures the hose to the filter. Unscrew the filter assembly from the carburetor and disconnect it from the hose. Discard the old filter assembly and install a new one of the correct type. The hose is attached to the filter and then secured by crimping a new clamp over it.

On all models the fuel filter is renewed at intervals of 35,000 miles.

Air cleaner:

Only on the earlier models the element should be cleaned, by tapping it onto a firm surface or gently blowing through with an air-line. On earlier models clean the element at intervals of 6000 miles and renew it after 36,000 miles. On the later models the element is renewed at intervals of 24,000 miles. Under no circumstances should the element be washed in any form of solvent. The distances given are those for normal use and if the vehicle is used in very dirty or dusty conditions the servicing intervals should be reduced.

Before removing the element, remove the complete air cleaner assembly so as to ensure that dirt does not fall down into the carburetor. Disconnect all hoses to the unit before removing it. Undo the wing-type screws that

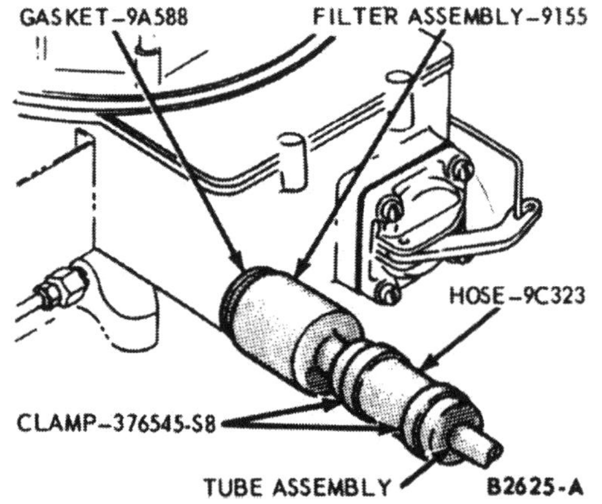

FIG 2:3 The line filter fitted to later models

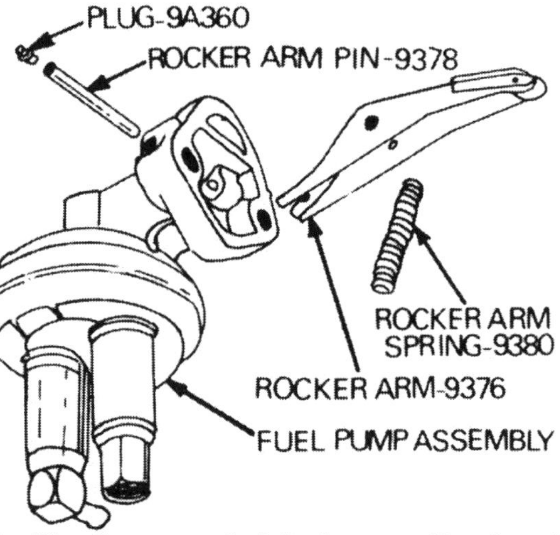

FIG 2:4 The Carter sealed fuel pump fitted to later models

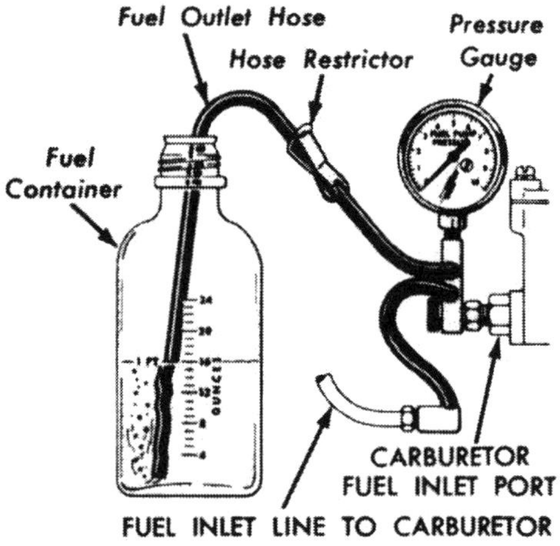

FIG 2:5 Checking the fuel pump performance

secure the assembly to the carburetor (and thermostatically controlled duct if fitted). Remove the cover and take out the element. Wipe out the container with a cloth moistened with gasoline to remove any dirt and oil inside it. Install the correct new element so that the marking FRONT faces forwards and the marking TOP is uppermost.

Crankcase ventilation:

At regular intervals check the system. On earlier models the regulator valve and filter in the oil filler cap can be cleaned by washing in gasoline or denatured alcohol but on later models they must be renewed without cleaning.

2:3 The fuel pump

The components of the fuel pump fitted to earlier models are shown in **FIG 2:2.** On later models a permanently sealed fuel pump, shown in **FIG 2:4,** is fitted in conjunction with a line filter.

Removal:

Free the clips and disconnect the inlet and outlet lines from the fuel pump. Plug the inlet pipe to prevent fuel siphoning through. Take out the screws that secure the pump to the crankcase, discarding the old gasket.

Install the pump in the reverse order of removal, using a new gasket lightly smeared on both sides with sealing compound. Be sure the fuel pump rocker arm is riding on the camshaft eccentric and hold the pump firmly against the crankcase while alternately tightening the attachment screws. Finally tighten both screws to the correct torque

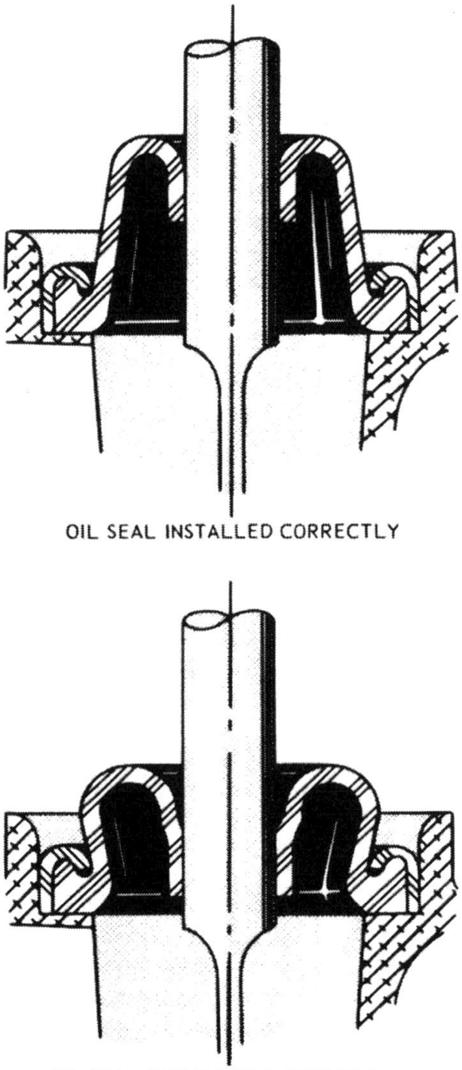

FIG 2:6 The fuel pump oil seal installation

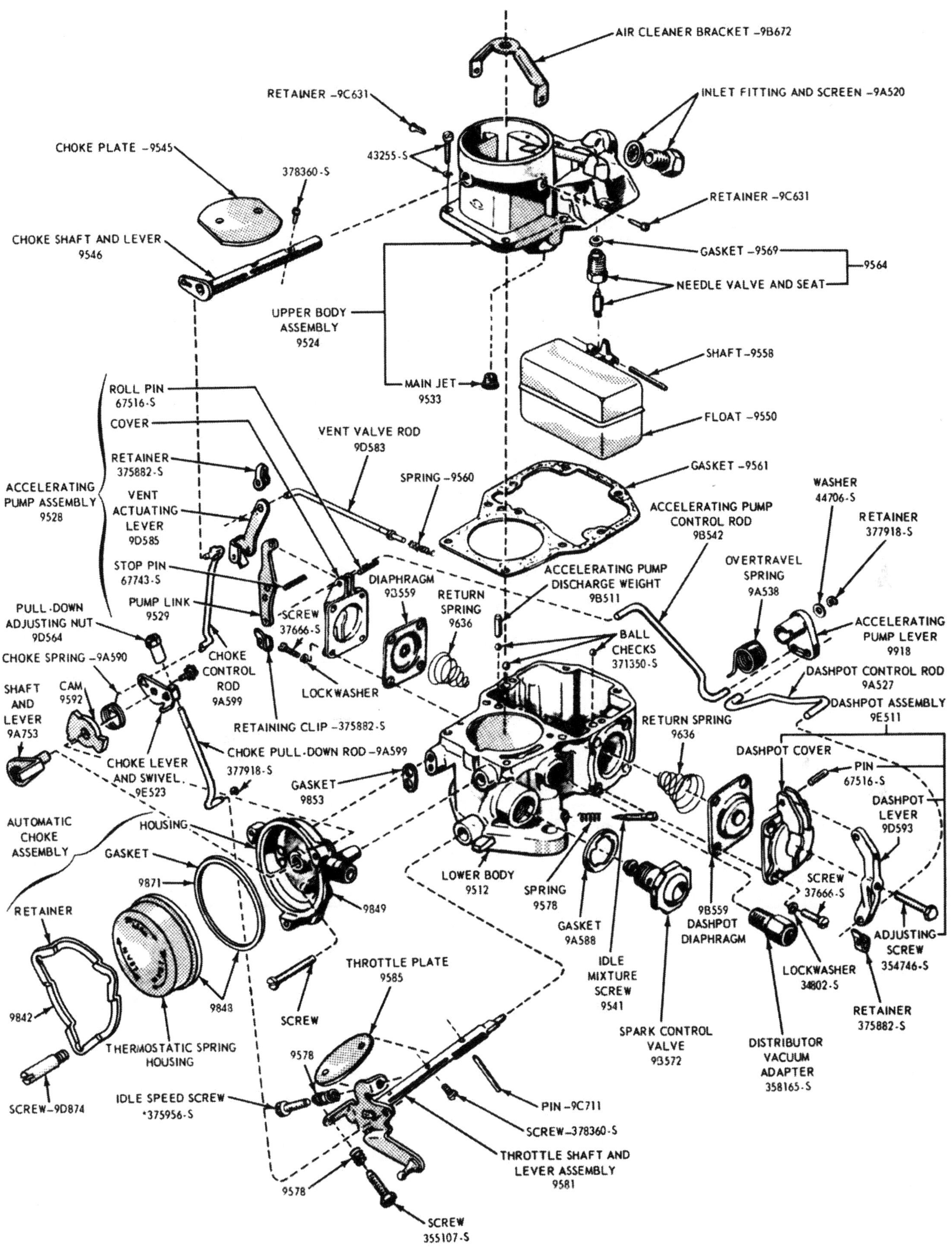

FIG 2:7 The components of the earlier carburetors

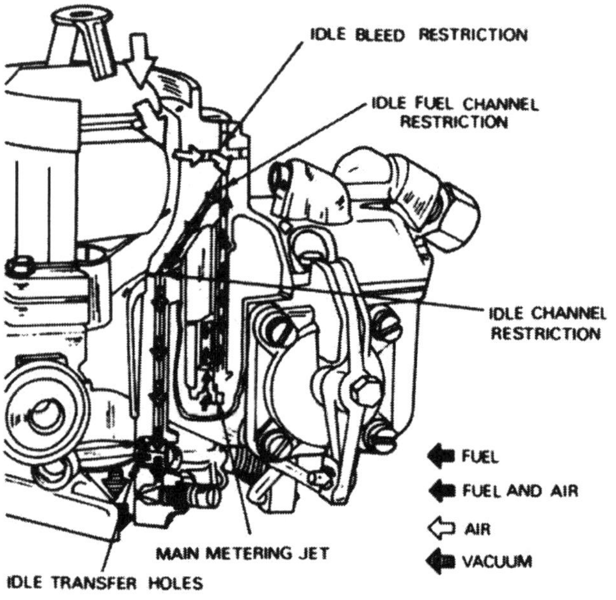

FIG 2:8 Typical idle fuel system

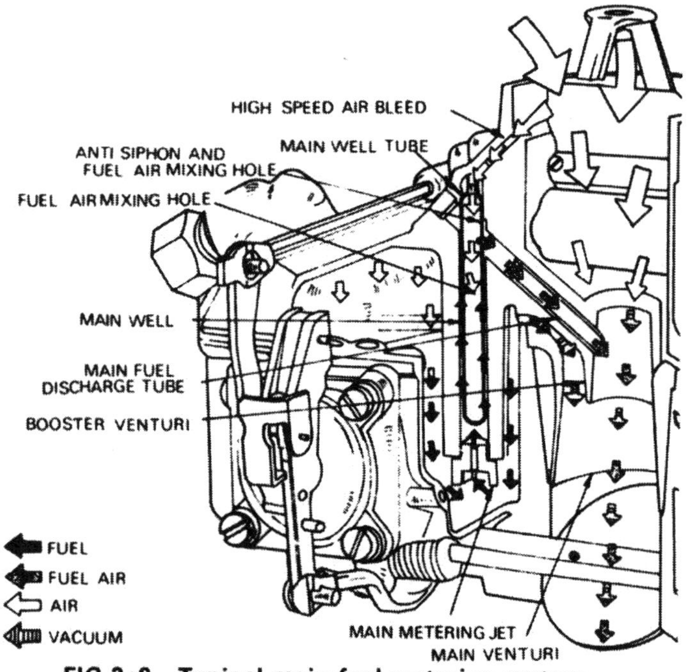

FIG 2:9 Typical main fuel metering system

load. Reconnect the fuel lines, using new clamps and crimping them as required. Start the engine and check for fuel and oil leaks from the pump. If the vehicle has stood for some time the engine will have to be cranked over sufficiently for the carburetor to prime before the engine will fire.

Testing:

The operation of the pump can be checked by cranking the engine over with the fuel line disconnected from the carburetor. Point the fuel line into a suitable container or well away from the engine. At every other turn of the engine there should be a good spurt of fuel from the line. Lack of fuel can be caused by a defective pump, blocked lines or strainer, or air leaks on the suction side of the pump.

For full and accurate tests a pressure gauge and graduated container must be fitted to the system, as shown in **FIG 2:5**. The fuel lines must be carefully and tightly connected to prevent leakage of fuel which could lead to a fire. Run the engine until it has reached its normal operating temperature. If the static pressure is low, renew the fuel filter and repeat the test. Both the pressure and flow tests are carried out with the engine at idling speed, and the system should be vented of air by briefly opening the restrictor on the hose before taking readings of pressure.

If the pressure is satisfactory, open the hose restrictor and note the time taken to deliver one pint of fuel. If the time taken to deliver one pint of fuel is excessive, stop the engine and disconnect the inlet line to the pump. Connect another line to the pump (suitable rubber tubing will do) and dip the free end into a clean container of fuel. Start the engine and vent air by opening the restrictor. Take great care not to allow the free fuel to slosh or spill because of the danger of fire. If the time taken to deliver one pint of fuel is now within limits, there is a restriction in the fuel lines of the vehicle. Check the strainer at the inlet in the tank and blow through the lines in a reverse direction with compressed air. If the pressure and volume flow are low in all cases, renew the pump.

Servicing:

On the later pumps, the plug can be levered out and the pin removed so that the rocker arm and its spring can be renewed. Pry out the old plug with a screwdriver and take the pressure of the rocker arm so that the pin can be shaken out. If the pin sticks, tap the casting gently against a firm surface to dislodge the pin. Reassemble the pump in the reverse order of disassembly, after cleaning and checking, using a new plug to hold the pin in place. This is all the servicing that can be carried out on later pumps and if there are any other defects, the complete pump must be renewed.

The components of the early pump are shown in **FIG 2:2**. There is an intermediate stage where the pump can be dismantled but a line filter is fitted instead of the canister type shown.

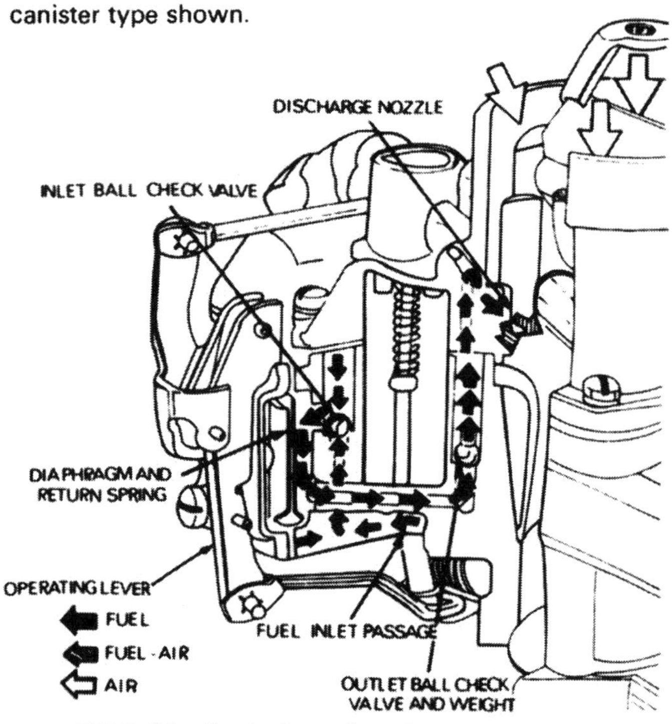

FIG 2:10 Typical accelerating pump system

The earlier type of pump can be dismantled by taking out the screws that secure the upper casting to the lower and driving out the pivot pin for the rocker. Before disassembling the pump, clean it externally and make light alignment marks across the castings. The valves and oil seal are held in place by staking, so if they are defective they should be removed and new parts fitted. Similarly renew the diaphragm if it has hardened or is split. If the oil seal has been removed, soak the new one in engine oil for two minutes and drive it back into place, preferably using the special drift No. 9350C. Install the diaphragm assembly and check that the oil seal is correctly positioned, as shown in **FIG 2:6**, and make sure that the diaphragm rod is correctly hooked onto the rocker arm. Install the top cover, making sure that the screws pass through the holes and do not tear the fabric. Tighten the screws in a diagonal sequence. Mask off all but one of the vent holes in the lower casting. Fit a rubber-tipped probe or hose to the remaining vent hole and apply a suction of 15 in Hg for a minimum of a minute. Release the vacuum and then apply it again for a further minute to ensure that the oil seal is correctly seated. If the valves have been removed, be sure that they are positioned as shown in the figure and then staked lightly back into place. Check that there is suction at the inlet and pressure at the outlet when the rocker arm is operated.

2:4 Carburetor operation

The operation of the Ford 1-V carburetor, fitted to earlier models, is described in this section. The carburetors fitted to later models operate on the same basic principles. The components of the Ford 1-V carburetor are shown in **FIG 2:7**.

Fuel from the fuel pump, under slight pressure, enters through the inlet in the cover and it passes into the float chamber through the needle valve assembly. When the correct level in the chamber is reached the float closes the needle valve, preventing any more fuel from entering.

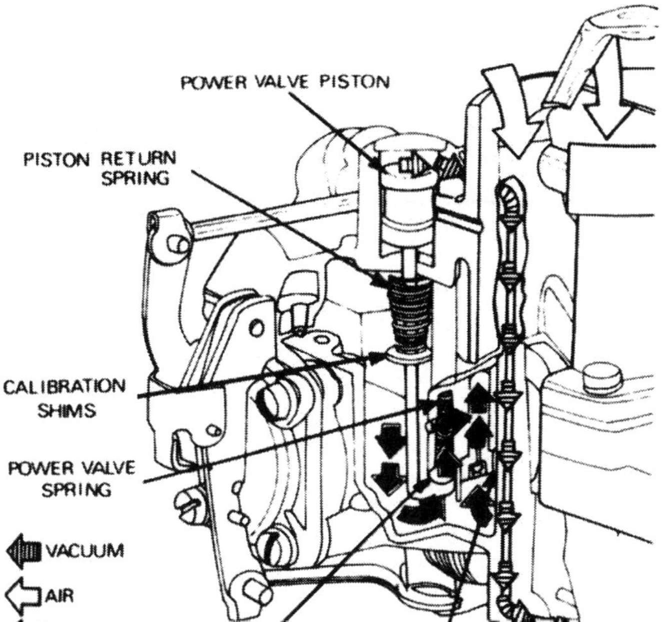

FIG 2:11 Power valve and enrichment device on earlier carburetors

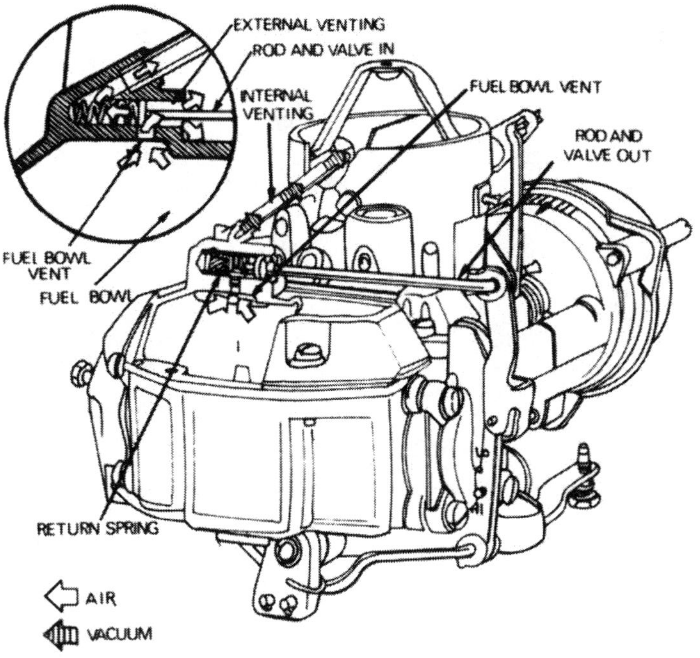

FIG 2:12 Autolite fuel bowl venting system

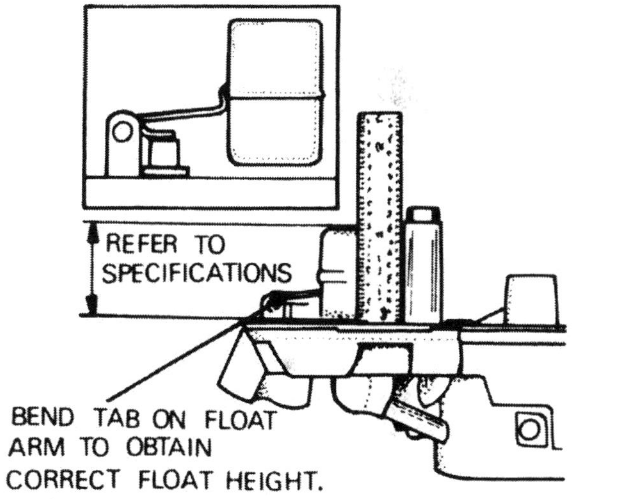

FIG 2:13 Typical method of setting fuel level. On some models dimension is measured between float and air horn

As fuel is used the float opens the needle valve, allowing fuel to enter to replenish the float chamber and keep the level constant within very narrow limits.

When an airflow is speeded up, as it passes through a restriction or venturi, its static pressure drops and use is made of this to draw the fuel through the calibrated jets.

At idle, the main suction is beside the throttle plate valve and fuel is drawn out through the ports beside the valve, as shown in **FIG 2:8**. When the throttle is fully closed all the mixture is drawn through the port controlled by the idle mixture screw and the ports above this one provide a smooth transition from idle to main jets.

As the throttle is opened the airflow through the booster venturi causes a drop in pressure at that point and fuel is then drawn through the main jet. The fuel is not drawn directly but air from the high speed air bleed is also drawn in and the two mix in the mixing tube to form a fuel/air emulsion which is then drawn in. This not only

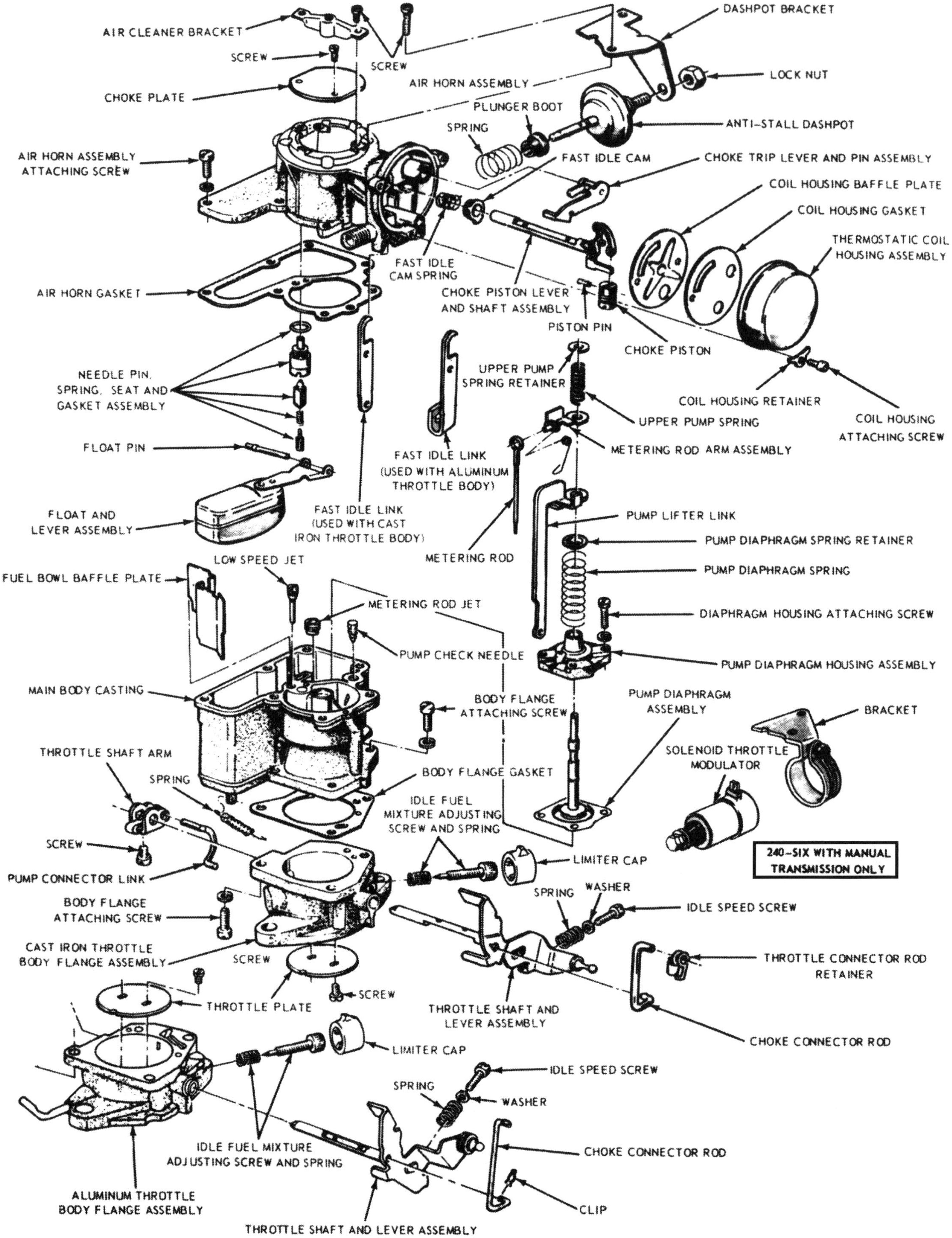

FIG 2:14 Carter carburetor components

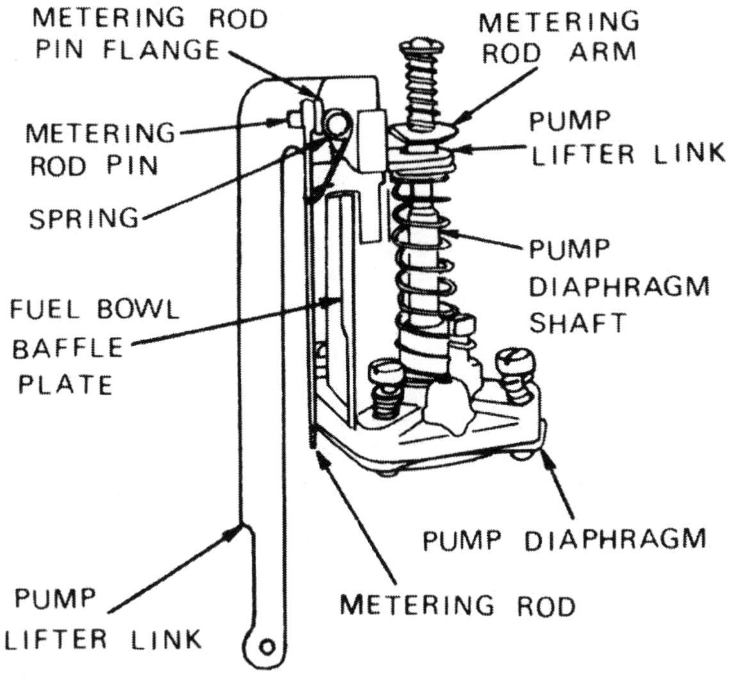

FIG 2:15 The Carter accelerating pump and lifter link assembly

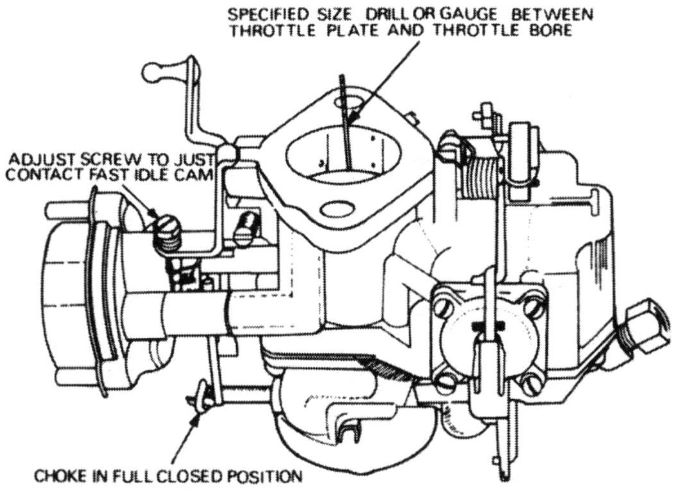

FIG 2:16 Autolite automatic choke fast idle cam adjustment

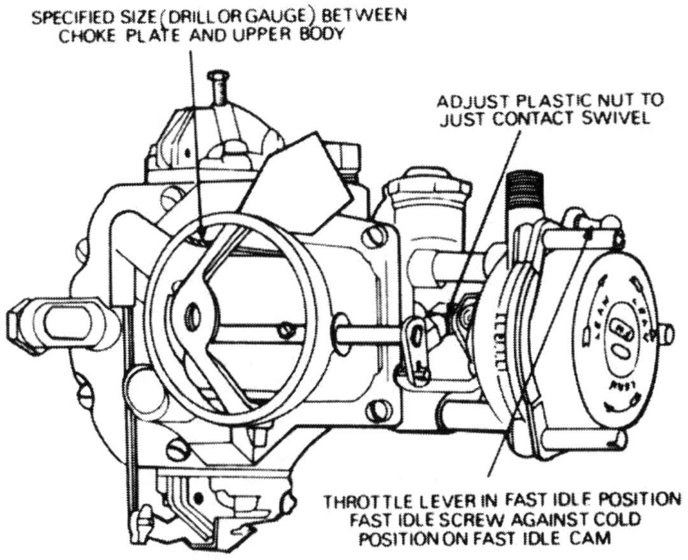

FIG 2:17 Autolite choke linkage adjustment

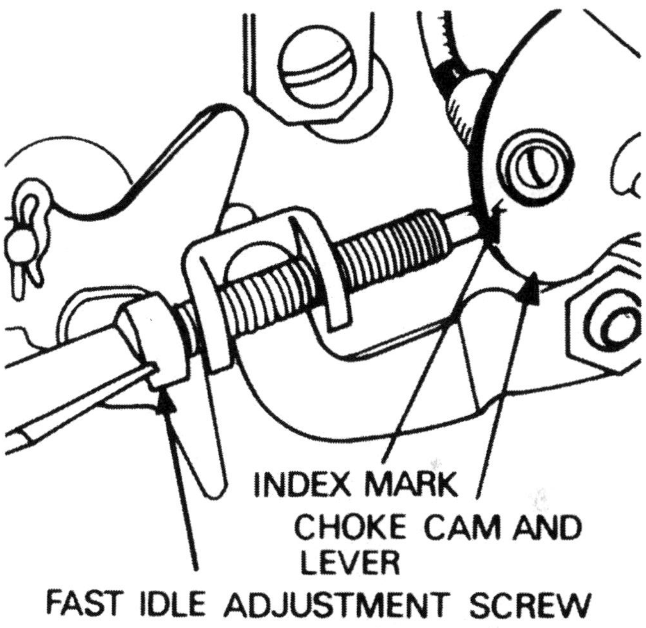

FIG 2:18 Setting the manual choke adjustment

ensures rapid mixing of the fuel with the air, but at high rates of fuel flow proportionally more air is drawn into the mixture tube reducing the overall mixture strength. The main metering system is shown in **FIG 2:9.**

On acceleration the fuel flow lags behind the airflow, because of the greater inertia of fuel, and if steps were not taken to correct this the mixture would become excessively weak. A diaphragm pump operated by the throttle linkage injects extra fuel as the throttle is opened. The system is shown in **FIG 2:10.** On the return stroke, fuel is drawn in from the float chamber through the inlet ball valve and the outlet ball check valve prevents air from being drawn in. The fuel is injected directly into the air stream.

For cold starts either a manually-operated or automatic choke can be fitted. A flap valve closes the inlet to the carburetor and as the engine cranks over a high suction is formed throughout the carburetor. This suction draws excess fuel through the jets to provide the very rich mixture required for starting. The choke flap valve is offset on its spindle, so that when the engine starts the extra air acts on the offset and partially opens the choke valve to reduce the very rich mixture to a rich mixture suitable for running a cold engine. The automatic choke is operated by a bi-metallic spring. When the spring is cold it exerts maximum torque and closes the choke valve. Air is drawn into the carburetor from the choke housing and this air is led from around the manifold. As the engine warms up so will the air being drawn through the choke housing and over the bi-metallic spring. The spring warms up and loses its tension, allowing the choke valve to open as the engine warms up. On later models a vacuum piston is fitted and when the engine is running this piston assists the choke plate valve to open. In case of hesitation or stall, the vacuum drops and the action of the vacuum piston is reduced so that the choke plate valve will automatically close slightly to enrich the mixture.

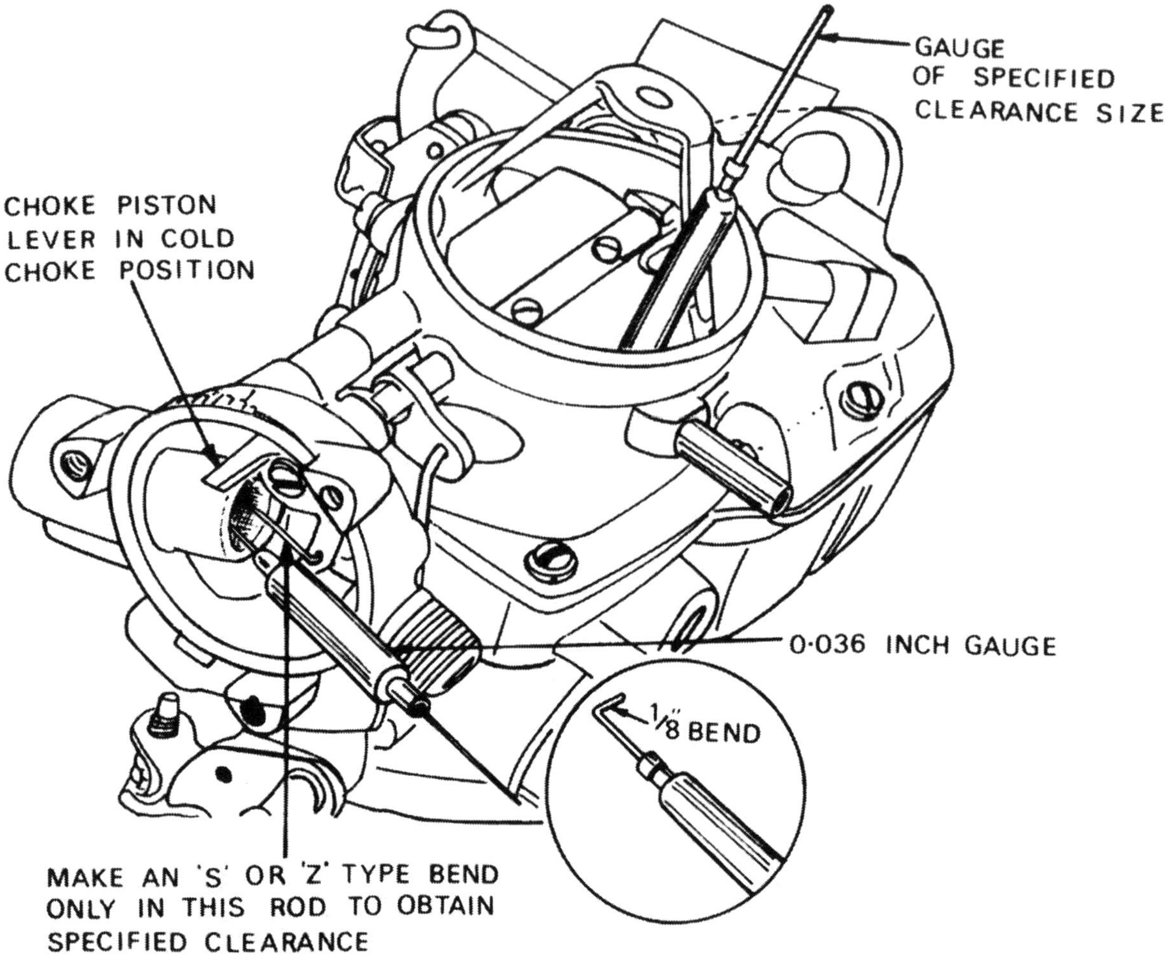

FIG 2:19 Adjusting the automatic choke on an Autolite 1101 carburetor

For full throttle operation, more fuel is required than is supplied by the main metering system. A power jet system, shown in **FIG 2:11**, is fitted to supply this extra fuel. Under part throttle operation the suction in the intake manifold is sufficient to hold up the piston of the system and prevent it from operating. At full throttle, when the vacuum is much reduced, the piston moves down and opens the power jet.

The float chamber vent is also operated by the throttle linkage, as shown in **FIG 2:12**. At idle the vent is open to the atmosphere and this ensures that there is no build-up of fumes when the engine is stationary. When the engine is running, any fumes are drawn into the intake and burnt with the mixture in the engine.

2:5 Carburetor faults

The most common cause of faults in the carburetor is dirt or gum blocking the jets or passages. Wear does not affect the carburetor excessively as there are few wear points. The most likely point of wear is throttle valve spindle or its bores in the body. Air leaking past at this point will cause defective idling. After very long service the accurately calibrated orifices of the jets will wear and alter the mixture but it should be noted that this wear can be rapidly speeded up if incorrect cleansing techniques are used.

If wear or dirt are causing faulty operation, the carburetor must be removed, taken apart and reassembled after cleaning and renewing worn parts. When reassembling carburetor use all new gaskets and seals—available in service kits.

No separate section on disassembling the carburetor is given but the faults are laid out in such an order that, by following this section through, the carburetor will be completely disassembled.

Faults that can be blamed on carburettion can also be caused by other components of the motor. Faulty ignition will cause misfiring and poor idle, so be sure that the ignition system is operating correctly and that the points and spark plugs are clean and correctly gapped (see **Chapter 3**). Faults in the crankcase breathing system and Thermactor emission control will also make it impossible to set a satisfactory idle. Check the air cleaner for cleanliness and correct operation. Check all vacuum systems and their hoses for leaks or poor connections. Similarly check the intake manifold joints for leaks. Any air leaking in through a vacuum system or leaking joint will not be passing through the carburetor and it will therefore greatly affect the idle.

Removal:

Take off the air cleaner to gain access to the carburetor. Disconnect the fuel line from the carburetor inlet and disconnect the vacuum line to the carburetor. On some models a water heated lower housing is fitted and if this is the case, the cooling system must be partially drained before disconnecting the hoses. Disconnect the automatic choke hoses, or operating cable if a manual choke is fitted. Free the return spring and disconnect the accelerator control rod. Evenly slacken the nuts and remove the carburetor.

Before installing the carburetor, check the carburetor intake manifold and spacer for damage to the flanges. High spots or nicks can be smoothed down by careful use of a fine file. Excessive distortion (usually caused by overtightening the attachment nuts) can be lapped down using a fine grade of emerycloth, or grinding paste, spread over plate glass.

Flooding or fuel starvation:

The appearance of flooding can be given by defective seals and gaskets which allow fuel to leak out, so check for this before taking further steps. The most likely cause of flooding or fuel starvation is a defective float needle valve assembly. Remove the air cleaner and disconnect the control rods for the choke and vent valve. Take out the screws which secure the upper body (air horn) and lift it off. Use a suitable wrench and unscrew the needle valve assembly. If the needle can be removed from the valve, do this and check the tapered seat of the needle. If the tapered seat is worn to a step, renew the complete needle valve assembly. Hold the needle in place with light finger pressure and try to blow through the valve. If the valve leaks or the needle sticks in the closed position the complete valve assembly must be renewed.

Remove the float from the fuel bowl. Check that the float is undamaged and that it does not contain any fuel.

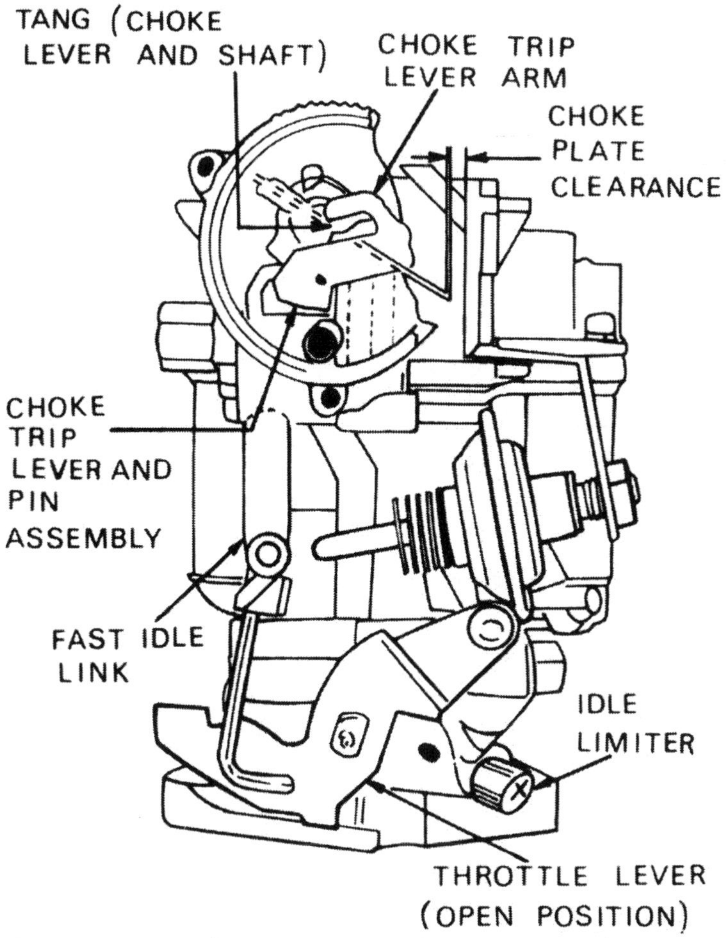

FIG 2:20 Adjusting the choke linkage on a Carter carburetor

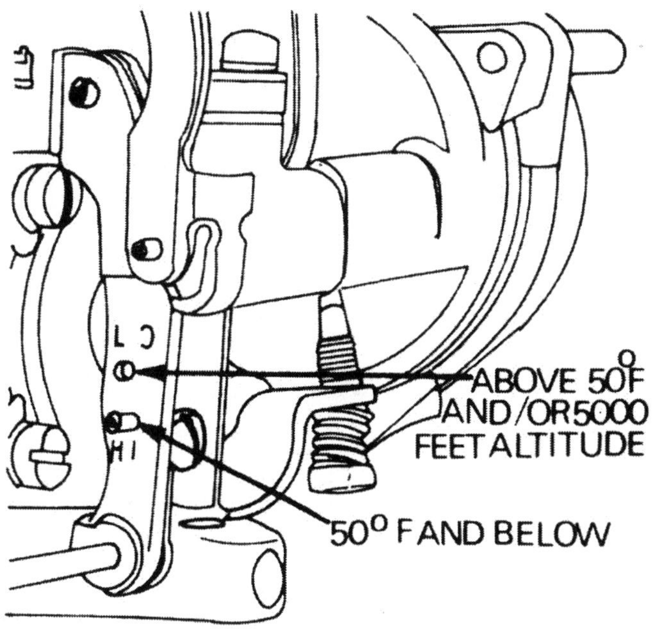

FIG 2:21 Accelerating pump settings

A defective float must be renewed. Only in an emergency should any fuel be driven out by alternate heating and cooling followed by soldering up of the leak.

Install the parts in the reverse order of removal, after checking the filter screen fitted to the inlet of the carburetor.

Fuel level:

The method of checking the fuel level is shown in **FIG 2:13.** On later models the distance between the float and air horn is checked and a gauge of suitable thickness can be made up to slip between float and air horn. If the setting is incorrect, carefully bend the float arm to adjust. Note that the air horn must be removed from the car-carburetor and then held in an inverted position.

Jet cleaning:

Dirt in the jets and bores will cause erratic running. The fuel is well filtered on its way to the carburetor, so dirt should not reach the jets if the fuel filters are cleaned and changed at the correct intervals. Remove the air horn and float assembly as described earlier. It is essential to use well-fitting tools when disassembling the carburetor as the jets and parts are made of soft metal which will easily become damaged by the use of poorly fitting tools. Remove each jet in turn. It is advisable to clean each jet in turn as several of them are very similar in appearance and if they are all removed there is a danger of fitting the jets back into the incorrect positions. If all the parts are being removed, store them in several small containers to prevent intermixing. Special cleansing fluids are available for cleaning the carburetor components but usually clean gasoline or denatured alcohol can be used to soften and remove deposits. Harsher solvents such as trichlorethylene or carbon tetrachloride can be used, but these must not be allowed to contact seals and gaskets. Swill the jets and parts in solvent and then blow through them, preferably with compressed air. The bores and ports in the castings should be blown through in a reverse direction to clear them. **Do not poke wire or**

bristles through the jets or ports as this will wear and deform the accurately calibrated orifices. Renew any jets that are damaged or have worn orifices.

Brush and scrape out any sediment from the fuel bowl and then flush with clean gasoline or suitable solvent. Reassemble the carburetor in the reverse order of dismantling.

If there are excessive amounts of dirt or water in the carburetor, check the fuel filter. If there is persistent dirt, disconnect the fuel lines after draining the fuel tank. Blow through the lines in a reverse direction with compressed air and fill the tank with clean gasoline after reconnecting the lines. In extreme cases it will be necessary to remove and swill out the tank or even renew it.

Poor acceleration or hesitation:

Remove the air cleaner and operate the throttle linkage. There should be a good spurt of fuel.

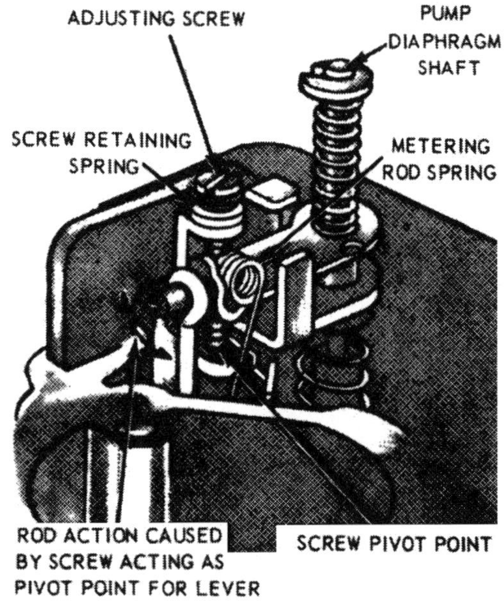

FIG 2:22 Adjusting the metering rod on a Carter carburetor

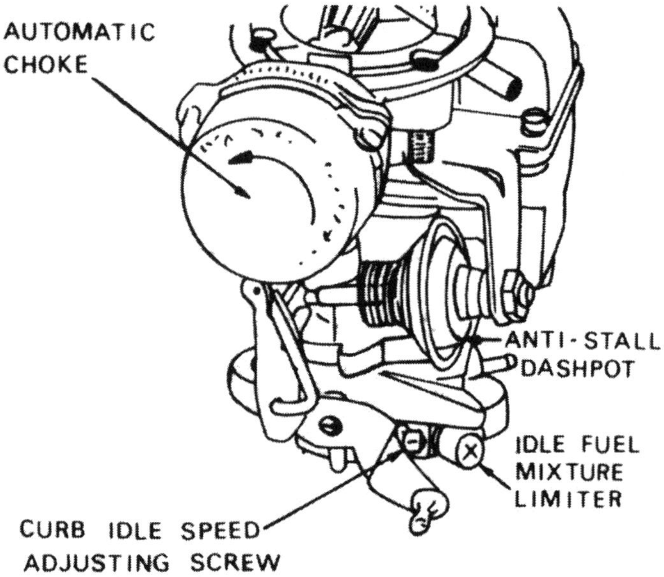

FIG 2:23 The adjustment points on a Carter carburetor

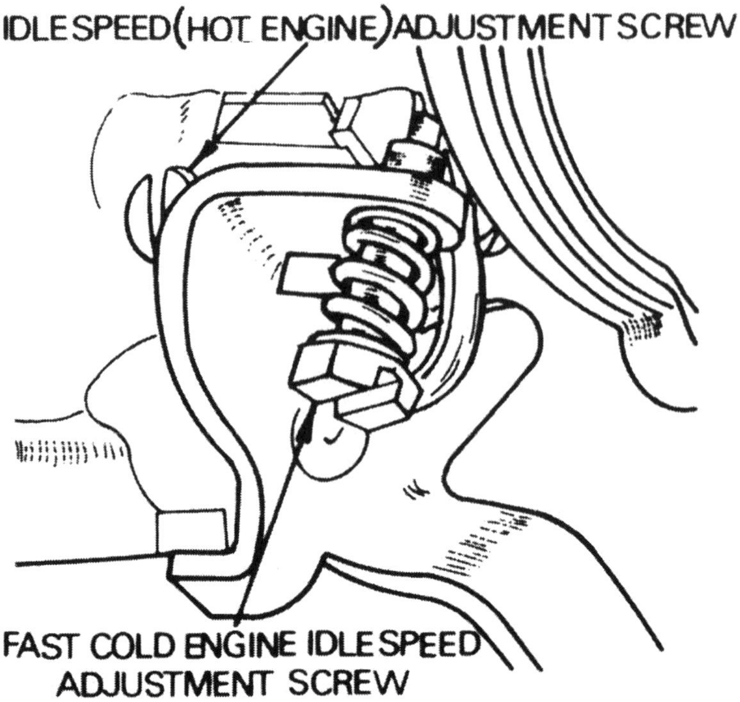

FIG 2:24 The idle speed adjustments on an Autolite carburetor

The components of the Carter YF 1-V carburetor fitted to later models are shown in **FIG 2:14.** Earlier carburetors of whatever designation are all very similar and their components are shown in **FIG 2:7.**

Remove the check valves and weight for the accelerator pump, after taking out the screw that holds the weight in place, and check that they are operating correctly and not blocked or damaged. Disconnect the linkage and remove the accelerating pump diaphragm by taking out the attachment screws. On the Carter carburetor the accelerating pump and metering needle are removed as a sub-assembly (shown in **FIG 2:15**). Check the diaphragm and all springs, renewing parts as required.

Reassemble the carburetor in the reverse order of dismantling.

Idle mixture screw:

On later models the range of adjustment is limited by a limiter so it is less likely that the taper on the needle will become damaged. On earlier models, if it is found difficult to set the idle mixture satisfactorily, with small movements of the screw having disproportionate effects, stop the engine and remove the idle mixture screw. If the tapered seat has worn to a step, discard the screw and install a new one. **Never tighten the idle mixture screw down firmly onto its seat as this is the main cause of wear on the tapered portion.** If an idle mixture screw fitted with a limiter is damaged, **do not attempt to renew the screw but take the car to a service agent so that the idle mixture strength can be checked with exhaust gas analyser equipment and the limiter correctly installed.**

Throttle valves and shafts:

Wear at this point will seriously affect the idling characteristics. On the Carter carburetors a separate housing is used for the throttle valve and throttle shaft.

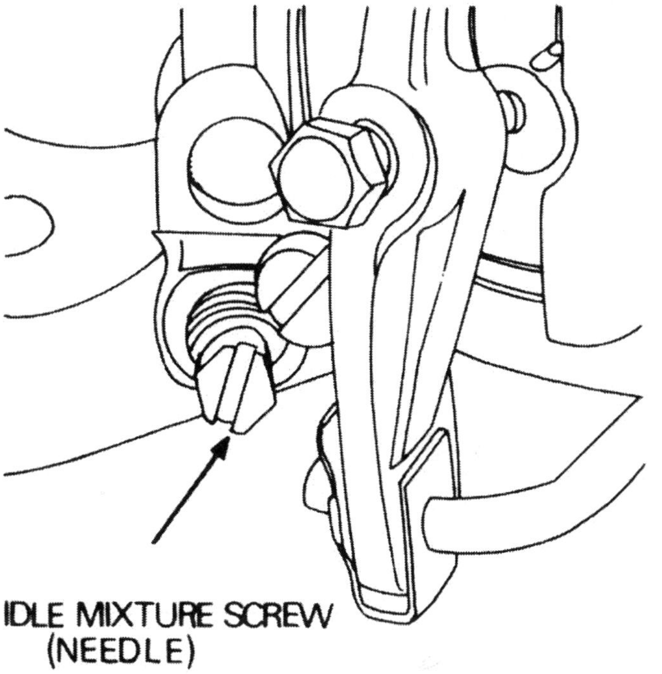

FIG 2:25 The idle speed adjustments on an Autolite carburetor

Care must be taken in ordering the correct replacement part as some models have light-alloy housings while others have cast iron housings. On all models, free the linkage parts from the shaft and take out the screws that secure the throttle valve plate to the shaft. Slide the valve plate out of its shaft and then withdraw the shaft.

Try installing a new shaft only but if there is still excessively play the housing must be renewed. On Carter models only the throttle housing need be renewed but on Autolite and Ford models a complete new carburetor will be required.

Install the throttle shaft and slide the plate valve back into position, noting that the notch must align with the idle ports. Tighten the screws which secure the plate sufficiently to prevent the plate from moving. Operate the shaft and gently move the plate round until the valve opens and closes freely without binding in the bore of the housing. Fully tighten the screws and check that the valve still moves freely throughout its range. Lock the screws by staking them. When staking the screws, support the shaft on a block of wood or metal so that it does not become damaged or bent.

2:6 Carburetor static settings

These are set on manufacture and do not normally require altering. They should be checked and adjusted as required after dismantling the carburetor or if a fault is found in operation.

Ford 1-V and early Autolite choke adjustments:

Be sure that the correct index marks align if an automatic choke is installed.

Insert a twist drill or gauge of the correct diameter (see **Technical Data**) between the throttle plate and the side of the bore, as shown in **FIG 2:16.** Hold the throttle valve closed and adjust the fast idle screw until it just touches the fast idle cam. Once this adjustment has been set, remove the original gauge and insert a gauge of the correct thickness between the choke valve and bore, holding the choke valve closed against the gauge as shown in **FIG 2:17.** Close the throttle valve until the fast idle screw just touches the fast idle cam and adjust the plastic nut until it just contacts the swivel on the choke lever assembly.

On models equipped with a manually-operated choke, run the engine until it has reached its normal operating temperature. Manually rotate the choke cam and lever until the fast idle screw is opposite to the index mark, shown in **FIG 2:18,** and then adjust the fast idle screw to give the correct fast idle speed.

Autolite 1101 choke plate adjustment:

Remove the air cleaner and choke housing so that the carburetor appears as shown in **FIG 2:19.** The thicknesses of the gauges required are given in **Technical Data.** Bend one gauge as shown in the insert in the figure, and insert it between the lower edge of the piston slot and the upper edge of the righthand slot in the housing. Pull the choke piston lever counterclockwise until the gauge is held in place and keep light pressure on the linkage to hold the gauge in place. Open the throttle slightly to ensure that the fast idle screw is not

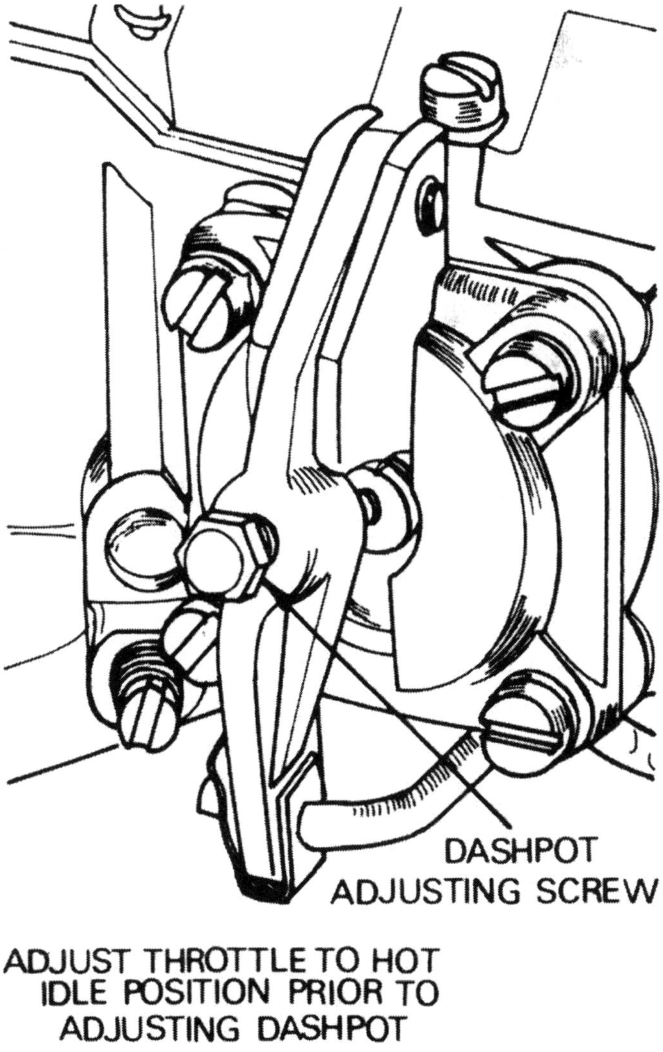

FIG 2:26 The anti-stall dashpot adjustment point on an Autolite carburetor

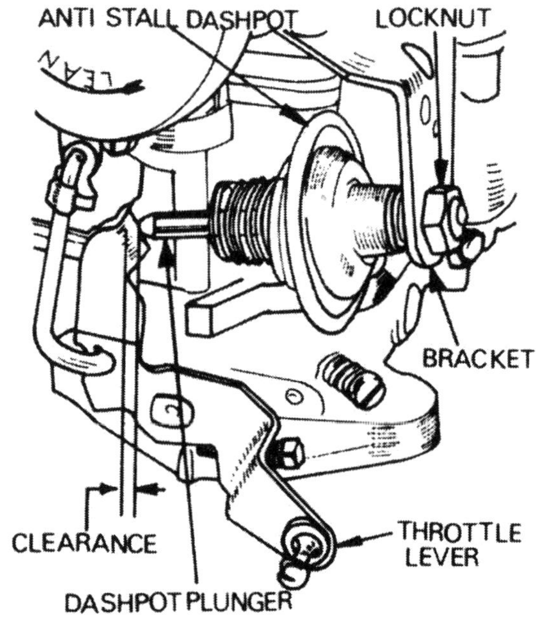

FIG 2:27 Anti-stall dashpot adjustments on a Carter carburetor

contacting the cam. Make a careful S- or Z-type bend in the link between the choke piston and piston lever, to set the correct clearance of the choke plate valve. Be sure that the bends in the link are not made so that the link fouls during operation.

Remove the gauges and install the choke housing. Be sure that the eye of the bimetallic spring picks up on the tang of the lever. Turn the housing until the correct index marks align and tighten the screws which secure it. Once these adjustments have been completed, the running adjustments must also be reset.

Carter YF 1-V choke adjustments:

Remove the air cleaner and the choke spring housing. Hold the throttle plate valve fully open and close the choke plate valve as far as possible without using force. Measure the clearance between the choke plate valve and the bore of the carburetor with a suitable gauge or drill shank. If the clearance is not within specification, adjust by bending the arm on the choke trip lever shown in **FIG 2:20.** Bending the arm downwards will increase the clearance and bending upward will decrease it.

Install the thermostatic housing and set it to the correct index mark before fully tightening the screws. Check the running adjustments.

Accelerating pump:

The two positions in which the pump can be set are shown in **FIG 2:21** and the figure is self-explanatory.

Autolite fuel chamber vent:

This should only be adjusted after the accelerating pump has been adjusted and the idle settings made. With the throttle closed to curb idle, the notch on the vent operating rod should be even with the open end of the vent. If the adjustment is incorrect, carefully bend the end of the operating rod where it contacts the accelerating pump lever until the setting is correct.

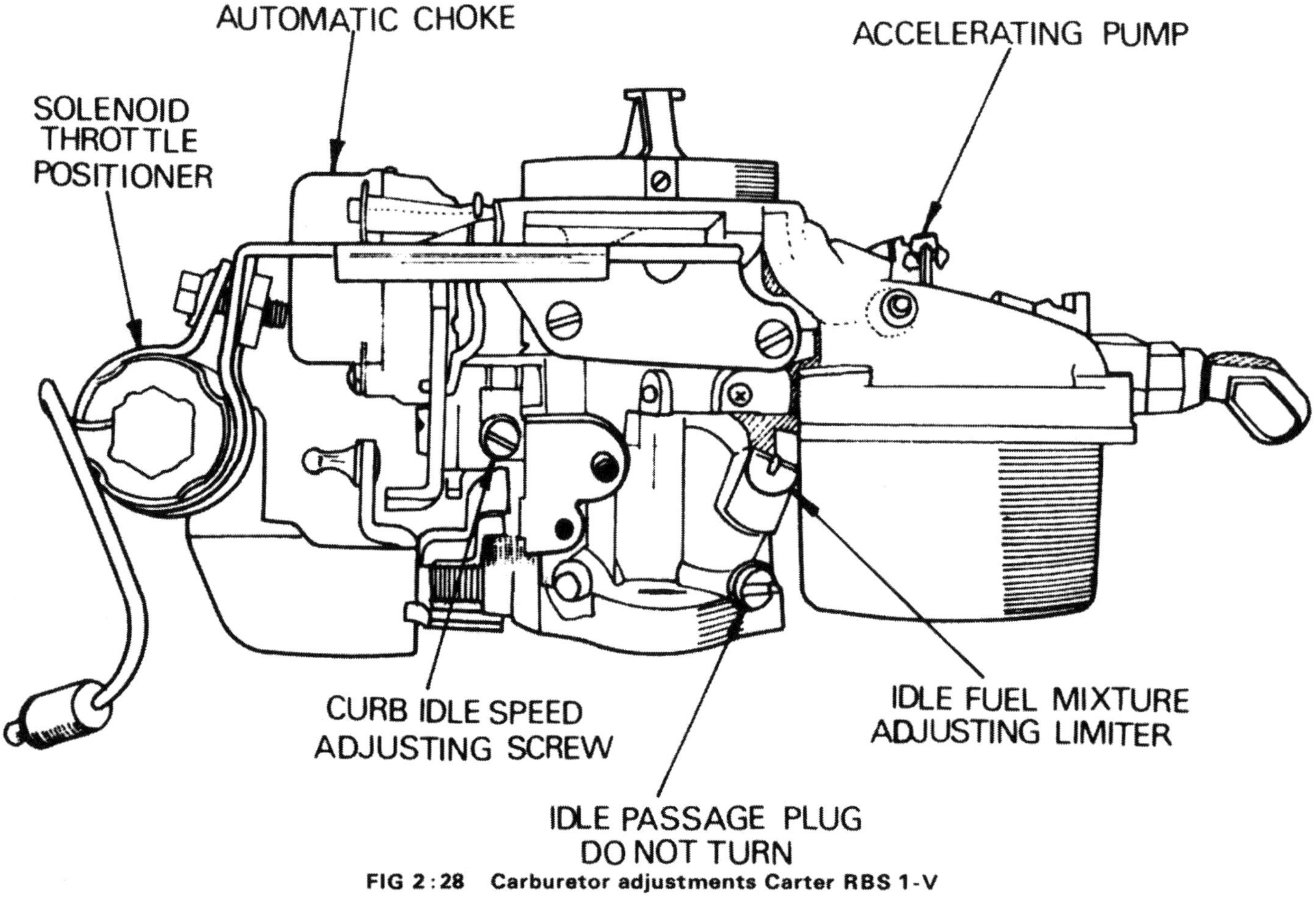

FIG 2:28 Carburetor adjustments Carter RBS 1-V

Carter metering needle:

The method of adjustment is shown in **FIG 2:22.** Remove the air horn assembly. Back out the curb idle adjustment screw until the throttle plate valve is fully closed in the bore. Press down on the diaphragm shaft until the metering rod arm contacts the lifter link at the diaphragm stem, and then turn the adjusting screw until the metering rod just bottoms in the casting. Release the diaphragm shaft and turn the adjusting screw one further turn clockwise for the correct setting. Install the air horn.

Dashpot:

This is only installed on models fitted with automatic transmission and it damps the rate of closure of the throttle near idling to prevent the engine from stalling. Adjustments are made after the idling adjustment.

2:7 Idling adjustments

Before attempting to set the idle, check that all vacuum systems are operating correctly with no leaks in the lines. Similarly check that the ignition settings are correct.

On later carburetors, limiter stops are installed to ensure that the idle mixture screw cannot be set beyond its range of minimum engine emission. **If a limiter is damaged or comes off, it must be renewed by a service agent using electronic equipment to ensure that it is set within the correct limits.**

On models fitted with limiter caps, turn the idle mixture screw as far counterclockwise as the limiter will allow. On other models gently tighten the screw until it just contacts its seat and then back it out by $1\frac{1}{2}$ turns. Unscrew the curb idle speed adjusting screw until it is clear of its stop and the throttle plate valve fully shut. Tighten the curb idle speed screw until it just touches the stop and then turn it a further $1\frac{1}{2}$ turns so that the throttle plate valve opens slightly. These settings will allow the engine to start and run. The adjustment points on the Carter carburetor are shown in **FIG 2:23** while the curb idle screw on an Autolite is shown in **FIG 2:24** and the idle mixture screw in **FIG 2:25.**

Start the engine and run it until it has reached its normal operating temperature. Check the choke plate valve to see that it is fully open. Install the air cleaner again after checking the choke. Running at 1500 rev/min for 20 minutes will allow all temperatures to stabilize.

Connect a tachometer to the engine and use the curb idle adjusting screw to set the correct idle speed (see **Technical Data**). On vehicles with less than 50 miles running (or a new engine) the curb idle speed should be set 25 rev/min below specification to allow for the increase as the engine 'loosens'.

Once the correct idle speed has been set, adjust the idle mixture screw to give the smoothest possible idle. If a vacuum gauge is connected to the intake manifold the idle mixture screw should be set to give the highest vacuum reading. Turn on the headlamps to load the alternator and make small final adjustments to the idle adjustment screw to set the correct curb idle speed. It is most advisable to have the settings checked by a service agent using exhaust gas analyser equipment to ensure that the emissions are within legal limits.

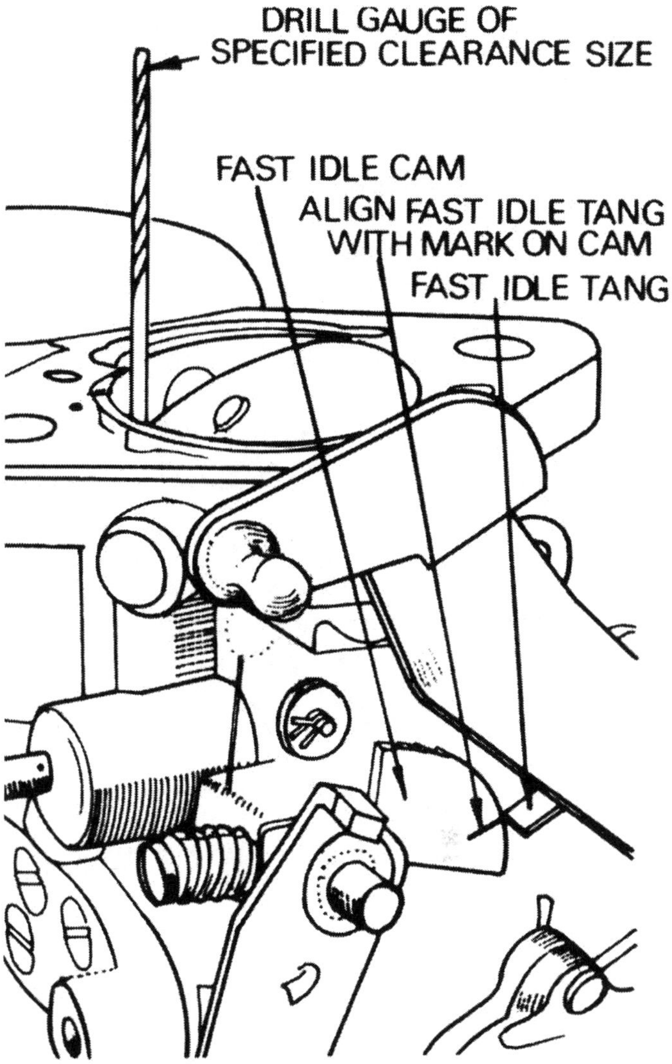

FIG 2:29 Throttle plate clearance

Autolite dashpot adjustment:

The adjustment position is shown in **FIG 2:26.** Turn the screw until it is clear of the dashpot and then tighten it until it just touches the dashpot. Turn it clockwise the correct number of turns (see **Technical Data**) after the initial contact.

Carter dashpot adjustment:

The typical adjustment method is shown in **FIG 2:27.** Hold the throttle closed and depress the plunger of the dashpot with a screwdriver. Measure the clearance shown. If the clearance is incorrect, slacken the dashpot locknut and turn the dashpot in or out until the clearance is correct. Tighten the locknut when the setting is correct.

2:8 Carter carburetor type RBS 1-V

This carburetor is fitted with 250 cu inch engines and has a number of differences from the types described earlier in this chapter. A diaphragm controlled step-up type metering rod controls the fuel supply and the accelerator pump is spring actuated. A vacuum piston type automatic choke is used and a pneumatic dashpot is used to control the rate of throttle return to idle. In some installations a solenoid throttle positioner is fitted.

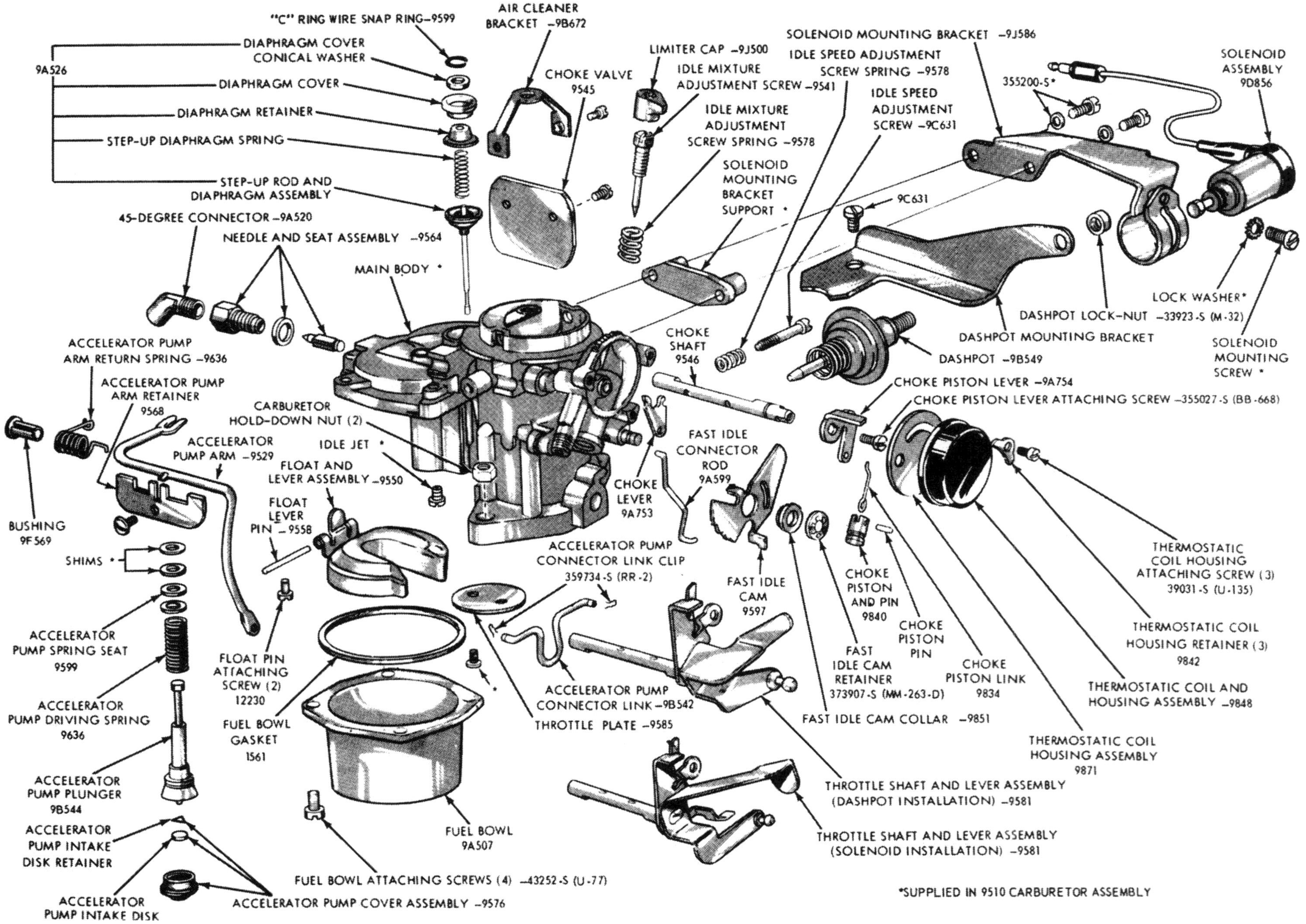

FIG 2:30 Components of the Carter carburetor type RBS 1-V

Idle speed and fuel mixture:

As with other carburetors for use in the U.S. a limiting device is fitted to the idle fuel mixture control to prevent unauthorized adjustment with the possibility of infringing Federal regulations on pollution.

A plastic idle limiter cap is fitted to the head of the idle fuel mixture control screw shown in **FIG 2:28**, and any adjustment must be within the range of this limiter. Provided that all other systems are operating correctly, satisfactory results should be obtained as follows:

Set the idle fuel mixture screw and limiter cap as far anticlockwise as possible. In the case of those vehicles not equipped with EEC, turn the screw in until it is **lightly** seated and then out for $1\frac{1}{2}$ turns.

Back off the curb idle speed adjusting screw until the throttle plate seats in the throttle bore, ensuring that the dashpot or solenoid positioner (when fitted) does not interfere with the throttle lever. It may be necessary to loosen the dashpot or solenoid to allow the throttle plate to seat correctly.

Turn the idle speed screw in to the point where it just contacts the screwstop on the throttle shaft and lever assembly and then a further $1\frac{1}{2}$ turns to establish a basic idle setting.

The parking brake must be set while making adjustments with the engine running and where a vacuum release brake is fitted the vacuum line must be removed from the vacuum release power unit, the line plugged and then the parking brake set. The vacuum power unit must be deactivated in order to keep the brake engaged when running in DRIVE as is necessary with automatic transmissions.

Bring the engine up to its full working temperature, ensure that the choke plate is fully open and the hot-idle compensator (when fitted) is fully seated. Turn the headlights to main beam and the air conditioner to OFF.

Adjust the curb idle speed to 500-600 rev/min and note that tachometer readings must be taken with the air cleaner in position, then turn the idle mixture screw inwards to obtain the smoothest idle possible within the range of the idle limiter.

Fast idle:

This adjustment which depends upon the setting of the fast idle linkage and a specified throttle plate clearance, requires the removal of the carburetor.

Fully close the choke plate and, referring to **FIG 2:29**, align the fast idle tang on the throttle lever with the scribed line on the cam.

The choke connector rod end should be at the top end of the slot in the fast idle cam and, if necessary, the choke connector rod should be bent at the offset position to obtain this setting.

Still with the tang aligned with the scribed index mark insert a .04 inch (1 mm) drill between the throttle plate and the bore as shown. This should be a close fit and, if necessary, the tang on the throttle lever should be bent to obtain the correct clearance.

The components of the RBS 1-V carburetor are shown in the exploded view of **FIG 2:30** to assist those who might wish to dismantle it.

2:9 Trouble shooting guide

(a) Leakage or insufficient fuel delivery

1 Air vent to fuel tank blocked
2 Fuel lines partially blocked
3 Fuel filters partially blocked
4 Defective fuel pump
5 Defective float in fuel bowl
6 Defective needle valve assembly in fuel bowl
7 Loose hose connections
8 Loose carburetor parts or defective gaskets

(b) High fuel consumption

1 Carburetor incorrectly adjusted
2 Dirty air cleaner
3 Automatic choke incorrectly set or failing in operation
4 Fuel leakage
5 Fuel level in fuel bowl too high
6 Loose or worn jets
7 Incorrect jets installed
8 Brakes binding
9 Tires under-inflated
10 Engine running too hot
11 Vehicle regularly overloaded

(c) Idling speed too high or difficult to set

1 Defective air cleaner
2 Defective Thermactor
3 Air leaks at manifold
4 Air leaks through vacuum operated systems
5 Carburetor incorrectly set
6 Throttle linkage sticking
7 Defective automatic choke
8 Worn carburetor throttle plate or shaft
9 Worn idle mixture screw tapered seat

(d) No fuel delivery

1 Fuel bowl needle valve jammed closed
2 Vent in fuel tank blocked
3 Fuel filters, strainer or screen blocked
4 Fuel lines blocked
5 Defective fuel pump
6 Bad air leak on suction side of pump
7 No gasoline in tank

Inches	Decimals	Millimetres	Inches to Millimetres: Inches	Inches to Millimetres: mm	Millimetres to Inches: mm	Millimetres to Inches: Inches
1/64	.015625	.3969	.001	.0254	.01	.00039
1/32	.03125	.7937	.002	.0508	.02	.00079
3/64	.046875	1.1906	.003	.0762	.03	.00118
1/16	.0625	1.5875	.004	.1016	.04	.00157
5/64	.078125	1.9844	.005	.1270	.05	.00197
3/32	.09375	2.3812	.006	.1524	.06	.00236
7/64	.109375	2.7781	.007	.1778	.07	.00276
1/8	.125	3.1750	.008	.2032	.08	.00315
9/64	.140625	3.5719	.009	.2286	.09	.00354
5/32	.15625	3.9687	.01	.254	.1	.00394
11/64	.171875	4.3656	.02	.508	.2	.00787
3/16	.1875	4.7625	.03	.762	.3	.01181
13/64	.203125	5·1594	.04	1.016	.4	.01575
7/32	.21875	5.5562	.05	1.270	.5	.01969
15/64	.234375	5.9531	.06	1.524	.6	.02362
1/4	.25	6.3500	.07	1.778	.7	.02756
17/64	.265625	6.7469	.08	2.032	.8	.03150
9/32	.28125	7.1437	.09	2.286	.9	.03543
19/64	.296875	7.5406	.1	2.54	1	.03937
5/16	.3125	7.9375	.2	5.08	2	.07874
21/64	.328125	8.3344	.3	7.62	3	.11811
11/32	.34375	8.7312	.4	10.16	4	.15748
23/64	.359375	9.1281	.5	12.70	5	.19685
3/8	.375	9.5250	.6	15.24	6	.23622
25/64	.390625	9.9219	.7	17.78	7	.27559
13/32	.40625	10.3187	.8	20.32	8	.31496
27/64	.421875	10.7156	.9	22.86	9	.35433
7/16	.4375	11.1125	1	25.4	10	.39370
29/64	.453125	11.5094	2	50.8	11	.43307
15/32	.46875	11.9062	3	76.2	12	.47244
31/64	.484375	12.3031	4	101.6	13	.51181
1/2	.5	12.7000	5	127.0	14	.55118
33/64	.515625	13.0969	6	152.4	15	.59055
17/32	.53125	13.4937	7	177.8	16	.62992
35/64	.546875	13.8906	8	203.2	17	.66929
9/16	.5625	14.2875	9	228.6	18	.70866
37/64	.578125	14.6844	10	254.0	19	.74803
19/32	.59375	15.0812	11	279.4	20	.78740
39/64	.609375	15.4781	12	304.8	21	.82677
5/8	.625	15.8750	13	330.2	22	.86614
41/64	.640625	16.2719	14	355.6	23	.90551
21/32	.65625	16.6687	15	381.0	24	.94488
43/64	.671875	17.0656	16	406.4	25	.98425
11/16	.6875	17.4625	17	431.8	26	1.02362
45/64	.703125	17.8594	18	457.2	27	1.06299
23/32	.71875	18.2562	19	482.6	28	1.10236
47/64	.734375	18.6531	20	508.0	29	1.14173
3/4	.75	19.0500	21	533.4	30	1.18110
49/64	.765625	19.4469	22	558.8	31	1.22047
25/32	.78125	19.8437	23	584.2	32	1.25984
51/64	.796875	20.2406	24	609.6	33	1.29921
13/16	.8125	20.6375	25	635.0	34	1.33858
53/64	.828125	21.0344	26	660.4	35	1.37795
27/32	.84375	21.4312	27	685.8	36	1.41732
55/64	.859375	21.8281	28	711.2	37	1.4567
7/8	.875	22.2250	29	736.6	38	1.4961
57/64	.890625	22.6219	30	762.0	39	1.5354
29/32	.90625	23.0187	31	787.4	40	1.5748
59/64	.921875	23.4156	32	812.8	41	1.6142
15/16	.9375	23.8125	33	838.2	42	1.6535
61/64	.953125	24.2094	34	863.6	43	1.6929
31/32	.96875	24.6062	35	889.0	44	1.7323
63/64	.984375	25.0031	36	914.4	45	1.7717

UNITS	Pints to Litres	Gallons to Litres	Litres to Pints	Litres to Gallons	Miles to Kilometres	Kilometres to Miles	Lbs. per sq. In. to Kg. per sq. Cm.	Kg. per sq. Cm. to Lbs. per sq. In.
1	.57	4.55	1.76	.22	1.61	.62	.07	14.22
2	1.14	9.09	3.52	.44	3.22	1.24	.14	28.50
3	1.70	13.64	5.28	.66	4.83	1.86	.21	42.67
4	2.27	18.18	7.04	.88	6.44	2.49	.28	56.89
5	2.84	22.73	8.80	1.10	8.05	3.11	.35	71.12
6	3.41	27.28	10.56	1.32	9.66	3.73	.42	85.34
7	3.98	31.82	12.32	1.54	11.27	4.35	.49	99.56
8	4.55	36.37	14.08	1.76	12.88	4.97	.56	113.79
9		40.91	15.84	1.98	14.48	5.59	.63	128.00
10		45.46	17.60	2.20	16.09	6.21	.70	142.23
20				4.40	32.19	12.43	1.41	284.47
30				6.60	48.28	18.64	2.11	426.70
40				8.80	64.37	24.85		
50					80.47	31.07		
60					96.56	37.28		
70					112.65	43.50		
80					128.75	49.71		
90					144.84	55.92		
100					160.93	62.14		

UNITS	Lb ft to kgm	Kgm to lb ft	UNITS	Lb ft to kgm	Kgm to lb ft
1	.138	7.233	7	.967	50.631
2	.276	14.466	8	1.106	57.864
3	.414	21.699	9	1.244	65.097
4	.553	28.932	10	1.382	72.330
5	.691	36.165	20	2.765	144.660
6	.829	43.398	30	4.147	216.990

CHAPTER 3

THE IGNITION SYSTEM

3:1 Description

A schematic layout of the ignition system is shown in **FIG 3:1**. The system shown is actually for a V-8 engine, as can be seen by counting the electrodes on the distributor cap, but the six-cylinder system is very similar. The distributor is fitted with a low-tension make-and-break as well as a rotor arm, and the distributor shaft is synchronized with the engine. The body of the distributor is set so that the contact points are just opening, and the rotor arm is pointing towards the correct electrode in the distributor cap, as a piston approached the firing point.

When the contacts are closed, current flows through the primary windings of the ignition coil, inducing a magnetic field around them. When the contacts open, the current is sharply cut off. The condenser in the distributor ensures that the cut-off is rapid and the magnetic field around the primary coils also collapses rapidly. The collapse of the magnetic field induces a very high voltage across the many turns of the secondary coil in the ignition coil. This high voltage is led to the distributor cap centre electrode and from there by a a carbon brush to the rotor arm. The voltage jumps the small gap between rotor arm electrode and is then led to the appropriate spark plug, where it jumps across the electrodes and the spark ignites the mixture in the cylinder.

When the engine is running, the current from the battery passes along a resistance wire which drops the voltage at the ignition coil. The ignition coil is designed to operate at full efficiency on this reduced voltage. When the starter motor is being operated, a terminal on the starter relay is live and at full battery voltage. A wire connects this terminal to the ignition coil so that on start-up the coil is receiving full battery voltage. The battery voltage will drop because of the current drain of the starter motor but as the ignition coil is designed to run on reduced voltage, efficiency is not lost.

The actual ignition point for the most efficient combustion will vary with the speed and load on the engine. The distributor shaft is fitted with a centrifugal advance mechanism to advance the ignition with increasing engine speed. Weights are fitted between the distributor shaft and cam in such a way that they rotate the cam in an advance direction as they swing outwards. The weights are restrained by special springs. As the

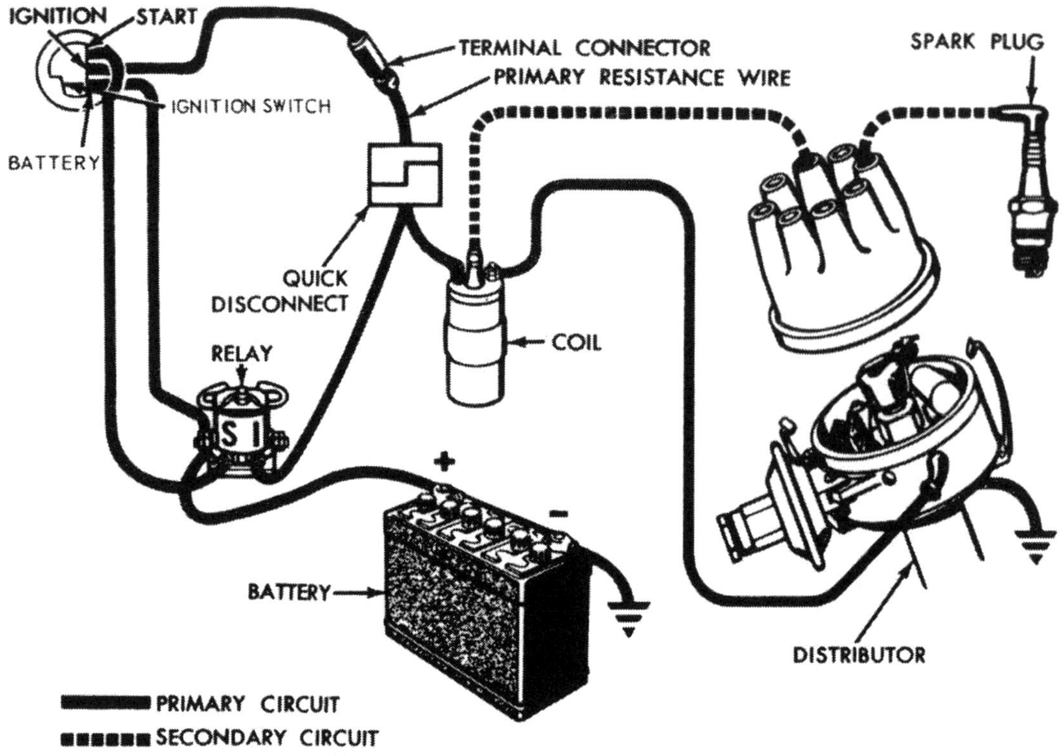

FIG 3:1 Typical ignition system circuit

engine speeds up the weights move outwards against the calibrated restraint of the springs under the action of centrifugal force, and this movement advances the ignition.

The vacuum in the intake manifold is proportional to the load on the engine and the throttle opening, the wider open the throttles the less the vacuum. A small-bore tube connects the intake manifold to the vacuum unit on the distributor. The vacuum unit contains a diaphragm which is acted on by the vacuum to rotate the movable breaker plate about the cam to alter the ignition point.

On later models the vacuum unit operates as just described during normal running except that carburetor vacuum is used for the outer diaphragm and manifold vacuum for the inner. The unit is fitted with two diaphragms and at idle or deceleration the inner diaphragm takes over to retard the ignition and therefore prevent excessive emission. Sectioned views of this dual type vacuum unit in operation are shown in **FIG 3:2**. A distributor vacuum valve is fitted into the system and this unit senses the temperature of the coolant. The unit contains a valve which connects the ports of the unit in such a way that when the coolant temperature becomes excessive the ignition timing is altered to speed up the curb idle speed of the engine so that during prolonged periods of idling the engine will be speeded up if it starts to overheat.

In order to achieve additional spark advance during the actual period of deceleration, in order to allow as much time as possible for the complete combustion of the fuel mixture, some models are equipped with a distributor vacuum advance control valve. During deceleration periods when inlet manifold vacuum rises above a specified value, this valve directs the inlet manifold depression to the distributor outer diaphragm, which advances the ignition until the engine has returned to its idle speed and retarded timing. Typical vacuum systems are shown diagrammatically in **FIGS 3:3** and **3:4**.

3:2 Maintenance

Regularly wipe the HT leads, ignition coil and distributor cap clean with a soft dry cloth. A build-up of dirt or moisture will cause arcing and reduce efficiency in the system.

The distributor points should be checked and adjusted at intervals of 12,000 miles and at the same time the distributor should be lightly lubricated. Clean the spark plugs at 6000 miles and renew them at 12,000 miles.

Lubrication:

Free the clips and remove the distributor cap from the distributor, leaving it attached to the HT leads. Firmly but squarely pull the rotor arm off from the cam. Pour a few drops of oil onto the felt pad in the top of the cam. With a clean finger smear a little high temperature lubricant around the cam face. Pour a little SAE.10W oil into the oil cup on the distributor body. **Wipe away all surplus lubricant and make sure that none goes onto the actual contact points.** Wipe the rotor arm and distributor cap all over with a clean soft cloth and install them on the distributor, making sure that the lug on the rotor arm fits into the slot in the cam.

If the distributor cap and rotor arm are very dirty, clean them with a stiff brush and denatured alcohol as a solvent. Be sure to clean out all crevices in the distributor cap.

3:3 Breaker points

The installation of the breaker plate is shown in **FIG 3:5**.

Cleaning:

The points should have a frosted slate grey color for efficient operation. Oil or grease on the contact points will blacken them and may also show up as a thin smudgy line on the base plate under the contacts. Cleaning in such cases should be carried out using a small stiff brush and chloroform as a solvent. If the points are badly blackened they should be renewed.

Badly blued or burnt points should also be renewed. The most likely causes of such points are defective condensers, incorrectly installed radio supressor (condensor), incorrect alignment of the points, incorrectly set charging system regulator or extended operation at high speeds.

In normal use there will be a transfer of metal from one contact to the other, causing a build-up on one and a pit on the other. No action need be taken until the build-up reaches the same dimension as the correct gap between the contacts when they are fully open. When this point is reached the contacts should be renewed as a set. Only in an emergency should the contacts be cleaned with a fine file or ground down.

Renewing contact breaker points:

Remove the distributor cap and rotor arm. Disconnect the primary and condenser wires from the breaker plate assembly. Free the diaphragm link and take out the plate retaining screws so that the breaker assembly can be removed from the distributor, complete with condensor. Remove the breaker contacts and condenser from the breaker plate.

Install the new parts in the reverse order of disassembly. Crank the engine until the breaker contacts are fully closed. It is essential that the vented contacts meet squarely and centrally as shown in **FIG 3:6,** otherwise they will rapidly burn and fail. Align the contact breakers by carefully bending the stationary breaker contact bracket as shown in **FIG 3:7.** Do not bend the moving breaker arm.

Spring tension:

The tension of the moving contact spring should be checked using a suitable spring gauge. Apply the load so that it is at right angles to the contact faces and pull until the contacts just part, noting the reading. A limited variation in the tension can be made by slackening the nut that secures the spring and moving it as shown in **FIG 3:8.** If the tension cannot be brought to specifications a new contact breaker set must be fitted.

Adjusting gap:

The most accurate method is to use a dwell meter and measure the dwell of the contacts, adjusting them as necessary to bring them within specifications. If feeler gauges are used on worn contacts there is a danger that the gauge will bridge the cap caused by the build-up of material and so give a false reading. If feeler gauges have to be used on worn contacts, make sure that the gauge is inserted between the unworn portions of the contacts. The method of adjusting using feeler gauges is shown in **FIG 3:9.**

The gap should be .027 inch and the dwell angle 35 to 40 deg. at idle speed.

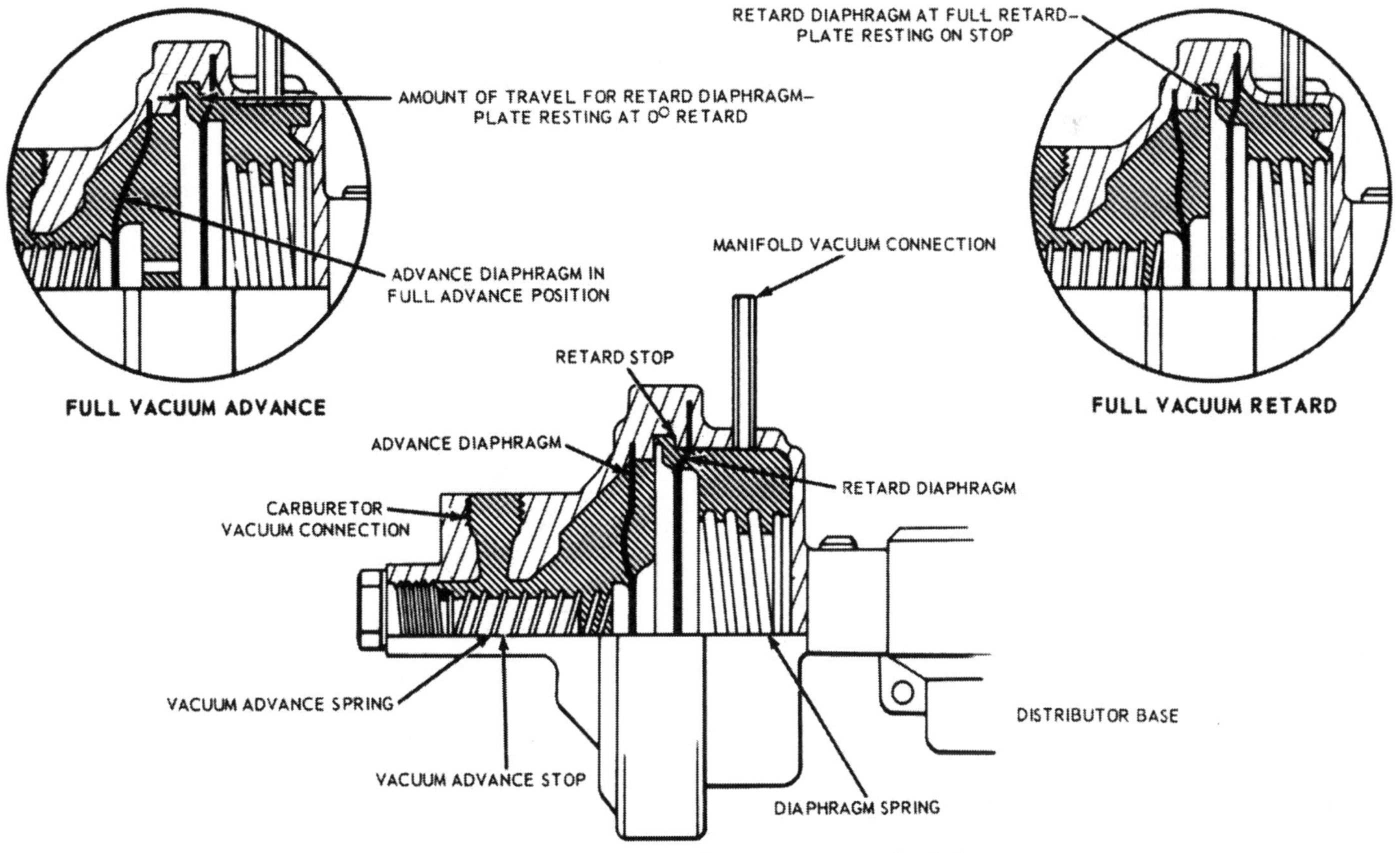

FIG 3:2 The operation of the dual diaphragm advance mechanism

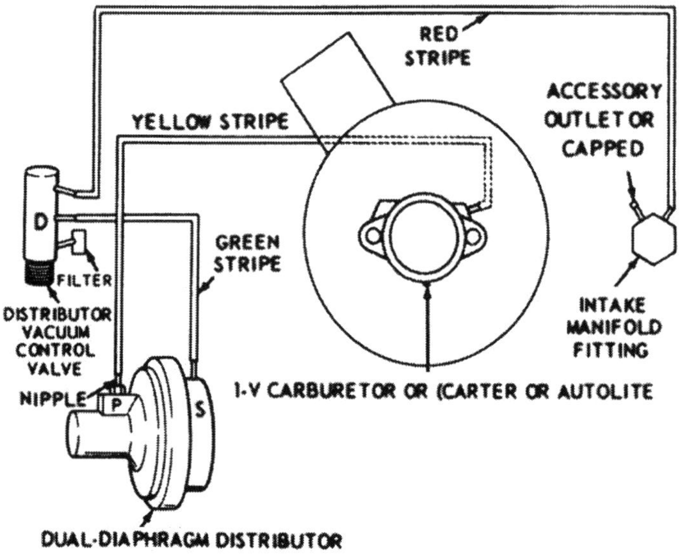

FIG 3:3 Schematic of the vacuum system for the 170 engine with Imco emission control and automatic transmission as well as all models of 200 engine engine with Imco emission control

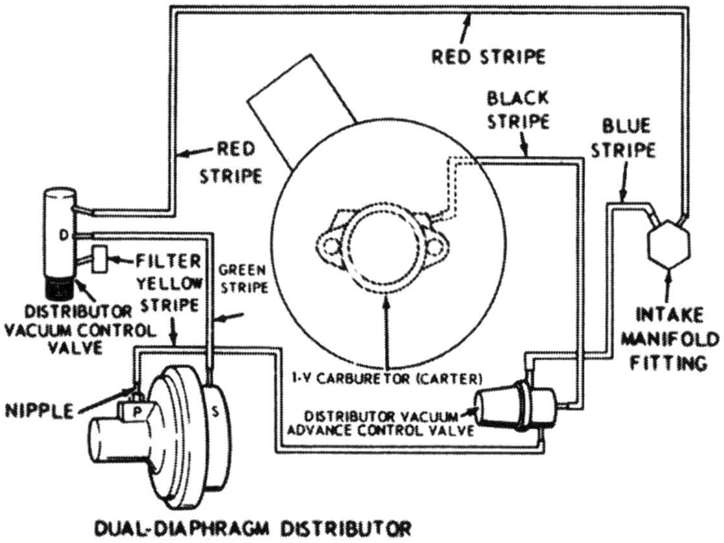

FIG 3:4 Schematic of the vacuum system for the 170 engine with Imco emission control and standard transmission

3:4 Ignition faults

A persistent misfire in the engine can be caused by defective ignition. The quickest way of determining faults is to connect the vehicle to electronic test equipment and the oscilloscope will then quickly pinpoint the fault.

If test equipment is not available, check the fuel system for faults which can be the cause of misfiring (see **Chapter 2**). Remove the distributor cap and check the contact breaker points for damage or incorrect gap as well as making sure that they are clean.

When likely faults have been checked for, start the engine and set it running at a fast idle speed. Either block the accelerator linkage open or adjust the curb idle screw on the carburetor.

Use rags, dry gloves or special insulated tongs to disconnect each plug lead from its plug in turn while listening carefully to the note of the engine. When disconnecting the plug leads, twist and pull directly on the moulded cap and not on the lead itself. If cylinder is not firing, the note of the engine will not change but if a cylinder is firing there will be a drop in speed. Listen carefully as the other cylinders can easily mask one cylinder not firing.

When the defective cylinder(s) has been found, stop the engine. Insert an old spark plug or length of metal rod into the moulded cap of the suspect lead and restart the engine. Hold the rod or spark plug approximately $\frac{3}{16}$ inch away from a convenient grounded metal part of the vehicle. Do not allow the lead near the fuel system because of the fire risk. If the lead and ignition are satisfactory there will be a regular blue spark. If it is found that the sparks are weak on all cylinders, disconnect the HT lead from the central electrode of the distributor cap and repeat the test using the lead while cranking the engine on the starter motor. The sparks will be faster than for a single cylinder but they should still be fat, healthy and blue.

If the sparks are weak, irregular or thin and reddish in color, check the HT lead and renew it if it appears burnt, cracked or perished. Repeat the test to see if there is an improvement.

If the sparks are still weak, remove the distributor cap and rotor arm. Thoroughly clean both of them, removing the HT leads if necessary, using a small stiff brush and chloroform or denatured alcohol as a solvent. Check the rotor arm for damage and renew if it is defective. Check the distributor cap carefully for hairline cracks or thin black lines between electrodes or from an electrode to the edge of the cap. The thin black lines are 'tracking' where the high voltage has burnt a path. In either case renew the distributor cap as no repairs are possible.

If the sparks are good, then the spark plug is at fault. Remove the spark plug so that it can be cleaned and tested or fit a new spark plug into its place. If the sparks from the central HT lead are poor and the lead is in good condition then the fault lies with the ignition coil.

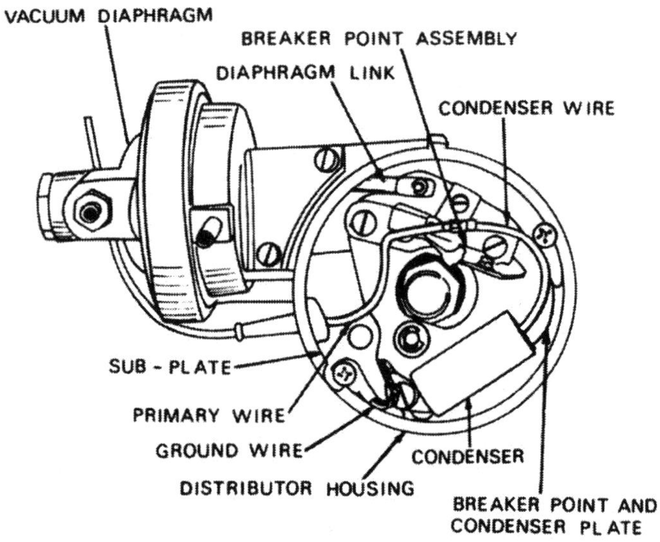

FIG 3:5 Typical breaker point installation

Testing the primary circuit:

Crank the engine until the contact points are closed, with the distributor cap removed. Switch on the ignition and flick the points apart with a fingernail. If there is current flowing there will be a small low voltage spark across the contacts.

A more accurate test is to disconnect the primary wire from between the terminal on the ignition coil and the low-tension terminal on the distributor. Reconnect it using a low-wattage 12-volt test lamp in series. Remove the distributor cap so that the contacts can be observed. Slowly crank the engine over with the ignition switched on. If the system is satisfactory the lamp will come on as the points close and then go out again as they open. It is advisable to remove all the spark plugs so that the engine can be cranked over by hand, rotating the crankshaft damper.

If the lamp fails to come on at all, trace back through the wiring until the fault is found. Use a voltmeter or test lamp to find the live points in the circuit and then work forwards until the break or fault is found. Note that the resistance wire will lower the voltage, so do not expect full battery voltage beyond the resistance wire. Stuck open contacts can also be a cause.

If the clamp comes on and does not go out at all there is a shortcircuit in the distributor. Recheck the contacts to make sure that they are opening, and that they are correctly installed. Check the wires in the distributor for faulty connection or frayed insulation. If the points and wires are satisfactory, repeat the test with the condenser disconnected. If the lamp now operates correctly there is a shortcircuit in the condenser.

Resistance wire:

This can be checked using a suitable resistance meter, and the resistance should be 1.3 to 1.4 ohms. The wire can also be checked using a voltmeter. Disconnect the leads from the ignition coil and connect the DIST terminal (normally connected to the distributor) to a good ground using a jumper lead. Disconnect the quick disconnect connector and connect the voltmeter between

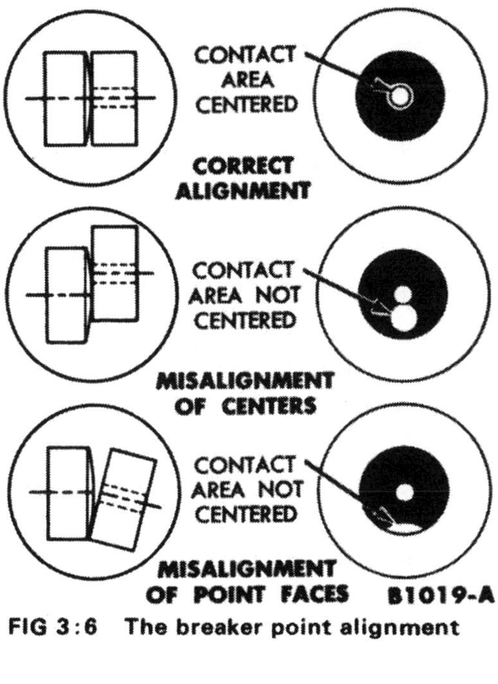

FIG 3:6 The breaker point alignment

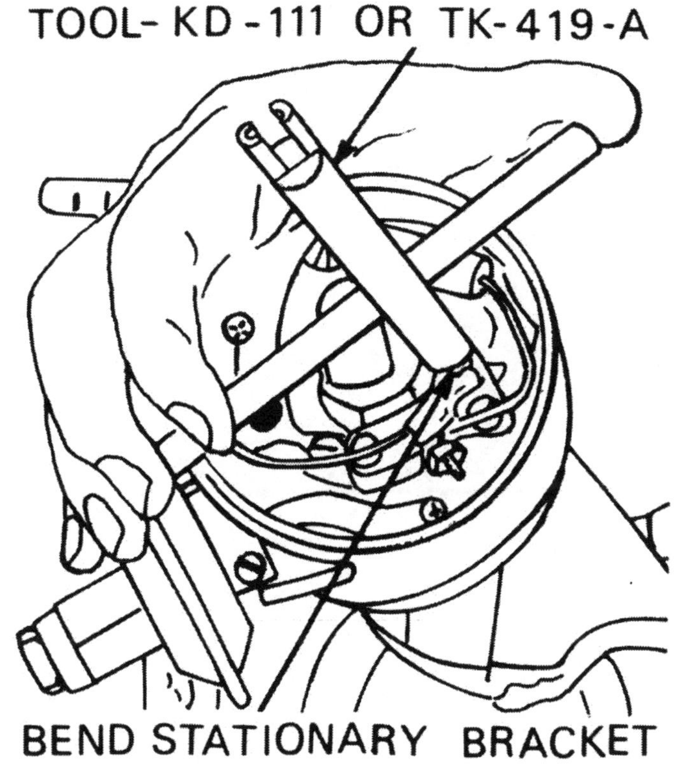

FIG 3:7 Setting the breaker point alignment

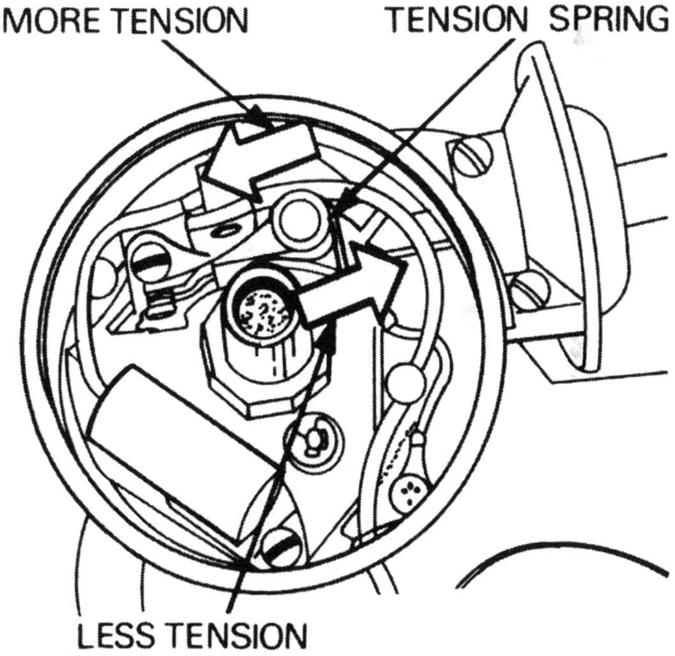

FIG 3:8 Adjusting the breaker spring tension

the end of the resistance wire and the other terminal on ignition coil. Switch on the ignition and if the voltage reading lies between 4.5 and 6.6 volts the resistance wire is satisfactory. Renew the wire if the voltage is outside specifications.

Condenser:

Condenser failure is not a common fault. If the paper insulation in the unit fails usually the spark erodes away the foil in the area so that the condenser is self-healing. A shortcircuited condenser is easily detected as described earlier.

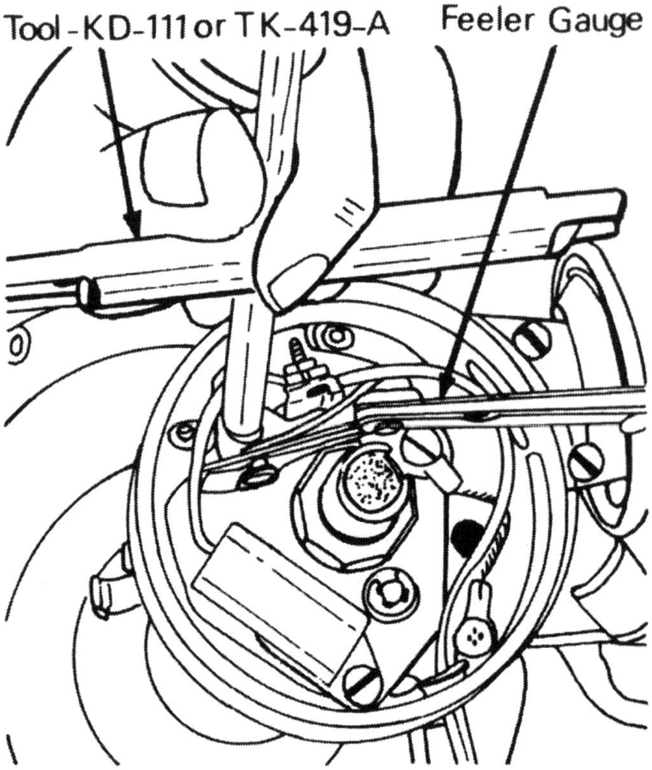

FIG 3:9 Checking the contact gap with feeler gauges

Open circuit is more difficult to detect without special equipment though it may be suspected if starting is difficult and the contacts are excessively 'blued' or burnt. The best check is to fit a new condenser and see if this cures the fault.

3:5 Removing and installing the distributor

Disconnect the primary wire at the distributor body. Disconnect the vacuum hoses from the vacuum unit, labelling them if the colors have faded. Remove the distributor cap, leaving it attached to the HT leads.

Make two aligning marks, one on the distributor body and the other on the engine. Make a mark on the top edge of the distributor case by the brass end of the rotor arm. Provided that the engine is not turned and the distributor is installed so that all the marks are again in alignment, the ignition timing will not be lost. Remove the distributor clamp hold down bolt and withdraw the distributor from the engine.

Install the distributor in the reverse order of removal. When pressing the distributor back into place, allow for the rotation of the gears as they mesh to ensure that the rotor arm is still pointing at the mark when the distributor is fully in place. Be sure that the distributor shaft engages fully and correctly with the oil pump intermediate shaft. If the crankshaft has been rotated since the distributor was removed it will be necessary to set the ignition timing from scratch.

If the HT leads have been removed, install them in the correct order into the distributor cap. When removing or installing the leads, pull and twist directly on the moulded cap itself, as pulling on the wire can damage it. The correct installation of the leads is shown in **FIG 3:10.** The ends of the leads must be clean and a little silicone grease may be used as waterproofing.

3:6 Setting the ignition timing

If the timing has been completely lost, leave the distributor out until the engine has been set. Turn the engine until it is at TDC after the compression stroke on No. 1 (front cylinder). The timing marks are shown in **FIG 3:11.** The compression stroke of the engine can be found after the spark plug has been removed. Crank the engine slowly with a thumb blocking the spark plug hole and there will be a pressure rise felt in the cylinder as the piston moves up on the compression stroke. The position can also be found if the rocker cover is off the engine, as both valves on No. 1 cylinder will be fully closed when the timing marks are at the correct TDC position.

Insert the distributor back into the engine. The distributor casing must be installed so that the vacuum unit is pointing in the original direction otherwise it may be impossible to rotate the case to set the timings accurately. Use the original engine and case aligning marks if the old distributor is being reinstalled. The distributor should be installed so that when it is fully in position the rotor arm is pointing to the position of the electrode in the cap connected to No. 1 spark plug. Leave the distributor clamp just tight enough to prevent the distributor from being moved accidentally.

Crank the engine forward by two turns. Just before the end of the second revolution, turn the engine very slowly and stop turning when the correct timing marks are in line. If the marks are overshot, turn the engine a further two revolutions, or else turn it back far enough to ensure that all backlash will be taken up when the engine is again rotated forwards to the correct timing position. The correct timing for all models is 6 deg. BTDC.

Disconnect the primary wire from between the distributor and DIST terminal on the ignition coil and reconnect it with a low-wattage test lamp in series. Switch on the ignition. Apply light backwards pressure on the rotor arm to take up all backlash. Rotate the distributor body backwards or forwards slightly until the point is found where the test lamp just goes out, indicating that the contacts have just opened. Tighten the distributor clamp in this position. Crank the engine forwards for another two turns, slowing the rate of turning at the end of the second revolution, and stop turning the instant that the test lamp goes out (noting that it will have come on and gone out in during the revolutions). If the timing is correct and the engine was stopped immediately that the lamp went out, the timing marks will again be in exact alignment.

The timing may also be checked with a stroboscopic light when the engine is running. Connect the light as instructed by the maker of the instrument. Mark the timing points on the crankshaft damper and the edge of the pointer with thin white lines so that they will be more easily seen. Start the engine and run it until it has reached its normal operating temperature. Disconnect the vacuum lines from the distributor, and plug them to prevent the air from altering the idling. Slacken the distributor clamp slightly so that the body can be rotated. Start the engine and shine the light onto the timing marks. **Take great care not to wear loose clothing, and avoid the moving drive belts, keeping tools and hands well away from them.** With the engines at idle, rotate the distributor body

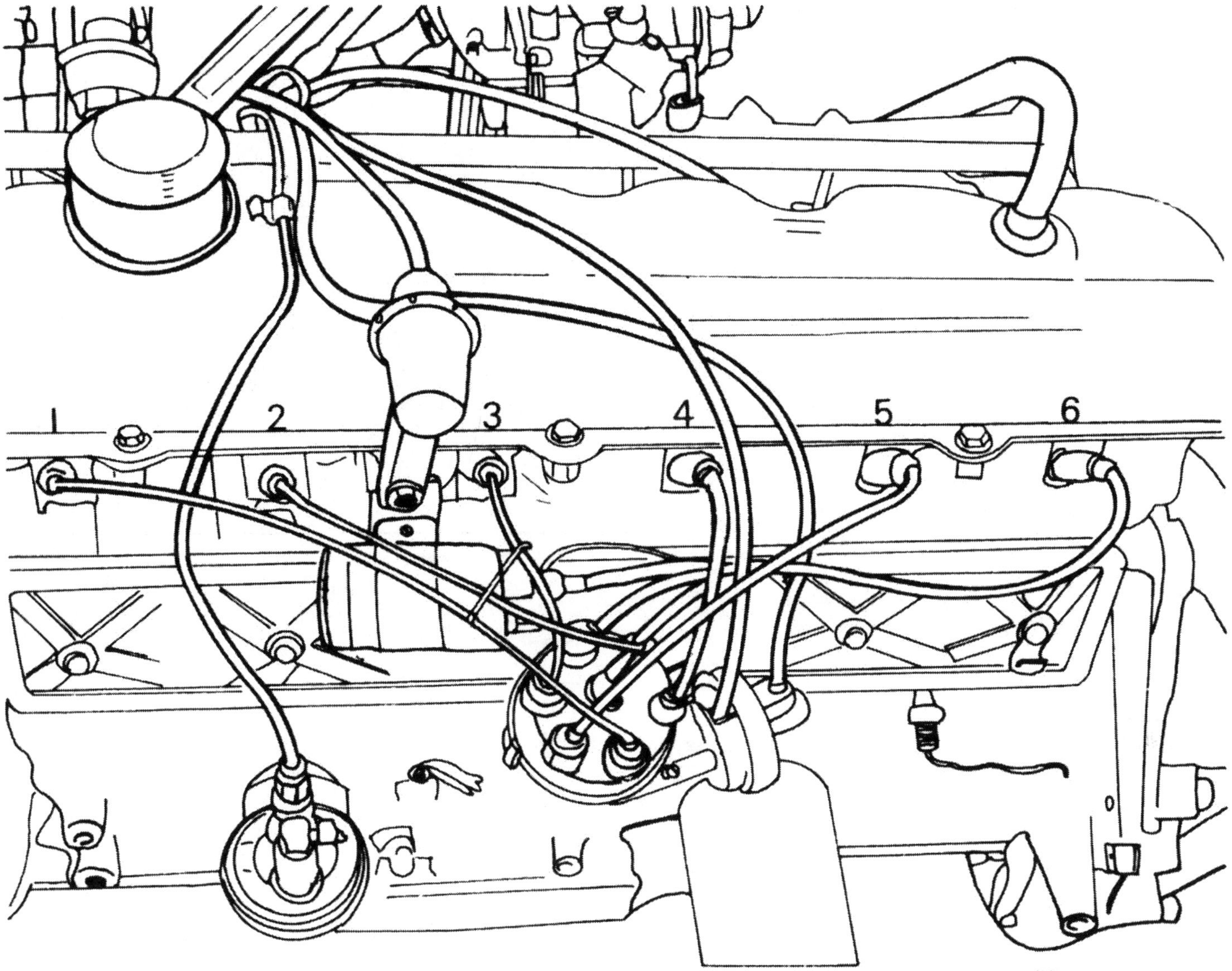

FIG 3:10 Typical HT lead installation

until the 6 deg. BTDC and pointer are apparently exactly in line. Tighten the clamp that secures the distributor. Remove the stroboscopic light and re-connect the vacuum lines.

3:7 The distributor

The components of the distributor are shown in **FIG 3:12**. A press and special tools will be required to disassemble and service the distributor, so the best solution for the owner is to fit a new tested distributor if the old one is excessively worn.

The end play of the distributor shaft can be checked with a dial gauge after the cap and rotor arm have been removed, without removing the distributor from the engine.

If the distributor is dismantled, it is essential to mark the weight springs and drive gear so that they are installed into their original positions.

3:8 Temperature sensing valve

This is fitted into the vacuum system and measures the temperature of the coolant. To check the valve, connect a tachometer to the ignition. Check that all vacuum hoses are correctly routed and in good condition. Start the engine and run it until it has reached its normal operating temperature, with the choke plate valve fully open. At this stage make sure that the engine does not overheat. Note the idle speed with the transmission in neutral and the throttle valve fully closed to the curb idle stop.

Disconnect the vacuum hose that connects the valve to the intake manifold at the valve. Plug the hose. If the valve is satisfactory the idle speed will not alter. If the idle speed drops by 100 rev/min or more, the valve is defective and must be renewed.

Cover the radiator and allow the engine to run until the overheat warning light comes on, or the needle on the gauge moves into the overheat sector. Before carrying out this part of the test it is essential that the cooling system is checked and the level in the system also checked (see **Chapter 4**). If the valve is satisfactory the idle speed will increase by 100 rev/min when the engine overheats. If the idle speed does not increase, remove and discard the old valve, fitting a new one in its place.

Uncover the radiator and allow the engine to run at a fast idle to cool before switching off.

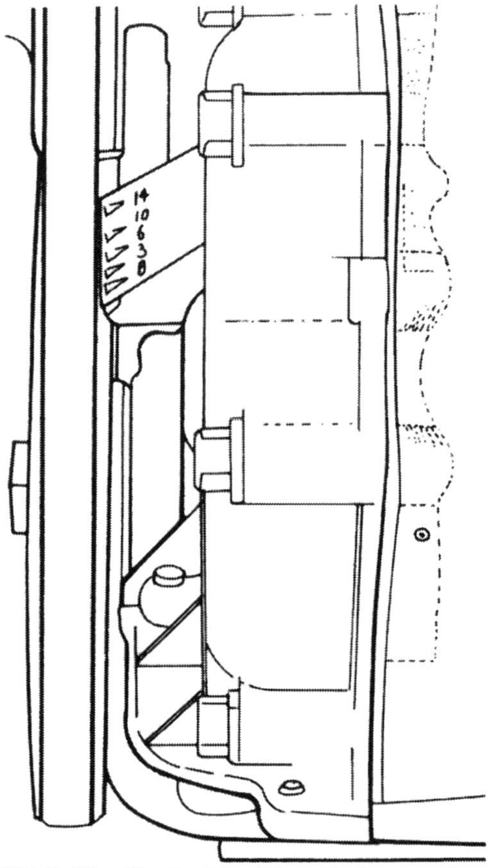

FIG 3:11 Typical ignition timing marks

3:9 The vacuum advance control valve

The valve with its cover removed is shown in **FIG 3:13.** Make sure that the engine idle speed is correctly set (see **Chapter 2**). Turn off the headlights and note the engine speed. Remove the plastic cover from the valve and turn the adjusting screw in a counterclockwise direction, without exerting excessive inward pressure. After 5 to 6 turns the idle speed should suddenly increase to approximately 1000 rev/min. Do not unscrew the adjusting screw further than this point otherwise the compressed spring in the valve will be released. If, after the sixth turn, the idle speed does not increase, press in on the adjusting screw and release it to trigger the engine to 1000 rev/min. Turn the adjusting screw clockwise until the idle speed returns to the original level and then make one further turn clockwise. Increase the engine speed to 2000 rev/min using the throttle, hold this speed for approximately 5 seconds, and release the throttle. If the engine takes longer than 4 seconds to return to the idle speed, free the anti-stall dashpot and repeat the test from speed. If, with the dashpot released, the engine still takes longer than 3 seconds to return from 2000 rev/min to idle, turn the adjusting screw a further quarter turn clockwise and repeat the test. Up to one further full turn may be made from the original setting, taking it in quarter turn increments, to bring the time taken to return from 2000 rev/min within specifications. If the valve requires more than one full turn clockwise from the original setting, the valve is defective and a new one must be fitted.

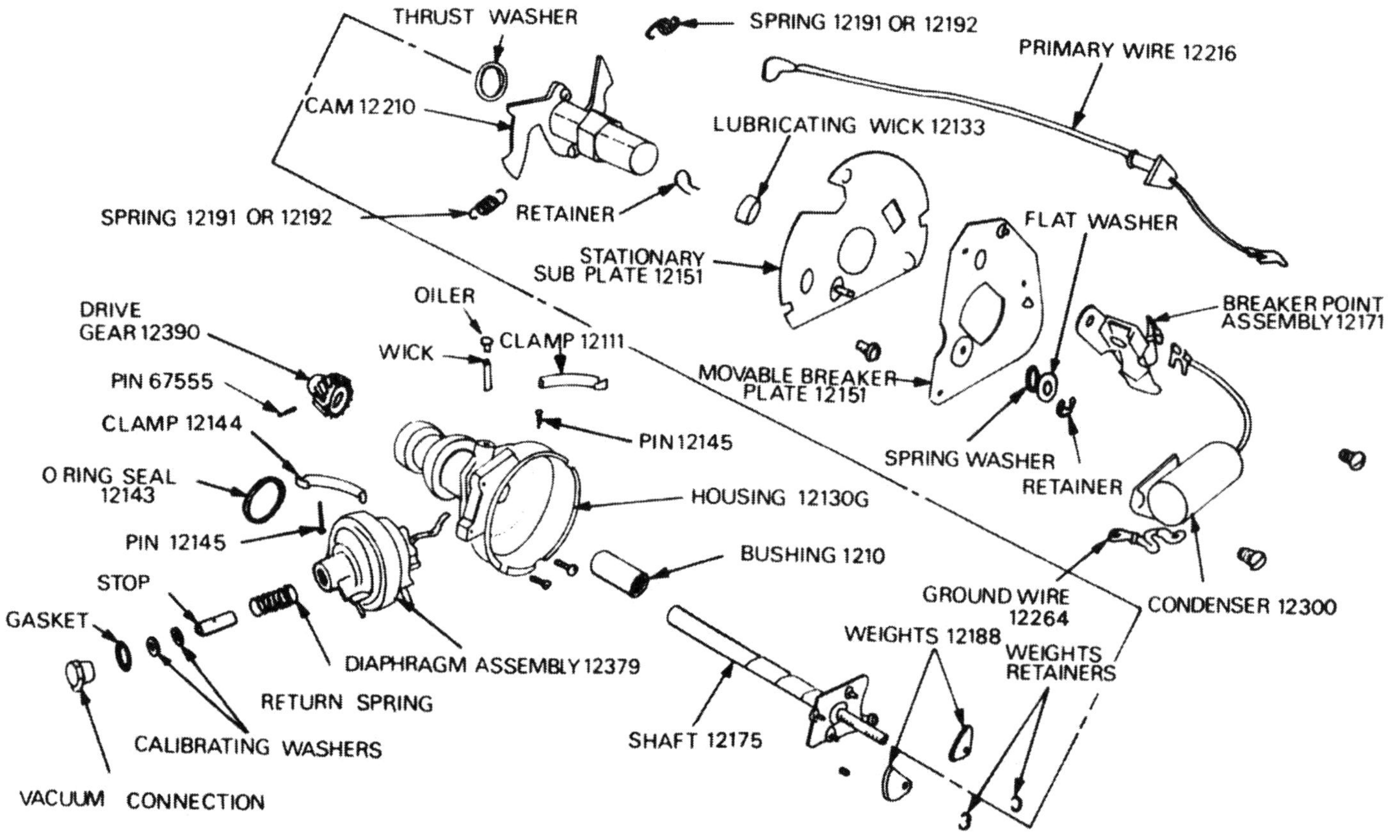

FIG 3:12 The distributor components

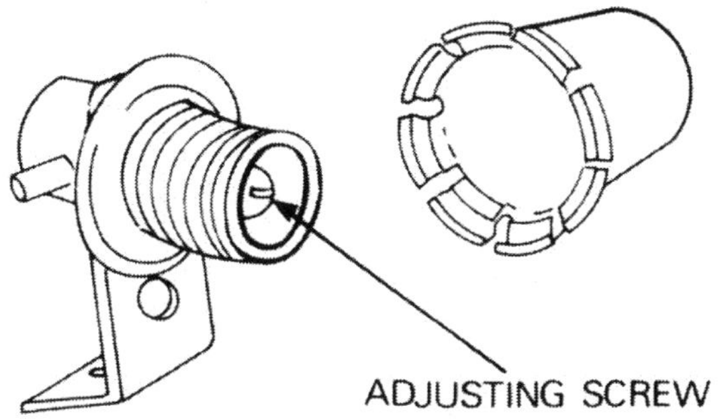

FIG 3:13 The vacuum advance control valve

3:10 The spark plugs

The spark plugs should be cleaned and tested at intervals of 6000 miles and renewed at intervals of 12,000 miles.

Removal:

Slacken the spark plug using a well fitting box wrench or plug wrench. Blow away all loose dust and dirt from around the base of the plug with an air-line or tire pump. Unscrew the spark plug by hand, unless it is too stiff. If the spark plug is stiff to remove, use the wrench to take it out as far as possible without exerting heavy pressure and then turn the plug in and out to this point a few times before attempting to unthread it further. In extreme cases, wrap a piece of rag around the base, soak the rag with kerosene or penetrating oil and leave it overnight to free the threads.

As the spark plugs are removed they should be stored for examination of their firing ends.

Examination:

The color and condition of the deposits on the firing ends of the plugs will give a good guide to the conditions inside the combustion chambers.

After examination, throw away any plugs that have cracked insulators or badly burnt electrodes, without bothering to clean or test them.

A light powdery deposit ranging in color from brown to greyish tan, coupled with light wear on the electrodes, indicates that the conditions are normal. Much city (or constant-speed driving) will leave the deposits white or yellowish. Cleaning and testing the spark plugs is all that is required.

If the deposits are wet and black they are caused by excess oil leaking into the combustion chamber. The only cure is an engine overhaul but fitting a hotter running grade of plug may help to alleviate the problem.

Black but dry and fluffy deposits are caused by incomplete combustion. This may be caused by excessive idling (particularly if all the spark plugs are similarly affected) but it may also be caused by defective ignition or spark plug or running with too rich a mixture.

Overheated spark plugs have a white blistered look about the central electrode. If lead-based fuels are used there may be glints of metal on the central electrode. The electrodes will also be excessively burnt away. Some possible causes are engine overheating, very weak mixture, incorrectly set ignition timing, incorrect grade of plug or running at high speeds with the vehicle overloaded.

Cleaning, adjusting and testing:

Wash oily plugs in gasoline before cleaning them further. Have the plugs cleaned on an abrasive-blasting machine and then pressure tested after attention to the electrodes. Use a steel-wire brush to clean the threads, and spark plug in an emergency.

Trim the electrodes square with a fine file. Adjust to the correct gap of .032 to .036 inch by bending the side electrode only. Never bend the central electrode as this will crack the insulator.

Check that the plugs spark under pressure. While still applying air pressure to the plug, drip a little thin engine oil around the joint of the insulator and body. If air bubbles are seen in the oil, reject the plug.

Clean the external portion of the insulator with denatured alcohol, unleaded gasoline or similar suitable solvent.

Installing:

If the threads in the cylinder head are dirty, clean them using a well-greased tap. Failing a tap, use an old spark plug with crosscuts down the threads. Graphite grease only may be used on the threads as any other grease will bake hard and lock the plug to the head.

Thread the plugs in by hand until they bottom. Tighten them to a torque of 15 to 20 lb ft. If a torque wrench is not available, tighten the plug by at most a further half-turn from the handtight position. Overtightening is likely to cause damage and defects.

3:11 Trouble shooting guide

(a) Engine will not fire

1 Battery discharged or terminals dirty
2 Breaker contact points dirty or incorrectly set
3 Distributor cap dirty, cracked or tracking
4 Rotor arm not making contact with carbon button in cap
5 Distributor rotor arm cracked, dirty or omitted
6 Defective ignition coil
7 Defective HT lead between coil and distributor
8 Broken or stuck contact breaker spring
9 Contact points stuck open
10 Internal shortcircuit in distributor

(b) Engine misfires

1 Check 2, 3 and 6 in (a)
2 Weak contact breaker spring
3 Defective HT lead
4 Spark plug(s) loose
5 Spark plug insulation cracked or dirty
6 Spark plug gap incorrectly set
7 Ignition timing too far advanced
8 HT leads installed in wrong order

Inches	Decimals	Milli-metres	Inches to Millimetres		Millimetres to Inches	
			Inches	mm	mm	Inches
1/64	.015625	.3969	.001	.0254	.01	.00039
1/32	.03125	.7937	.002	.0508	.02	.00079
3/64	.046875	1.1906	.003	.0762	.03	.00118
1/16	.0625	1.5875	.004	.1016	.04	.00157
5/64	.078125	1.9844	.005	.1270	.05	.00197
3/32	.09375	2.3812	.006	.1524	.06	.00236
7/64	.109375	2.7781	.007	.1778	.07	.00276
1/8	.125	3.1750	.008	.2032	.08	.00315
9/64	.140625	3.5719	.009	.2286	.09	.00354
5/32	.15625	3.9687	.01	.254	.1	.00394
11/64	.171875	4.3656	.02	.508	.2	.00787
3/16	.1875	4.7625	.03	.762	.3	.01181
13/64	.203125	5·1594	.04	1.016	.4	.01575
7/32	.21875	5.5562	.05	1.270	.5	.01969
15/64	.234375	5.9531	.06	1.524	.6	.02362
1/4	.25	6.3500	.07	1.778	.7	.02756
17/64	.265625	6.7469	.08	2.032	.8	.03150
9/32	.28125	7.1437	.09	2.286	.9	.03543
19/64	.296875	7.5406	.1	2.54	1	.03937
5/16	.3125	7.9375	.2	5.08	2	.07874
21/64	.328125	8.3344	.3	7.62	3	.11811
11/32	.34375	8.7312	.4	10.16	4	.15748
23/64	.359375	9.1281	.5	12.70	5	.19685
3/8	.375	9.5250	.6	15.24	6	.23622
25/64	.390625	9.9219	.7	17.78	7	.27559
13/32	.40625	10.3187	.8	20.32	8	.31496
27/64	.421875	10.7156	.9	22.86	9	.35433
7/16	.4375	11.1125	1	25.4	10	.39370
29/64	.453125	11.5094	2	50.8	11	.43307
15/32	.46875	11.9062	3	76.2	12	.47244
31/64	.484375	12.3031	4	101.6	13	.51181
1/2	.5	12.7000	5	127.0	14	.55118
33/64	.515625	13.0969	6	152.4	15	.59055
17/32	.53125	13.4937	7	177.8	16	.62992
35/64	.546875	13.8906	8	203.2	17	.66929
9/16	.5625	14.2875	9	228.6	18	.70866
37/64	.578125	14.6844	10	254.0	19	.74803
19/32	.59375	15.0812	11	279.4	20	.78740
39/64	.609375	15.4781	12	304.8	21	.82677
5/8	.625	15.8750	13	330.2	22	.86614
41/64	.640625	16.2719	14	355.6	23	.90551
21/32	.65625	16.6687	15	381.0	24	.94488
43/64	.671875	17.0656	16	406.4	25	.98425
11/16	.6875	17.4625	17	431.8	26	1.02362
45/64	.703125	17.8594	18	457.2	27	1.06299
23/32	.71875	18.2562	19	482.6	28	1.10236
47/64	.734375	18.6531	20	508.0	29	1.14173
3/4	.75	19.0500	21	533.4	30	1.18110
49/64	.765625	19.4469	22	558.8	31	1.22047
25/32	.78125	19.8437	23	584.2	32	1.25984
51/64	.796875	20.2406	24	609.6	33	1.29921
13/16	.8125	20.6375	25	635.0	34	1.33858
53/64	.828125	21.0344	26	660.4	35	1.37795
27/32	.84375	21.4312	27	685.8	36	1.41732
55/64	.859375	21.8281	28	711.2	37	1.4567
7/8	.875	22.2250	29	736.6	38	1.4961
57/64	.890625	22.6219	30	762.0	39	1.5354
29/32	.90625	23.0187	31	787.4	40	1.5748
59/64	.921875	23.4156	32	812.8	41	1.6142
15/16	.9375	23.8125	33	838.2	42	1.6535
61/64	.953125	24.2094	34	863.6	43	1.6929
31/32	.96875	24.6062	35	889.0	44	1.7323
63/64	.984375	25.0031	36	914.4	45	1.7717

UNITS	Pints to Litres	Gallons to Litres	Litres to Pints	Litres to Gallons	Miles to Kilometres	Kilometres to Miles	Lbs. per sq. In. to Kg. per sq. Cm.	Kg. per sq. Cm. to Lbs. per sq. In.
1	.57	4.55	1.76	.22	1.61	.62	.07	14.22
2	1.14	9.09	3.52	.44	3.22	1.24	.14	28.50
3	1.70	13.64	5.28	.66	4.83	1.86	.21	42.67
4	2.27	18.18	7.04	.88	6.44	2.49	.28	56.89
5	2.84	22.73	8.80	1.10	8.05	3.11	.35	71.12
6	3.41	27.28	10.56	1.32	9.66	3.73	.42	85.34
7	3.98	31.82	12.32	1.54	11.27	4.35	.49	99.56
8	4.55	36.37	14.08	1.76	12.88	4.97	.56	113.79
9		40.91	15.84	1.98	14.48	5.59	.63	128.00
10		45.46	17.60	2.20	16.09	6.21	.70	142.23
20				4.40	32.19	12.43	1.41	284.47
30				6.60	48.28	18.64	2.11	426.70
40				8.80	64.37	24.85		
50					80.47	31.07		
60					96.56	37.28		
70					112.65	43.50		
80					128.75	49.71		
90					144.84	55.92		
100					160.93	62.14		

UNITS	Lb ft to kgm	Kgm to lb ft	UNITS	Lb ft to kgm	Kgm to lb ft
1	.138	7.233	7	.967	50.631
2	.276	14.466	8	1.106	57.864
3	.414	21.699	9	1.244	65.097
4	.553	28.932	10	1.382	72.330
5	.691	36.165	20	2.765	144.660
6	.829	43.398	30	4.147	216.990

CHAPTER 4

THE COOLING SYSTEM

4:1 Description

A sectional view of the engine showing the coolant flow is shown in **FIG 4:1**. The coolant circulates around the outside of the cylinder bores and through the passages in the cylinder heads, taking away surplus heat. Hot water normally tends to rise, because of expansion, and the heated coolant therefore passes out through the elbow outlet on the engine and from this elbow through a hose to the header tank of the radiator. The coolant passing through the radiator core is cooled by the passage of air over the radiator fins and then returns to the engine. To ensure a rapid and positive flow of coolant at all times, the thermo-syphon action of the coolant is assisted by a centrifugal engine driven water pump. The pump draws the coolant from the radiator and passes it into the engine.

A cooling fan is fitted to the water pump hub to ensure that there is an adequate flow of air through the radiator core when the vehicle is stationary or travelling slowly. The fan is designed to operate only when required. On some models a special temperature-sensitive clutch is fitted to the hub so that the fan is only rotating at full speed when the engine is hot and in the low speed range. On other models the fan is fitted with special blades which flex and reduce the airflow and load at high engine speeds.

On vehicles fitted with automatic transmission, a heat exchanger is fitted into the cooler radiator tank to cool the transmission fluid. On downflow radiators the heat exchanger is fitted in the lower tank but on crossflow radiators the unit is fitted in the lefthand chamber.

A thermostat valve is fitted to the elbow outlet of the engine. The valve ensures that the engine temperature stays constant and also gives a shorter warm-up period. When the coolant is cold the valve stays shut and the coolant is bypassed directly back into the engine. The amount of coolant to be heated is therefore reduced and there are no heat losses from the radiator so the engine warms up rapidly. When the coolant heats, the valve opens and allows the coolant to pass through the radiator where it is cooled normally.

The filler cap to the radiator contains two concentric valves. The outer valve seals against the neck of the filler and only opens when the pressure of the system exceeds specification. Surplus coolant or air are then vented overboard. Pressurizing the system raises the boiling point of the coolant so that the engine can be run hotter without danger of boiling in localized hot spots. If the filler cap is removed while the engine is overheated the pressure will force scalding water over the operator's hand and the subsequent drop in pressure will allow the coolant to boil, forcing out even more scalding

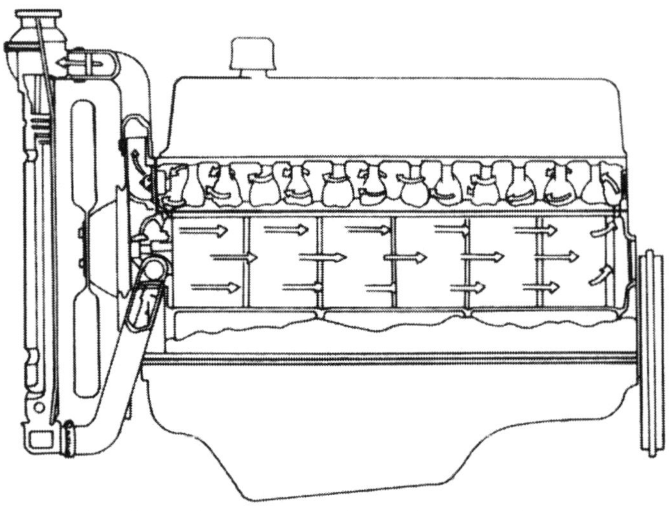

FIG 4:1 The coolant flow through the system

coolant. The cap has two positions, one of which is a lock that holds the cap in place but allows excess pressure to escape if the engine is warm. The inner valve of the cap opens as the system cools and allows air to enter and prevent a vacuum from forming in the cooling system.

A mixture of antifreeze and water should always be used. Even in hot climates the mixture should protect the system down to a temperature of 0°F. The antifreeze in these cases is not to protect against frost, but it does contain corrosion inhibitors as well as raising the boiling point of the water. In temperate climates a 50 per cent mixture of water and antifreeze should be used and this strength will protect the engine down to temperatures of —35°F and raise the boiling point to 250°F.

4:2 Maintenance

Regular maintenance is confined to checking that all the drive belts are in good condition and that their tension is correct, as well as checking the coolant level in the radiator. When checking the level make sure that the coolant is clean, and top up using a mixture of antifreeze and water. If water alone is used for topping up, the antifreeze in the system will be diluted.

The life of the antifreeze is two years. At yearly intervals have the strength of the mixture tested and add extra antifreeze as required. At the end of two years, drain the system and discard the coolant. Flush the system and refill with fresh coolant.

Coolant level:

The level should always be checked with the engine cold. On downflow radiators, top-up until the level is one inch below the bottom of the ring in the filler neck. On crossflow radiators the level is marked on the outside of the header tank.

Draining:

The radiator is fitted with a drain cock underneath it. If the system is being drained for any reason other than renewing the coolant, attach a hose to the cock and drain the coolant into a clean container so that it can be poured back into the system after the work has been carried out.

When completely draining the cooling system, remove the block drain plugs. A drain plug is fitted on each side of the block.

Flushing:

The use of Rotunda Cooling System Cleanser will remove scale and dirt if the instructions on the tin are followed.

The system can be partially cleaned by inserting a hose pipe into the filler neck, with all the drains open, and running water through until it comes out clean from the drains.

If the system is so clogged that Rotunda cleanser will not remove all the dirt and scale, have the system pressure flushed at a service garage. When pressure flushing, the thermostat must be removed from its housing.

In some cases it will help to remove the radiator, invert it, and flush through in a reverse direction so that water and gravity combined will remove scale trapped in the tubes.

Hoses:

Pinch the hoses lightly between the fingers and check them for perishing, hardening, softening, or cracking. Renew any hoses that are defective. Slacken off the hose clamps and pull the hoses free with a twisting motion.

Fit the hose clamp onto the hose so that it is at least $\frac{1}{8}$ inch from the end of the hose. Lubricate the inside of the hose with water, soft soap or an approved water-resistant sealer and slide the hose back into position. Make sure that the hose clamp is well past the bend on the pipe or connector and position it centrally on the portion of hose on the pipe. Tighten the hose clamps sufficiently to prevent leaks without cutting into the hoses.

Filling:

Make sure that all drains are closed and hoses connected. Fill the cooling system through the radiator filler cap until it is full, pouring it in slowly to allow air to escape. Disconnect the heater outlet hose at the water pump and wait until coolant flows out through it, with the heater set to hot, and reconnect the hose. Fill the engine to the correct level. Start the engine and check for coolant leaks. Leave the engine running at 2000 rev/min for 20 minutes and then switch it off and allow it to cool. When the system is cool, top up to the correct level. As much as a quart of coolant may be required for topping up to allow for the air trapped in the system. If the heater fails to operate on hot, disconnect the heater outlet hose and reconnect it when coolant flows out.

4:3 The radiator

Removal:

Drain the cooling system as described in the previous section. On models fitted with automatic transmission, disconnect the oil cooler lines at the radiator. Plug the hoses and ports on the coil cooler to prevent dirt from entering. Remove the radiator bolts and supports (if fitted) while supporting the radiator, and then lift out the radiator.

Install the radiator in the reverse order of removal. Fill the cooling system. On models fitted with automatic transmission, check the fluid level in the transmission (see **Chapter 7**).

Cleaning:

The internal flushing of the radiator has been dealt with in the previous section. Periodically use an air-line or hose pipe to blow through the radiator fins, from the back, so as to remove accumulations of mud and dead insects. In extreme cases, remove the radiator so that it can be soaked and scrubbed in a hot weak detergent solution.

Repairs:

If the radiator or transmission oil cooler are defective, the repairs should be left to a specialist firm.

4:4 Drive belts

Belts are used to transmit the drive to the water pump and alternator, as well as to the Thermactor air pump, air conditioning, and power steering pumps, if these are fitted.

All the belts are independently adjustable. If a belt is glazed or slack it will slip and drive will not be efficiently transmitted to the accessory while if the belt is too tight the bearings of the accessory may be damaged. The best method of checking the tension is to use the special gauge T63L-8620-A and this is the method to use wherever possible. A rough guide is that the tension is correct if the belt deflects by approximately $\frac{1}{4}$ inch from the straight line when moderate thumb pressure is applied at the center of a run. The settings using the gauge are given in Technical Data. Note that a belt is considered used when it has run for 10 minutes or more.

Water pump and alternator:

The alternator is pivoted about two bolts and secured by a third bolt to a slotted link. Slacken all three bolts and use a wooden lever behind the drive end bracket of the alternator to pull the alternator away from the engine until the correct belt tension is reached. Hold the alternator with the lever and tighten all three bolts.

Thermactor:

Slacken the Thermactor pump attachments and pull the pump away from the water pump until the correct belt tension is reached.

Hold the pump and tighten the attachments.

Power steering:

The pump must not be levered by the casing of the oil container otherwise the container may distort and leak. A boss is cast in with the pump and it is designed so that a wrench can be fitted over it. Slacken the pump attachments and use a wrench on the boss to move the pump so that the correct belt tension is set. Hold the pump and tighten its attachments.

Air conditioner:

An idler pulley is fitted to the belt. Position the idler so that the correct belt tension is set.

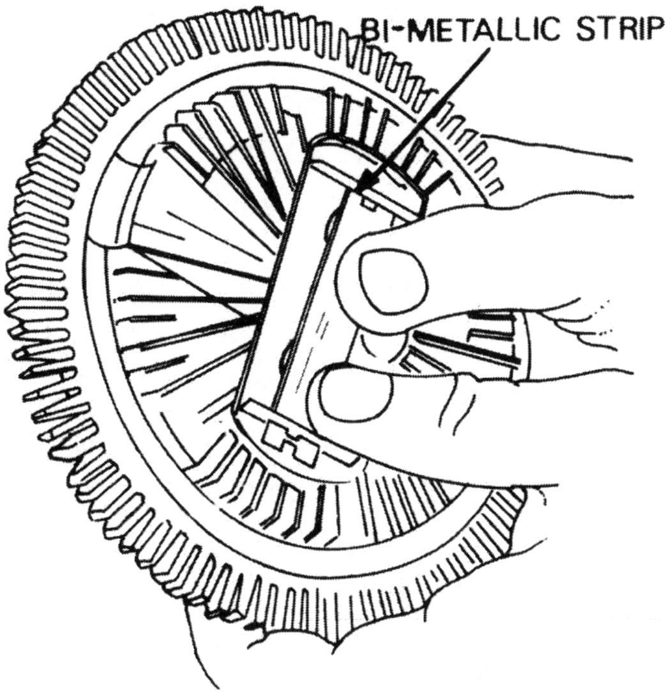

FIG 4:2 The cooling fan hub with flat bi-metallic spring

Belt removal:

Slacken off the belt as far as possible. Start removing the belt by rolling it forwards off the pulley and work in the direction of normal rotation. If the belt is too tight to remove fully, crank the engine to free it from the pulley. Once the belt is off one pulley it is easy to remove it from the vehicle.

Install the belt in the reverse order of removal, after checking it for defects or glaze on the sides, and set the correct tension. If two belts are fitted to drive the water pump and alternator, they must be renewed as a pair and not individually.

4:5 The cooling fan

On some models a Ford Flex-fan is fitted. The fan blades are flexible so that they alter pitch when under heavy load. When the engine is running fast, and therefore the vehicle moving fast, the blades reduce pitch and the fan takes less power from the engine.

On other models a clutch is fitted between the fan blades and water pump pulley. The unit operates by utilizing the drag of a special silicone-based fluid between the inner and outer members. When drive is required, centrifugal force and the pumping action in the unit force the fluid outwards between the inner and outer member, so that drag drives the outer member. Immediately the engine is started there will be maximum drive. A bi-metallic spring, either flat as shown in **FIG 4:2** or spiral as shown in **FIG 4:3**, is fitted to the clutch unit. The spring senses the temperature of the air coming from the radiator and adjusts the flow of fluid in the clutch accordingly. During the warm-up cycle, when the air temperature from the radiator is less than 150 to 180°F, the spring moves the plunger to allow all the fluid to return to the central chamber and stay there. With no fluid in the outer chamber there is very little drag and the fan freewheels, running at a comparatively slow speed.

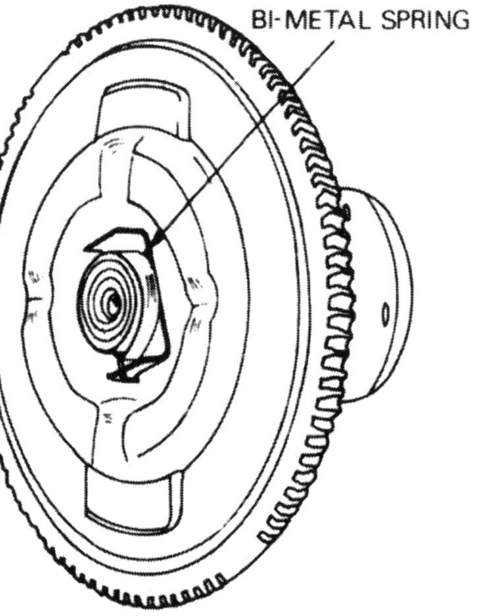

FIG 4:3 The cooling fan hub with spiral metallic spring

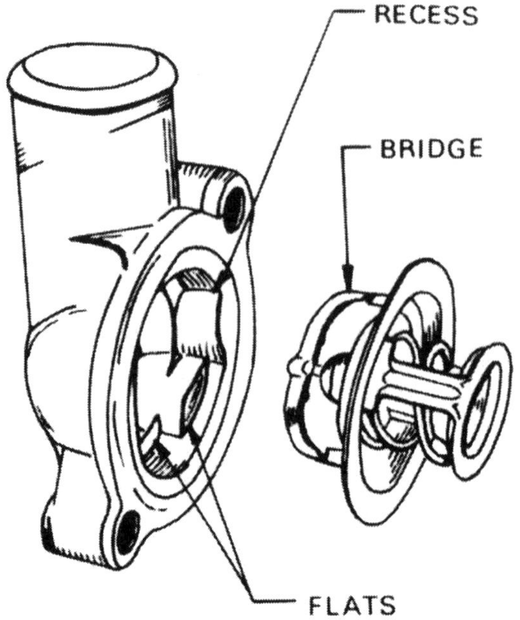

FIG 4:4 Later thermostat valve attachment

When the air temperature from the radiator rises, the spring senses this and moves the plunger so that fluid is allowed to return to the outer chamber. The fluid causes drag and the fan is then driven at engine speed. The spring will modulate the speed of the fan between the high and low settings, according to the road conditions and temperature of the engine. At high speed, the clutch action will limit the speed of the fan irrespective of the engine temperature.

To check the clutch, start the engine and allow it to run for a few minutes. Stop the engine and the fan should turn freely. Restart the engine and allow it to run until it has reached its normal operating temperature, covering the radiator to speed the process. Remove the blocking from the radiator and allow the engine to run for a minute so that the bi-metallic spring can soak to the temperature of the air from the radiator. Stop the engine. Protect the hands with rag and try to turn the fan blades. The blades should only turn with considerable pressure. If the fan turns easily the clutch is defective.

4:6 The thermostat

The attachment of the thermostat valve on later models is shown in **FIG 4:4** and the thermostat is installed and turned clockwise to lock it in the elbow. On earlier models a slightly different thermostat is fitted and the marking TOP must be uppermost otherwise the engine will overheat.

Drain the cooling system and take out the two bolts that secure the elbow in place, noting that the upper radiator hose is sufficiently flexible to leave it attached to the elbow. Pull off the elbow and remove the thermostat and gasket from it.

Before installing the thermostat, remove all traces of old gasket and sealing compound from the faces. Smear the faces of a new gasket with sealer and install it on the engine. Correctly fit the thermostat to the elbow and secure the elbow to the engine. Fill the cooling system and check for leaks.

Testing:

Suspend the thermostat in a container of water and gradually heat the water while stirring it with a thermometer. Note the temperature at which the valve starts to open, and less important, the temperature at which it is fully open. If the valve does not start to open within a few degrees of the correct temperature (see **Technical Data**), or the valve sticks in one position, the thermostat is defective and must be renewed.

Hold the thermostat up against a strong light and if there are gaps between valve and housing, the thermostat is again defective. A defective thermostat cannot be repaired and must be renewed.

4:7 The water pump

The pump has a bleed hole in the casting for the pump to allow coolant that has leaked past the seals to escape. This hole is not for lubrication purposes. If the pump leaks or is in any way defective it cannot be repaired and a new pump must be installed in its place.

Removal:

Drain the cooling system. Remove all the drive belts and take off the cooling fan with the pump hub and pulley assembly. Take out the bolts securing the pump and remove the pump, after disconnecting the radiator and heater hoses. On some models a bypass hose is fitted and this too must be disconnected from the pump. Discard the old gasket.

Clean the mounting faces free of old gasket or sealer. Lightly smear a new gasket with waterproof sealer and lay it back into position. Attach the pump, using the bolts, and make sure that the bypass hose (if fitted) connects to the pump. Install the fan, pulley and hoses in the reverse order of disassembly and reconnect the hoses. Set the correct belt tensions and fill the cooling system. Check for leaks both before and after starting the engine.

4:8 Trouble shooting guide

(a) Internal coolant leakage

1 Loose cylinder head bolts
2 Cracked cylinder head
3 Cracked cylinder block
4 Defective cylinder head gasket
5 Warped cylinder head
6 Loose intake manifold bolts

(b) Corrosion

1 Impurities in the coolant
2 Lack of inhibitor or antifreeze
3 Infrequent draining and flushing

(c) Poor circulation

1 Perished or collapsed coolant hoses
2 Defective thermostat
3 Low coolant level
4 Loose, slipping or broken drive belt
5 Engine water passages restricted
6 Radiator tubes or fins blocked

(d) Overheating

1 Check (c)
2 Defective cooling fan clutch
3 Sludge in oil pan or crankcase
4 Too low oil level in oil pan
5 Tight engine
6 Choked exhaust
7 Binding brakes
8 Slipping clutch (manual transmission only)
9 Incorrect valve timing
10 Retarded ignition
11 Mixture too weak
12 Automatic transmission overheating

Inches	Decimals	Millimetres	Inches to Millimetres Inches	mm	Millimetres to Inches mm	Inches
1/64	.015625	.3969	.001	.0254	.01	.00039
1/32	.03125	.7937	.002	.0508	.02	.00079
3/64	.046875	1.1906	.003	.0762	.03	.00118
1/16	.0625	1.5875	.004	.1016	.04	.00157
5/64	.078125	1.9844	.005	.1270	.05	.00197
3/32	.09375	2.3812	.006	.1524	.06	.00236
7/64	.109375	2.7781	.007	.1778	.07	.00276
1/8	.125	3.1750	.008	.2032	.08	.00315
9/64	.140625	3.5719	.009	.2286	.09	.00354
5/32	.15625	3.9687	.01	.254	.1	.00394
11/64	.171875	4.3656	.02	.508	.2	.00787
3/16	.1875	4.7625	.03	.762	.3	.01181
13/64	.203125	5·1594	.04	1.016	.4	.01575
7/32	.21875	5.5562	.05	1.270	.5	.01969
15/64	.234375	5.9531	.06	1.524	.6	.02362
1/4	.25	6.3500	.07	1.778	.7	.02756
17/64	.265625	6.7469	.08	2.032	.8	.03150
9/32	.28125	7.1437	.09	2.286	.9	.03543
19/64	.296875	7.5406	.1	2.54	1	.03937
5/16	.3125	7.9375	.2	5.08	2	.07874
21/64	.328125	8.3344	.3	7.62	3	.11811
11/32	.34375	8.7312	.4	10.16	4	.15748
23/64	.359375	9.1281	.5	12.70	5	.19685
3/8	.375	9.5250	.6	15.24	6	.23622
25/64	.390625	9.9219	.7	17.78	7	.27559
13/32	.40625	10.3187	.8	20.32	8	.31496
27/64	.421875	10.7156	.9	22.86	9	.35433
7/16	.4375	11.1125	1	25.4	10	.39370
29/64	.453125	11.5094	2	50.8	11	.43307
15/32	.46875	11.9062	3	76.2	12	.47244
31/64	.484375	12.3031	4	101.6	13	.51181
1/2	.5	12.7000	5	127.0	14	.55118
33/64	.515625	13.0969	6	152.4	15	.59055
17/32	.53125	13.4937	7	177.8	16	.62992
35/64	.546875	13.8906	8	203.2	17	.66929
9/16	.5625	14.2875	9	228.6	18	.70866
37/64	.578125	14.6844	10	254.0	19	.74803
19/32	.59375	15.0812	11	279.4	20	.78740
39/64	.609375	15.4781	12	304.8	21	.82677
5/8	.625	15.8750	13	330.2	22	.86614
41/64	.640625	16.2719	14	355.6	23	.90551
21/32	.65625	16.6687	15	381.0	24	.94488
43/64	.671875	17.0656	16	406.4	25	.98425
11/16	.6875	17.4625	17	431.8	26	1.02362
45/64	.703125	17.8594	18	457.2	27	1.06299
23/32	.71875	18.2562	19	482.6	28	1.10236
47/64	.734375	18.6531	20	508.0	29	1.14173
3/4	.75	19.0500	21	533.4	30	1.18110
49/64	.765625	19.4469	22	558.8	31	1.22047
25/32	.78125	19.8437	23	584.2	32	1.25984
51/64	.796875	20.2406	24	609.6	33	1.29921
13/16	.8125	20.6375	25	635.0	34	1.33858
53/64	.828125	21.0344	26	660.4	35	1.37795
27/32	.84375	21.4312	27	685.8	36	1.41732
55/64	.859375	21.8281	28	711.2	37	1.4567
7/8	.875	22.2250	29	736.6	38	1.4961
57/64	.890625	22.6219	30	762.0	39	1.5354
29/32	.90625	23.0187	31	787.4	40	1.5748
59/64	.921875	23.4156	32	812.8	41	1.6142
15/16	.9375	23.8125	33	838.2	42	1.6535
61/64	.953125	24.2094	34	863.6	43	1.6929
31/32	.96875	24.6062	35	889.0	44	1.7323
63/64	.984375	25.0031	36	914.4	45	1.7717

UNITS	Pints to Litres	Gallons to Litres	Litres to Pints	Litres to Gallons	Miles to Kilometres	Kilometres to Miles	Lbs. per sq. In. to Kg. per sq. Cm.	Kg. per sq. Cm. to Lbs. per sq. In.
1	.57	4.55	1.76	.22	1.61	.62	.07	14.22
2	1.14	9.09	3.52	.44	3.22	1.24	.14	28.50
3	1.70	13.64	5.28	.66	4.83	1.86	.21	42.67
4	2.27	18.18	7.04	.88	6.44	2.49	.28	56.89
5	2.84	22.73	8.80	1.10	8.05	3.11	.35	71.12
6	3.41	27.28	10.56	1.32	9.66	3.73	.42	85.34
7	3.98	31.82	12.32	1.54	11.27	4.35	.49	99.56
8	4.55	36.37	14.08	1.76	12.88	4.97	.56	113.79
9		40.91	15.84	1.98	14.48	5.59	.63	128.00
10		45.46	17.60	2.20	16.09	6.21	.70	142.23
20				4.40	32.19	12.43	1.41	284.47
30				6.60	48.28	18.64	2.11	426.70
40				8.80	64.37	24.85		
50					80.47	31.07		
60					96.56	37.28		
70					112.65	43.50		
80					128.75	49.71		
90					144.84	55.92		
100					160.93	62.14		

UNITS	Lb ft to kgm	Kgm to lb ft	UNITS	Lb ft to kgm	Kgm to lb ft
1	.138	7.233	7	.967	50.631
2	.276	14.466	8	1.106	57.864
3	.414	21.699	9	1.244	65.097
4	.553	28.932	10	1.382	72.330
5	.691	36.165	20	2.765	144.660
6	.829	43.398	30	4.147	216.990

CHAPTER 5

THE CLUTCH

5:1 Description

A mechanically operated clutch is fitted to all models covered by this manual. The 1970 model mechanical linkage is shown in **FIG 5:1** and this is fairly typical of all the models in the range.

A sectioned view of the clutch in the flywheel housing is shown in **FIG 5:2.** The mechanical linkage operates the clutch release lever. The clutch cover assembly is bolted to the engine flywheel and therefore rotates with the engine. The clutch disc has a hub with splines in it and it is fitted to the input shaft of the gearbox so that they revolve together.

The clutch casing assembly contains springs which act between the case and a pressure plate to force the pressure plate forwards. The clutch disc is gripped, by its friction linings, between the pressure plate and face of the engine flywheel so that the clutch disc is forced to rotate with the engine and transmit drive to the input shaft of the transmission.

When the clutch pedal is depressed, the linkage operates the clutch release lever. The lever moves about its pivot and presses the clutch release bearing against the clutch fingers. The fingers in turn pull the pressure plate back against the pressure of the clutch spring, releasing the clutch disc so that it and ιne transmission input shaft are free to revolve independently or even come to a stop with the engine running.

As the clutch pedal is released, there is a gradual increase of pressure on the clutch disc and at first there is slip between the clutch disc and pressure faces so that drive is smoothly taken up. Eventually the point is reached where the clutch disc is revolving at engine speed and the drive is again direct. Damper springs are fitted into the clutch disc to help absorb and shock of take-up of drive.

All the components of a 1964 model are shown in **FIG 5:3** and the action of the clutch can be most easily followed through using this figure.

5:2 Clutch adjustments

As the friction linings on the disc wear, the clutch pedal free travel will alter. **It is essential that there is always a clearance between the clutch release bearing and clutch release fingers when the clutch is engaged.** If the release bearing contacts the fingers at all times there will be rapid wear on the release bearing and in excessive cases there may be sufficient pressure on the fingers to partially withdraw the pressure plate and allow the clutch to slip.

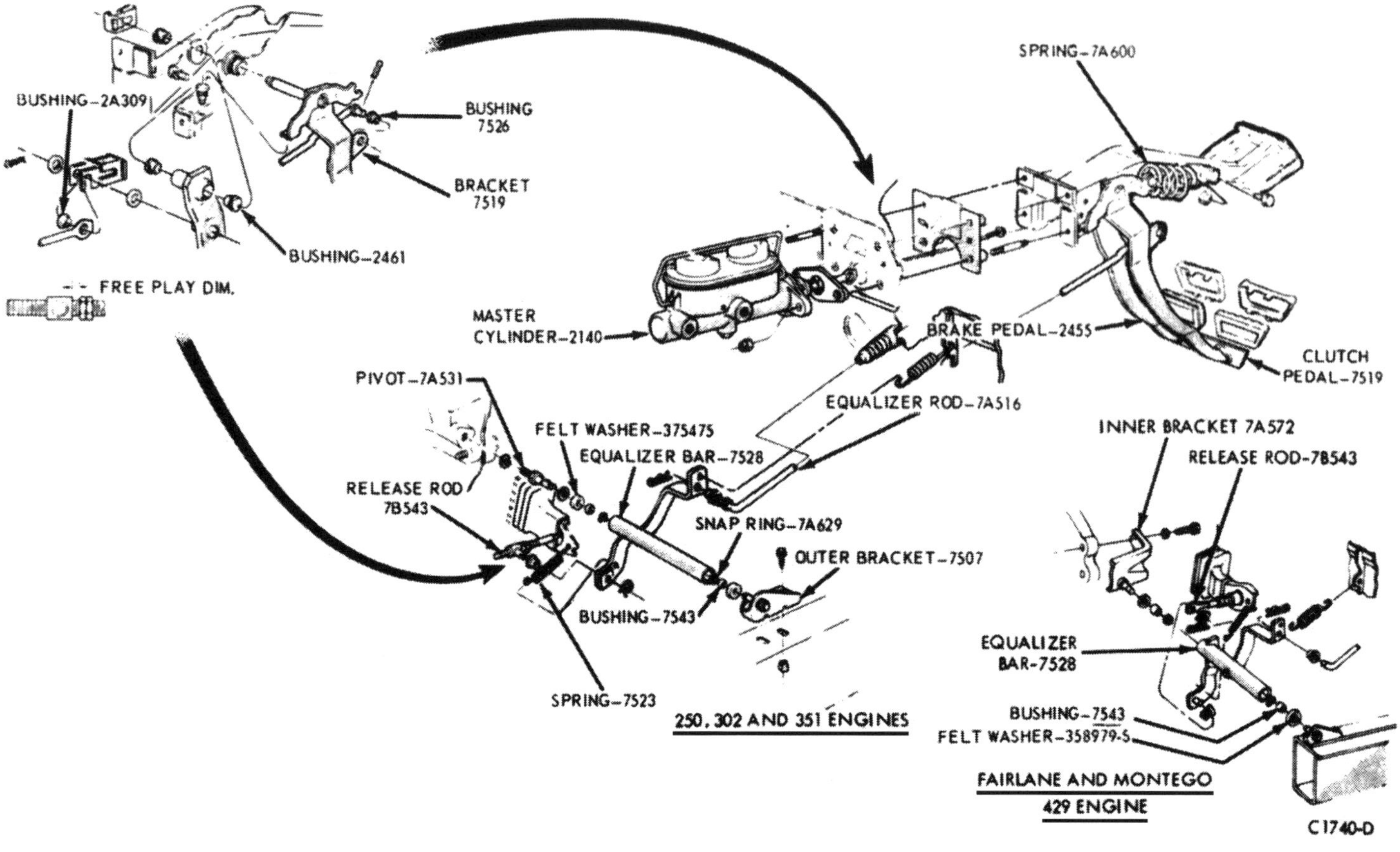

FIG 5:1 Late model clutch linkage

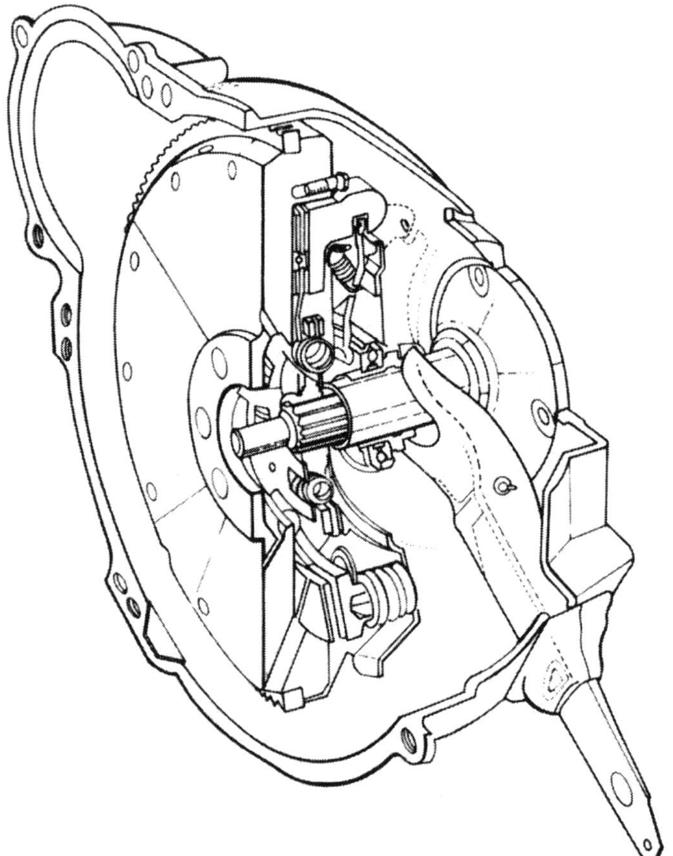

FIG 5:2 Sectional view of typical clutch

The method of checking on earlier models is shown in **FIG 5:4**. Press the clutch pedal right down and measure the total travel. If the total travel is incorrect, adjust the position of the bumper which the pedal hits when it is fully released to set the correct full travel. Start the engine and leave it running at idling. Press down the clutch pedal by hand until the release bearing can just be felt to contact the fingers in the clutch. This point is readily detectable by hand. On all models the adjustment is correct if the free travel of the clutch pedal is $\frac{7}{8}$ to $1\frac{1}{8}$ inch. The adjustment point on 1965 models is shown in **FIG 5:5**. Slacken the nuts and move them in the direction indicated to alter the free play. Tighten both nuts to the trunnion. When the free play at idling is correct, increase the engine speed to 3000 rev/min and check that the minimum pedal free travel is $\frac{1}{2}$ inch. If the free travel at speed is not sufficient, the free travel at idling must be increased.

On the later models (see **FIG 5:1**) the check is carried out in a similar manner. To adjust the free play, disconnect the return spring from the clutch release lever. Slacken the nuts on the release lever rod. Slacken off the adjusting nut and pull the release lever rearwards until the release bearing can just be felt to touch the fingers in the clutch. Press the release rod into the release lever so that they are in contact and hold them both in this position. Insert a .136 inch feeler gauge between the adjusting nut and swivel sleeve and tighten down the adjusting nut until it just nips onto the feeler gauge. Tighten the locknut to hold the adjusting nut in

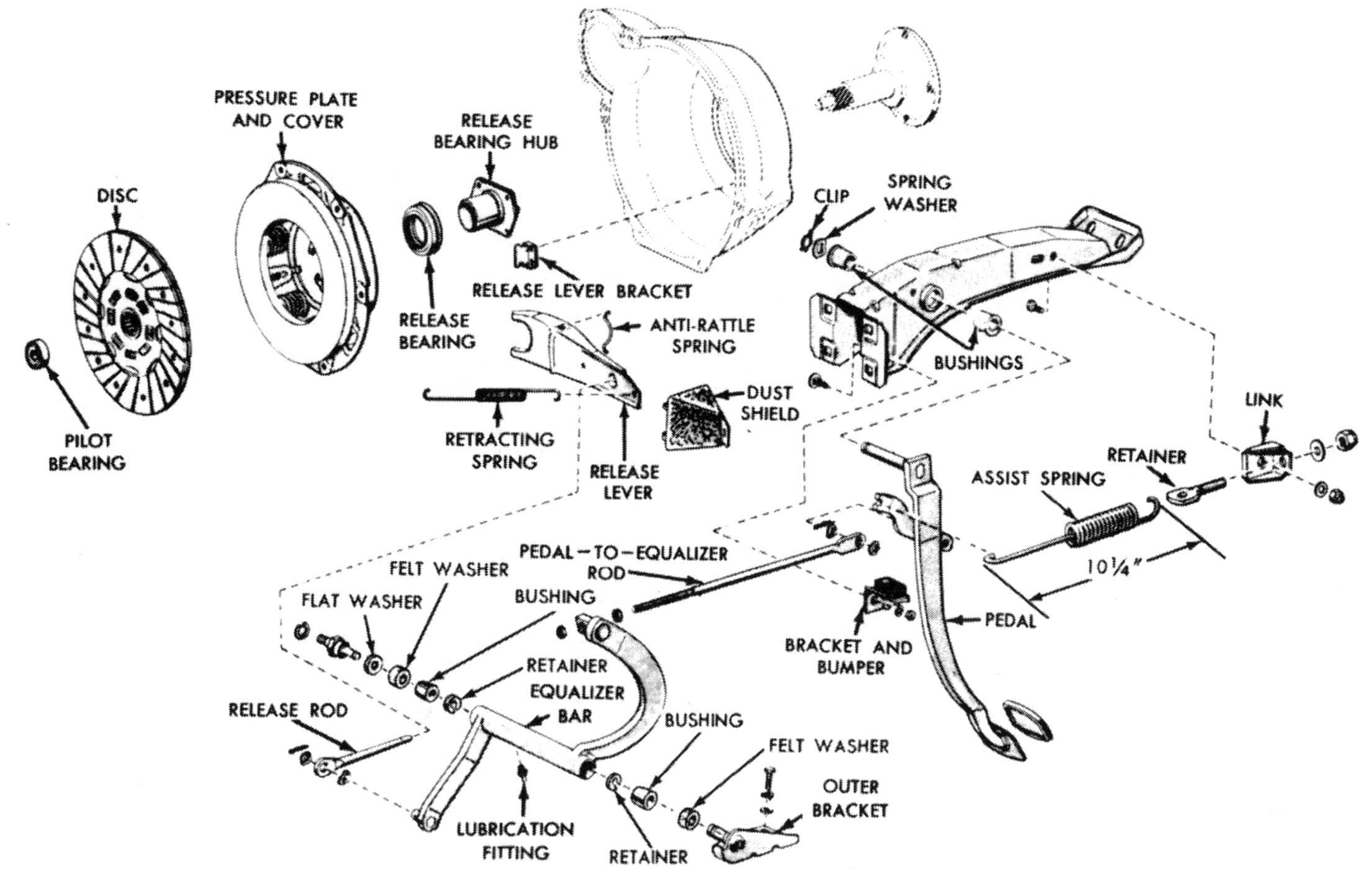

FIG 5:3 Early model clutch linkage

place and remove the feeler gauge. Reconnect the return spring to the release lever. Check that the pedal free travel is correct at idling and 3000 rev/min.

5:3 The flywheel housing

The removal of the flywheel housing, after the transmission has been removed, is dealt with in **Chapter 1, Section 1:11.** When removing the housing, carefully collect any shims fitted between it and the engine.

The shims are fitted to correct the alignment of the housing to the engine, though not more than .010 inch of shims may be used at any one point.

If it is suspected that the alignment of housing and engine is not correct, a dial indicator, pilot tool and Dia-L-Igner gauge will be required to check the alignment. The clutch release bearing and lever must be removed from the housing and the spark plugs removed from the engine. Full instructions are given on the Dia-L-Igner gauge, and shims are made up to the dimensions shown in **FIG 5:6.** If the Dia-L-Igner gauge gives a reading of greater than .014 inch then the housing is not acceptable.

A certain amount of misalignment can be corrected by a service garage, using new locating dowels and drilling faulty mounting holes oversize.

5:4 Release bearing

The release bearing and lever can be removed after the transmission and flywheel housing have been removed from the engine.

The bearing is sealed and no further lubrication can be added to it. **Do not wash the bearing in any solvent otherwise the lubricant will be washed out of it.**

Inspect the bearing for nicks or roughness and clean off any such damage with Crocus cloth. Apply pressure to the bearing and rotate it to check for roughness or noise in operation. If the bearing is rough or noisy, reject it and fit a new bearing in its place. The old bearing can be carefully driven off from the hub but when fitting a new bearing use the tools shown in **FIG 5:7.**

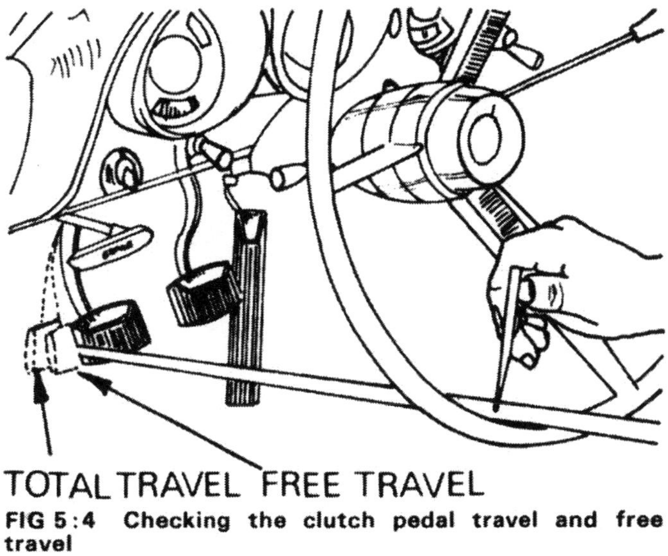

FIG 5:4 Checking the clutch pedal travel and free travel

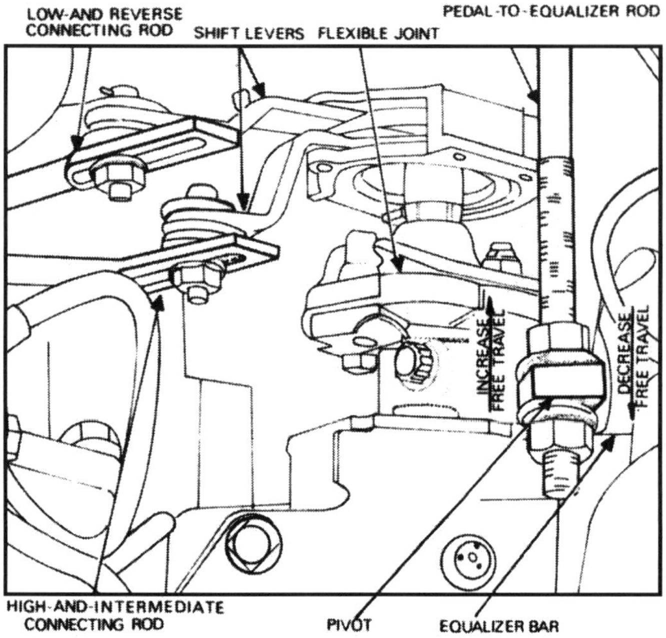

FIG 5:5 Adjusting the clutch linkage on early models

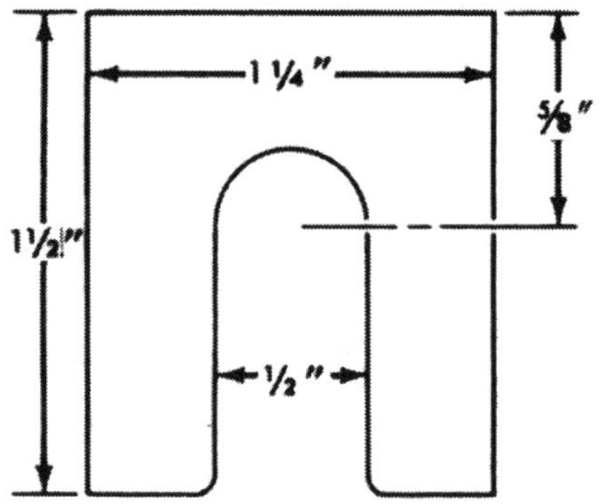

FIG 5:6 Shims to be made up for aligning the clutch housing

Install the parts in the reverse order of removal. The pivot and bearing points on the release lever should be lightly lubricated with Lubriplate and the operating face of the release bearing, where it contacts the release fingers, should be lightly coated with lithium-based grease. **Avoid overlubrication otherwise grease may find its way onto the clutch disc.**

5:5 Servicing the clutch

The removal and installation of the clutch is dealt with **in Chapter 1, Section 1:11. It is essential to use the arbor to centralize the clutch disc while installing the cover assembly,** otherwise the input shaft of the transmission cannot be inserted through the clutch disc into the pilot bushing in the end of the crankshaft.

Damage or wear to either the cover assembly or the clutch disc requires renewal of the component, as neither component can be disassembled.

Cover assembly:

Do not immerse the assembly in solvent. Cleaning should be confined to blowing and brushing out loose dirt and wiping oily surfaces with a cloth moistened with suitable solvent (such as denatured alcohol).

Examine the cover itself for cracks or distortion, paying particular attention to the areas around rivets and the attachment holes.

Inspect the surface of the pressure plate for burn marks, scores or hairline cracks. Light damage is removable, though not advisable, and if the damage is deep the cover assembly must be renewed.

Lay the assembly, pressure plate downwards, onto a clean floor or plank of wood. Check that the release fingers are all at approximately the same height and that no finger shows more wear than the others. Press each finger down in turn and release it smartly. If the finger does not return quickly there is a binding condition. If the fingers are at uneven heights or the binding condition cannot be cured, a new cover assembly must be fitted.

Before installing the clutch, lubricate the pivots, as shown in **FIG 5:8**, with a little lithium-based grease and pump the fingers to work the grease well in. Wipe away surplus grease.

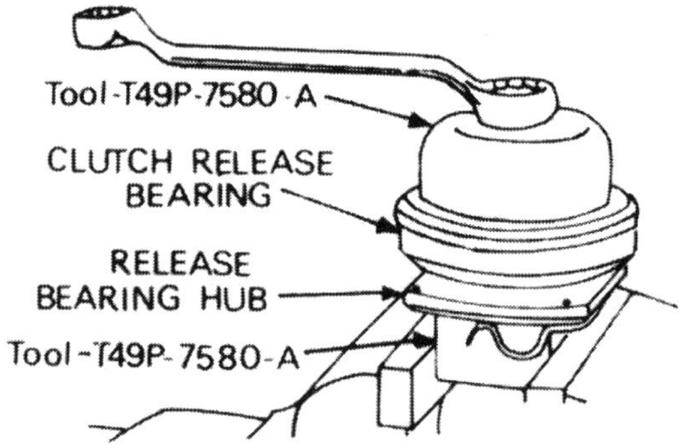

FIG 5:7 Installing a new clutch release bearing onto the hub

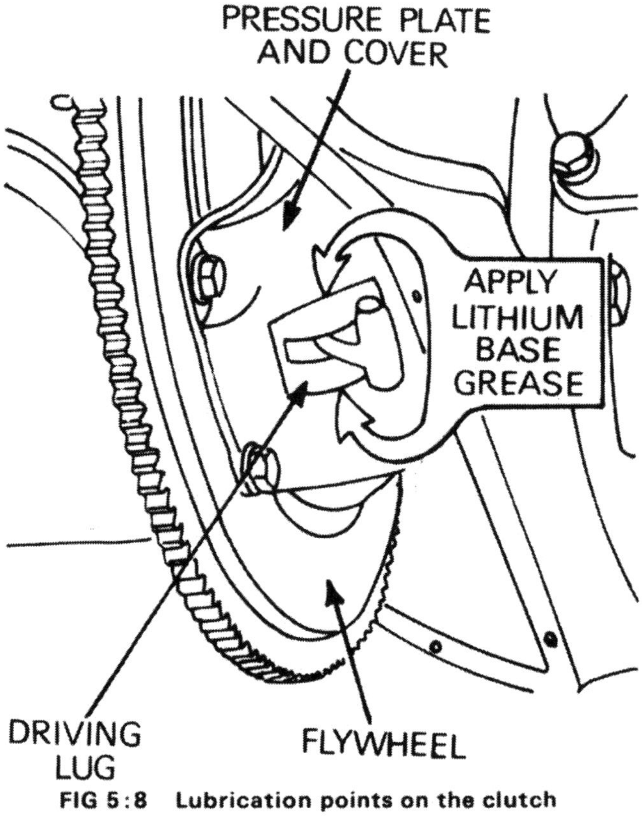

FIG 5:8 Lubrication points on the clutch

Clutch disc:

Take great care not to drop the assembly or to allow lubricant onto the friction linings.

Check the disc for wear, mechanical damage or contamination and renew it if it is defective.

The lining surface must be well clear of the rivet heads. Install a new disc if the linings are worn down nearly to the rivet heads as it will save carrying out the work again in a short time.

Check that all rivets are tight and secure. Check that the hub is securely in place and that none of the damper springs are loose or broken. Slide the disc back onto the transmission input shaft and check that it slides freely without excess rotational play or movement in the damper springs.

Any oil leaks into the flywheel housing must be traced and rectified before the clutch is installed. If gaskets and seals are satisfactory the cause can be overfilling of the transmission or a blocked transmission vent.

For maximum efficiency the friction linings should have a smooth polished finish through which the grain of the material is clearly visible. Small amounts of oil or grease will leave dark colored smears. Larger amounts of oil will leave a dark colored glaze over the friction surface, hiding the grain of the material, while excessive amounts will be obvious from the free oil in the housing and the oil-soaked appearance of the linings. The disc can be considered suitable for further use, providing that the linings are not excessively worn or mechanically defective, as long as the grain of the friction material can be seen. **Do not attempt to wash off grease or oil with solvents.**

5:6 Trouble shooting guide

(a) Drag or spin

1 Oil or grease on disc linings
2 Incorrectly adjusted clutch pedal free travel
3 Misalignment between engine and flywheel housing
4 Binding pilot bushing in crankshaft
5 Distorted clutch disc
6 Warped or damaged pressure plate
7 Broken disc linings

(b) Fierceness or snatch

1 Check 1, 3, 4, 5, 6 and 7 in (a)
2 Excessively worn disc linings
3 Worn or loose disc hub

(c) Slip

1 Check 1, 2 and 3 in (a) and 2 in (b)
2 Weak cover assembly springs

(d) Judder

1 Check 1, 2 and 7 in (a)
2 Contact area of friction linings not evenly distributed
3 Buckled disc
4 Faulty or loose engine mountings
5 Defective rear suspensions or drive shaft

(e) Rattles

1 Worn release linkage
2 Worn release lever and bearing
3 Broken spring in disc or cover assembly.

Inches	Decimals	Millimetres	Inches to Millimetres		Millimetres to Inches	
			Inches	mm	mm	Inches
1/64	.015625	.3969	.001	.0254	.01	.00039
1/32	.03125	.7937	.002	.0508	.02	.00079
3/64	.046875	1.1906	.003	.0762	.03	.00118
1/16	.0625	1.5875	.004	.1016	.04	.00157
5/64	.078125	1.9844	.005	.1270	.05	.00197
3/32	.09375	2.3812	.006	.1524	.06	.00236
7/64	.109375	2.7781	.007	.1778	.07	.00276
1/8	.125	3.1750	.008	.2032	.08	.00315
9/64	.140625	3.5719	.009	.2286	.09	.00354
5/32	.15625	3.9687	.01	.254	.1	.00394
11/64	.171875	4.3656	.02	.508	.2	.00787
3/16	.1875	4.7625	.03	.762	.3	.01181
13/64	.203125	5·1594	.04	1.016	.4	.01575
7/32	.21875	5.5562	.05	1.270	.5	.01969
15/64	.234375	5.9531	.06	1.524	.6	.02362
1/4	.25	6.3500	.07	1.778	.7	.02756
17/64	.265625	6.7469	.08	2.032	.8	.03150
9/32	.28125	7.1437	.09	2.286	.9	.03543
19/64	.296875	7.5406	.1	2.54	1	.03937
5/16	.3125	7.9375	.2	5.08	2	.07874
21/64	.328125	8.3344	.3	7.62	3	.11811
11/32	.34375	8.7312	.4	10.16	4	.15748
23/64	.359375	9.1281	.5	12.70	5	.19685
3/8	.375	9.5250	.6	15.24	6	.23622
25/64	.390625	9.9219	.7	17.78	7	.27559
13/32	.40625	10.3187	.8	20.32	8	.31496
27/64	.421875	10.7156	.9	22.86	9	.35433
7/16	.4375	11.1125	1	25.4	10	.39370
29/64	.453125	11.5094	2	50.8	11	.43307
15/32	.46875	11.9062	3	76.2	12	.47244
31/64	.484375	12.3031	4	101.6	13	.51181
1/2	.5	12.7000	5	127.0	14	.55118
33/64	.515625	13.0969	6	152.4	15	.59055
17/32	.53125	13.4937	7	177.8	16	.62992
35/64	.546875	13.8906	8	203.2	17	.66929
9/16	.5625	14.2875	9	228.6	18	.70866
37/64	.578125	14.6844	10	254.0	19	.74803
19/32	.59375	15.0812	11	279.4	20	.78740
39/64	.609375	15.4781	12	304.8	21	.82677
5/8	.625	15.8750	13	330.2	22	.86614
41/64	.640625	16.2719	14	355.6	23	.90551
21/32	.65625	16.6687	15	381.0	24	.94488
43/64	.671875	17.0656	16	406.4	25	.98425
11/16	.6875	17.4625	17	431.8	26	1.02362
45/64	.703125	17.8594	18	457.2	27	1.06299
23/32	.71875	18.2562	19	482.6	28	1.10236
47/64	.734375	18.6531	20	508.0	29	1.14173
3/4	.75	19.0500	21	533.4	30	1.18110
49/64	.765625	19.4469	22	558.8	31	1.22047
25/32	.78125	19.8437	23	584.2	32	1.25984
51/64	.796875	20.2406	24	609.6	33	1.29921
13/16	.8125	20.6375	25	635.0	34	1.33858
53/64	.828125	21.0344	26	660.4	35	1.37795
27/32	.84375	21.4312	27	685.8	36	1.41732
55/64	.859375	21.8281	28	711.2	37	1.4567
7/8	.875	22.2250	29	736.6	38	1.4961
57/64	.890625	22.6219	30	762.0	39	1.5354
29/32	.90625	23.0187	31	787.4	40	1.5748
59/64	.921875	23.4156	32	812.8	41	1.6142
15/16	.9375	23.8125	33	838.2	42	1.6535
61/64	.953125	24.2094	34	863.6	43	1.6929
31/32	.96875	24.6062	35	889.0	44	1.7323
63/64	.984375	25.0031	36	914.4	45	1.7717

UNITS	Pints to Litres	Gallons to Litres	Litres to Pints	Litres to Gallons	Miles to Kilometres	Kilometres to Miles	Lbs. per sq. In. to Kg. per sq. Cm.	Kg. per sq. Cm. to Lbs. per sq. In.
1	.57	4.55	1.76	.22	1.61	.62	.07	14.22
2	1.14	9.09	3.52	.44	3.22	1.24	.14	28.50
3	1.70	13.64	5.28	.66	4.83	1.86	.21	42.67
4	2.27	18.18	7.04	.88	6.44	2.49	.28	56.89
5	2.84	22.73	8.80	1.10	8.05	3.11	.35	71.12
6	3.41	27.28	10.56	1.32	9.66	3.73	.42	85.34
7	3.98	31.82	12.32	1.54	11.27	4.35	.49	99.56
8	4.55	36.37	14.08	1.76	12.88	4.97	.56	113.79
9		40.91	15.84	1.98	14.48	5.59	.63	128.00
10		45.46	17.60	2.20	16.09	6.21	.70	142.23
20				4.40	32.19	12.43	1.41	284.47
30				6.60	48.28	18.64	2.11	426.70
40				8.80	64.37	24.85		
50					80.47	31.07		
60					96.56	37.28		
70					112.65	43.50		
80					128.75	49.71		
90					144.84	55.92		
100					160.93	62.14		

UNITS	Lb ft to kgm	Kgm to lb ft	UNITS	Lb ft to kgm	Kgm to lb ft
1	.138	7.233	7	.967	50.631
2	.276	14.466	8	1.106	57.864
3	.414	21.699	9	1.244	65.097
4	.553	28.932	10	1.382	72.330
5	.691	36.165	20	2.765	144.660
6	.829	43.398	30	4.147	216.990

CHAPTER 6

THE TRANSMISSION

6:1 Description

A threespeed transmission is fitted as standard throughout the range covered by this manual. Fourspeed transmissions are fitted as standard to some Falcon models with V8 engines, so care must be taken to specify the type of gearbox when ordering parts. A column shift is fitted as standard to the models with in-line 6 engines. The forward speeds are all fitted with synchromesh engagement but a spur-cut idler gear is slid into mesh between mainshaft and countershaft gears to provide reverse.

The power flow in the various selections is shown in **FIG 6:1**. The input shaft and its gear are integral and they are splined to the disc of the clutch so that they revolve with the clutch. The input shaft rotates in a ballbearing in the casing and its rear end is supported on a needle roller bearing which fits over the spigot of the output shaft. The gear on the input shaft is in constant mesh with the front gear on the countershaft gear cluster so that the cluster is revolving in the opposite direction to the input shaft. The countershaft gear cluster rotates about needle roller bearings on the countershaft and the end float of the cluster is controlled by thrust washers at either end. The remaining gears on the countershaft cluster are in constant mesh with those on the output shaft. So far all the gears mentioned are of helical cut for quiet operation.

The gears on the output shaft are free to revolve about the shaft so that when a gear is not selected no drive is transmitted to the output shaft. A synchromesh unit connects the gear to the shaft to supply drive. The hub of the synchronizer is splined internally to the shaft and splined externally to the sleeve, the sleeve being moved by the appropriate selector fork. When the sleeve is first moved, pressure is applied to the blocking ring and friction on the conical surfaces brings the parts to the same speed. Further movement of the sleeve then meshes the dog teeth to make the drive positive.

The output shaft is supported by two ballbearings and the needle roller bearing in the input shaft. Thrust washers take the thrust of the helical cut gears and also control the end floats of the gears. The components of an early transmission are shown in **FIG 6:2**. Later transmissions differ slightly.

6:2 Maintenance

Seasonal changes in lubricant are not required. When removing the transmission from the vehicle it is advisable to drain out the old lubricant and discard it, filling with fresh lubricant after installation.

A drain plug is fitted underneath the casing, and the drain plug has a magnet in it to ensure that metallic particles are trapped and not allowed to circulate with the oil.

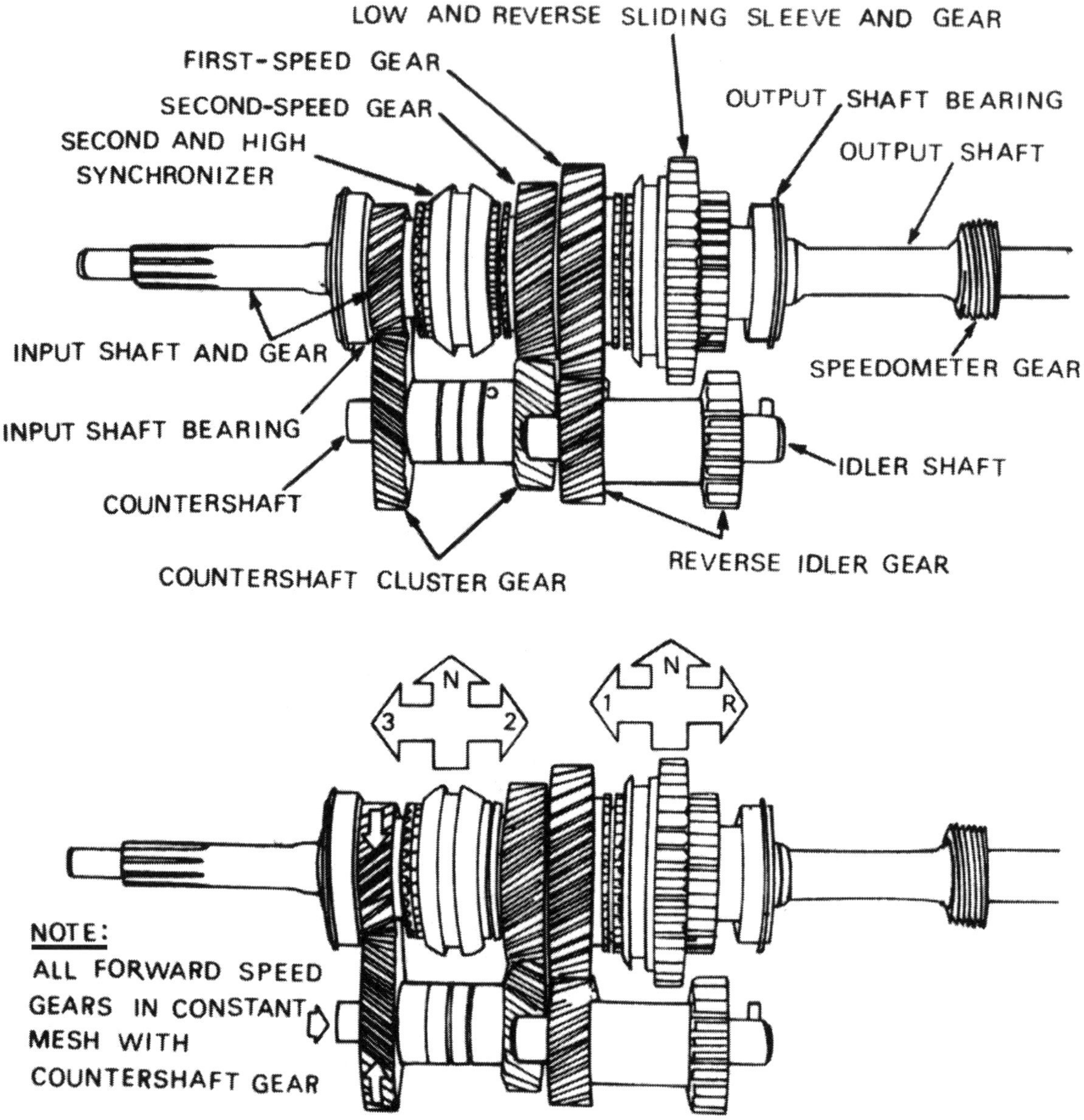

FIG 6:1 The power flow through the transmission

At regular intervals check the oil level in the transmission. Wipe the area around the combined filler and level plug clean and stand the vehicle on level ground. Remove the filler plug and check that the level is just up to the bottom of the filler orifice. If the level is low, fill using a recommended lubricant (Mild EP Gear Oil C3RZ 19547-B). Allow the surplus lubricant to drain out from the filler hole before installing the filler plug.

On some later models, no drain plug may be fitted and to drain the oil out remove the lower bolt that secures the rear extension housing.

6:3 Removing the transmission

There are only minor variations between the transmission attachments throughout the range. Full details of the drive shaft are given in **Chapter 8**, and if there is any difficulty in disconnecting the drive shaft this chapter should be consulted.

Removal and installation will be easiest if the car is raised on a hoist, or work carried out from a pit. If neither of these is available, support the vehicle securely on firm stands.

1 Drain the oil from the transmission. This is not absolutely essential but it will minimize the risk of spillage.

2 Disconnect the shift linkage rods (see next section for details).

3 Mark the flanges of the rear connection of the drive shaft and disconnect it from the rear axle. Wrap tape around the rear universal joint to prevent the loose bearings from falling off. Pull the drive shaft rearwards to disconnect it from the transmission. Blank off the aperture in the rear extension housing of the transmission to prevent oil from leaking out or dirt from entering.

4 Still underneath the vehicle, remove the capscrew and lockwasher that secure the speedometer drive cable to the transmission and withdraw the cable from the transmission.

5 Support the power unit with a jack and pad of wood under the engine oil pan. Remove the bolts that secure the rear crossmember to the frame. Raise the rear of the power unit slightly and remove the rear support crossmember. On some models the park

brake cable is attached to the crossmember and will therefore have to be freed. Again, on some models, it will be necessary to disconnect the park brake cable to allow the transmission to be removed.

6 Support the transmission on another jack and pad of wood or suitable support. Remove the four bolts that secure the transmission to the flywheel housing. It may be necessary to lower both jacks, keeping the transmission aligned with the engine, in order to draw the transmission back. Draw the transmission rearwards until the input shaft is well clear of the clutch and then lower the transmission out of the vehicle. **Do not allow the weight of the transmission to hang by the input shaft in the clutch or the clutch will be seriously damaged. Once the transmission has been removed, do not operate the clutch.** If the clutch is operated, the disc will fall and have to be realigned with the arbor.

Installation:

The transmission is installed in the reverse order of removal. The task will be made easier if two guide pins are fitted to the flywheel housing and the transmission slid along these to align the input shaft with the clutch. If the input shaft is difficult to enter into the clutch, select a gear and rotate the output shaft until the splines on the input shaft and clutch align. **Do not allow the weight of the transmission to hang by the input shaft in the clutch.**

When all connections have been made, fill the trans-transmission with fresh lubricant. Fill slowly as the oil is thick and will overflow before the unit is full if it is poured in rapidly.

Check the shift selector linkage for adjustment and road test the vehicle.

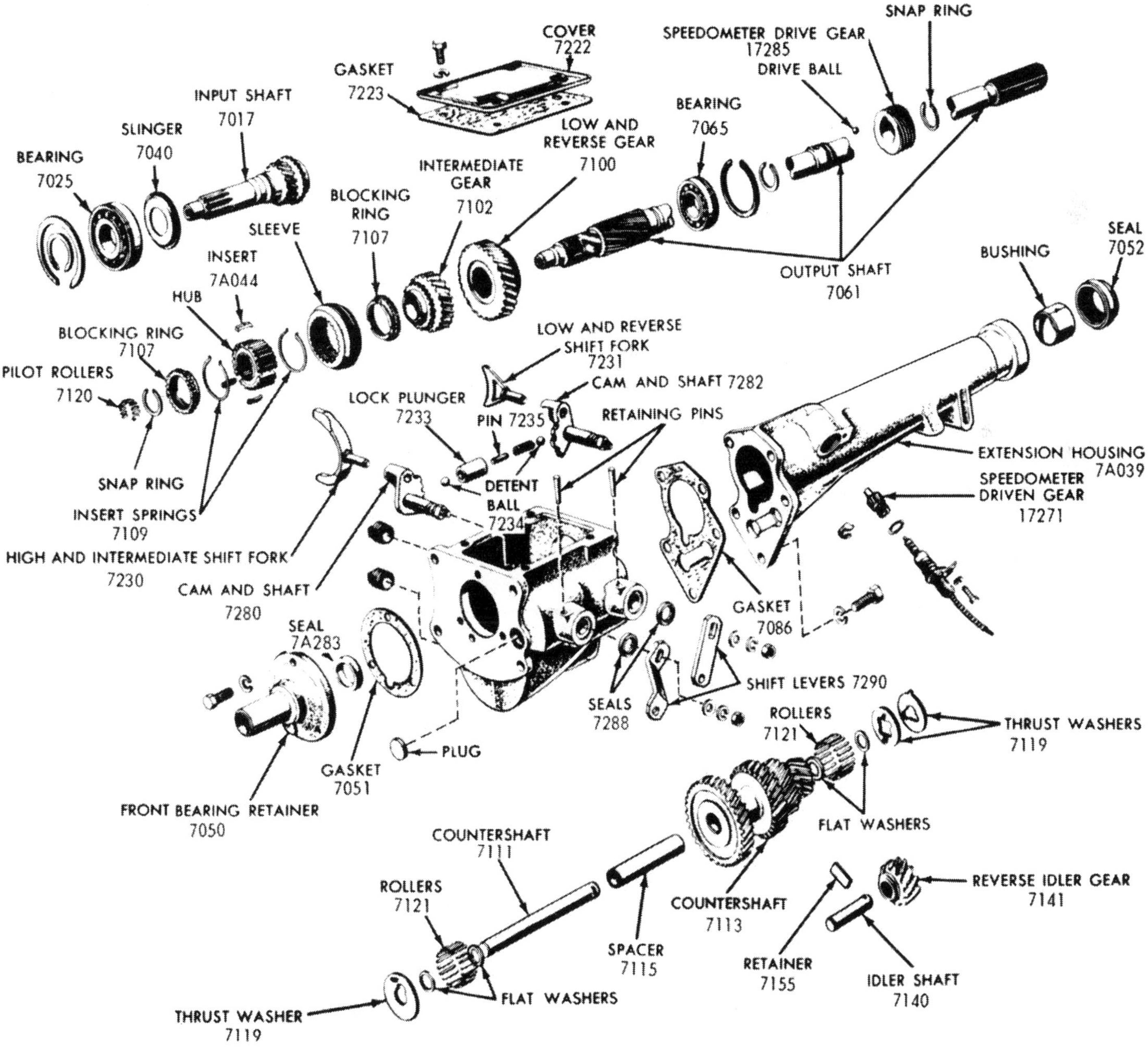

FIG 6:2 The components of the earlier transmission

6:4 Shift linkage adjustment

The method of adjusting the linkage up to and including 1967 models is shown in **FIG 6:3**. The gauge for the adjustment has to be fabricated and it is not available as a special tool. Move the gearshift lever into the Neutral position and slacken the adjusting nuts. Check that both the levers on the transmission are in the neutral position. Fit the fabricated gauge into position in the slot in the column and tighten the adjusting nuts. Remove the gauge and check that all gears engage freely and smoothly.

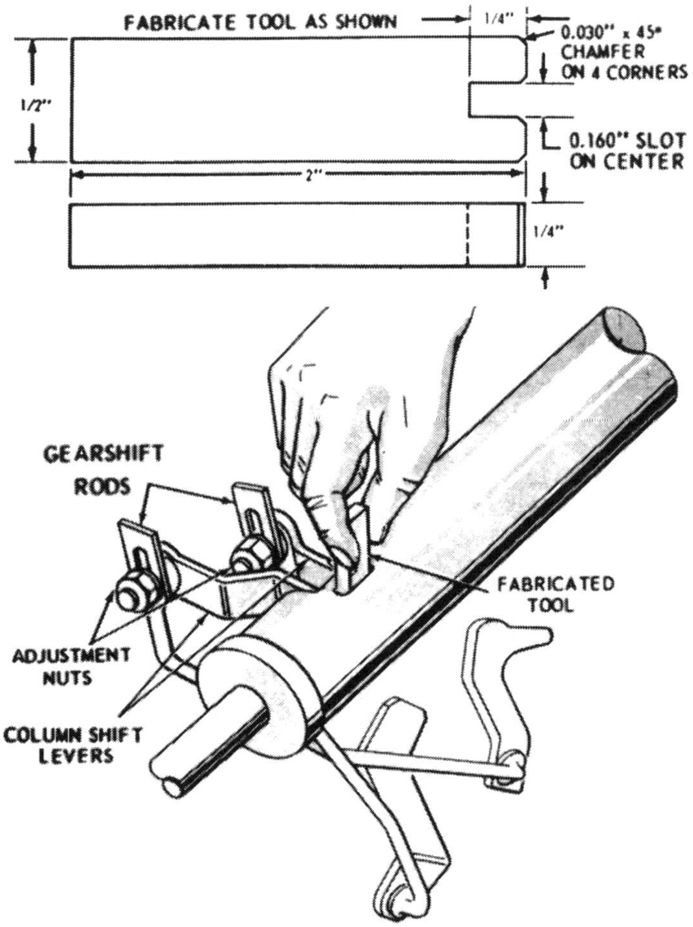

FIG 6:3 Adjusting the gearshift linkage on early models

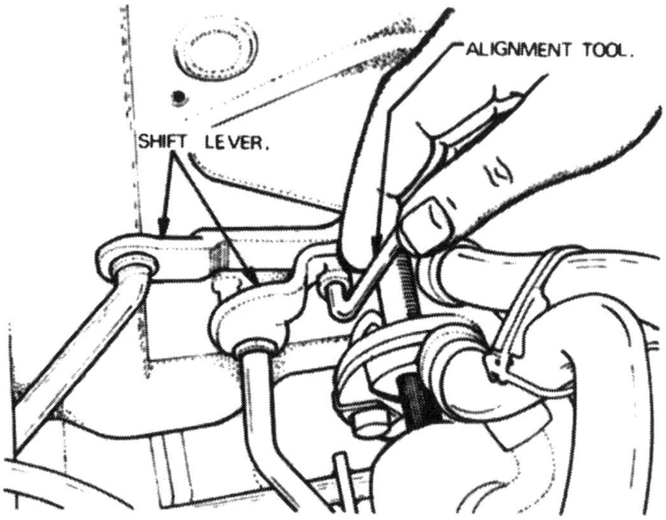

FIG 6:4 Adjusting the gearshift linkage on later models

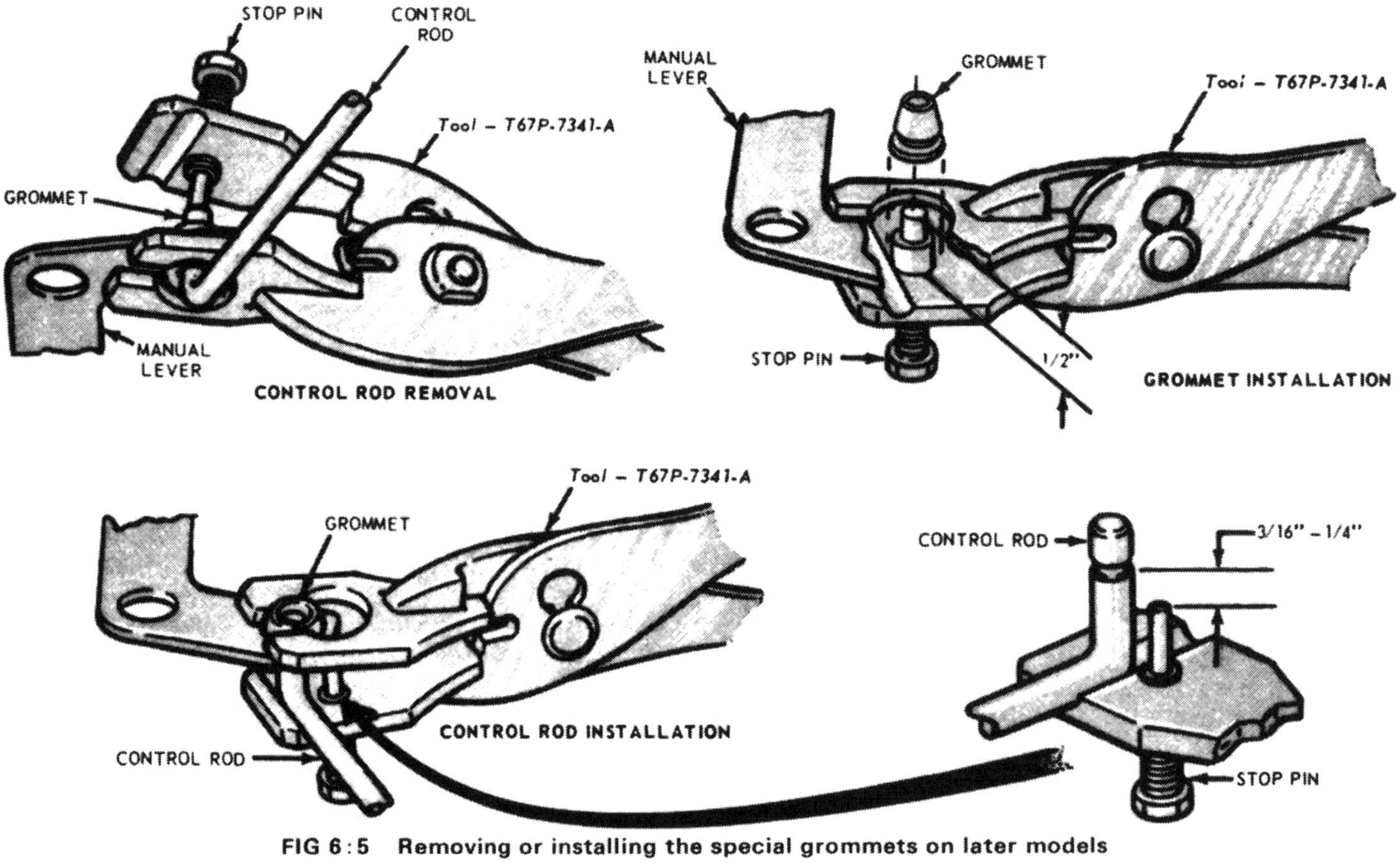

FIG 6:5 Removing or installing the special grommets on later models

On later models the method of adjustment is very similar and is shown in **FIG 6:4.** An alignment tool is made out of a piece of $\frac{3}{16}$ inch diameter rod bent to an L-shape with legs of approximately 1 and 2 inches. Set the gearshift lever to neutral and slacken the adjustment nuts by the transmission itself. Insert the short leg of the pin through the holes in the levers and holes in the casting. Move the selector levers on the transmission to their neutral positions and tighten the adjusting nuts. Remove the alignment tool, start the engine and check for ease of selection in all speeds.

On the later models, special grommets secure the rods to the levers at the base of the column. Special pliers are needed to remove and install the control rod and once the rod has been disconnected the old grommet must be discarded and a new one installed. The method of using the special pliers is shown in **FIG 6:5.** The stop pin must be adjusted to position the grommet, first in the lever and then on the rod.

6:5 Servicing the transmission

Special tools, extractors, and an arbor press are required for dismantling and reassembling the transmission. The parts must be installed with the correct end plays otherwise the unit will not operate quietly or efficiently. If the owner has any doubts on his ability to strip and reassemble the unit, he is advised to take it to a service garage.

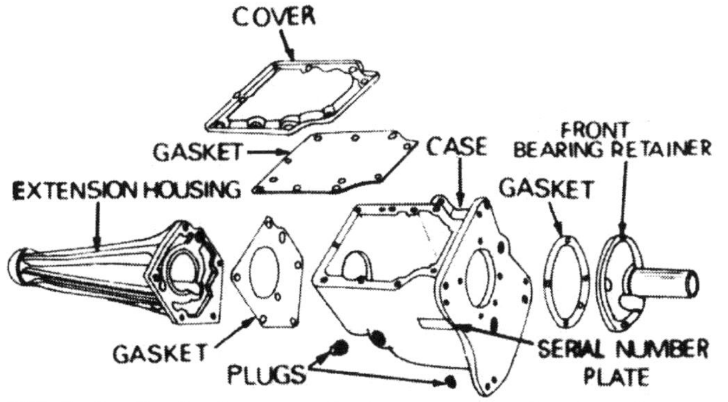

FIG 6:6 Typical transmission case and related parts

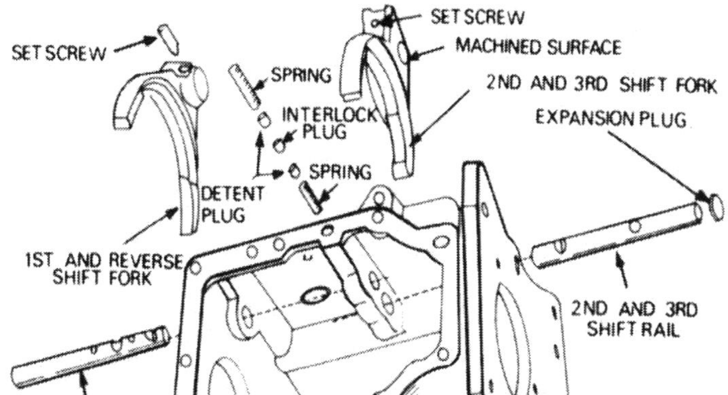

FIG 6:7 The shift rails and forks of the later transmission

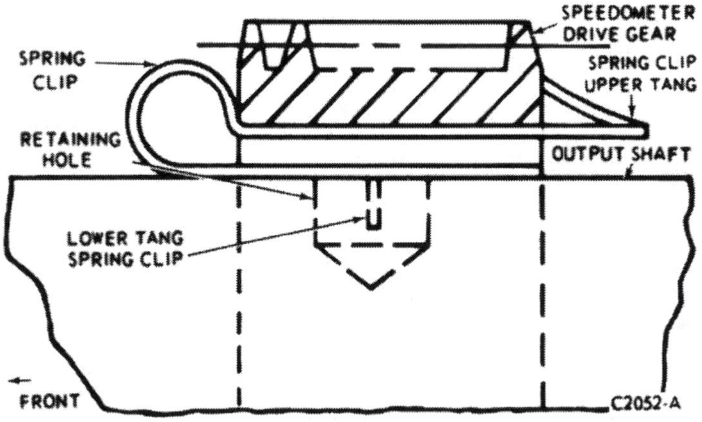

FIG 6:8 The latest method of securing the speedometer drive gear

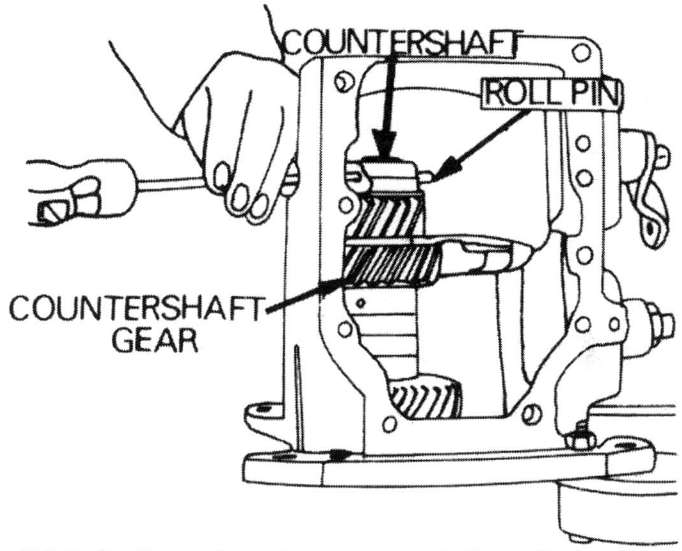

FIG 6:9 Removing the countershaft roll-pin from later models

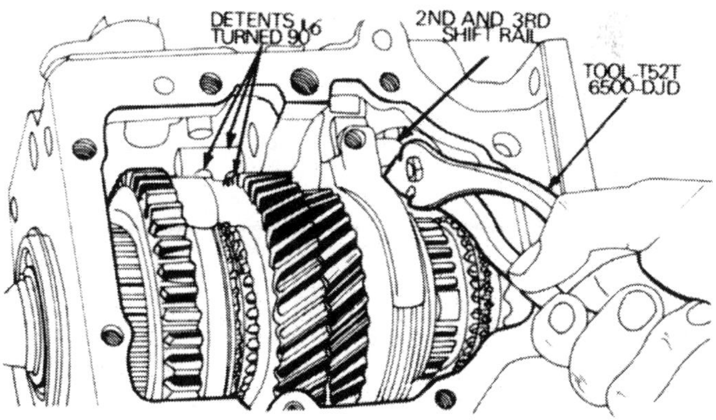

FIG 6:10 Rotating the shift rail on later models

Disassembly:

Typical transmission case and related parts are shown in **FIG 6:6.** Before starting work drain out the lubricant, if not already done, and refit the drain and filler plugs. Wash down the outside of the unit with kerosene, gasoline or similar suitable solvent.

Remove the cover and its gasket. Care must be taken in this operation as on the latest models the detent springs are held in place by the cover, as shown in **FIG 6:7.** On slightly earlier models a screw plug is fitted to hold the detent springs down in place.

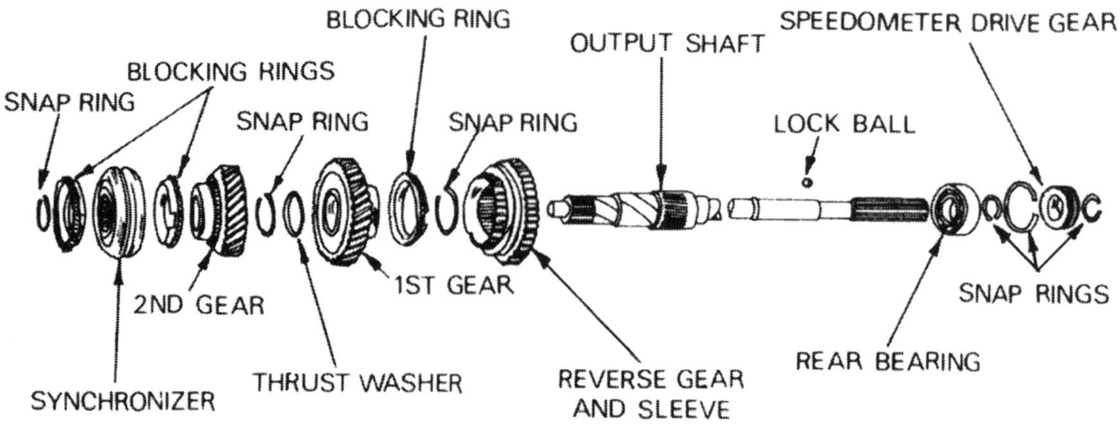

FIG 6:11 The output shaft components on later models

Take out the screws that secure the rear extension housing. Withdraw the extension housing, if necessary tapping the output shaft inwards to prevent it being moved with the extension housing and dislodging the needle rollers between output and input shafts. Remove the speedometer drive gear. On earlier models the speedometer drive gear is secured by a snap ring and driven by a ball. Remove the snap ring and draw off the gear, carefully collecting the drive ball. On the very latest models the drive is secured by a spring clip, as shown in **FIG 6:8.** Depress the upper tang and slide the gear off, removing the complete spring clip after the gear is off.

Take out the cap screws and remove the front bearing retainer complete with its gasket.

On the earlier models, remove the retainer (shown in **FIG 6:2**) which secures the countershaft in place. On later models, remove the filler plug and use a $\frac{1}{4}$ inch drift to drive out the roll-pin which secures the countershaft—as shown in **FIG 6:9.** Support the countershaft assembly with a hook made from wire. Press a dummy countershaft against the end of the countershaft and drive it through so that it displaces the countershaft. The dummy should be of the same diameter as the actual countershaft but should be shorter so that it is the same length as the countershaft gear cluster with thrust washers. When the dummy is in place, withdraw the countershaft and lower the countershaft gear cluster to the bottom of the casing. The dummy must be used, particularly on reassembly, otherwise the needle rollers of the countershaft will be displaced.

On the later models, now remove the selector and shift mechanism. It will be necessary to remove the snap ring that secures the bearing of the output shaft and remove the bearing with a suitable extractor acting in the snap ring annular groove. Take out the setscrew that secures the first/reverse shift fork to its rail. Slide out the rail through the rear end of the case, rotate the fork upwards and remove it from the case. Move the second/third speed fork to the second speed position, for access, and remove the setscrew that secures it to the rail. Rotate the shift rail through 90 deg., as shown in **FIG 6:10.** Remove the spring and lift out the detent plug and interlock plunger with a thin magnet. Tap on the inner end of the rail so that it drives out the expansion plug from the front of the case and withdraw the rail.

Remove the remaining plug and spring. On all models remove the input shaft assembly from the front of the casing, carefully collecting the needle rollers as they come free. If the bearing is defective, it will be necessary to remove the securing snap ring and then use a press to remove the bearing from the shaft. On the later models the second/third speed shift fork can now be rotated and removed upwards.

On earlier models, remove the snap ring that secures the front synchronizer assembly. Hold the synchronizer assembly together and withdraw the output shaft assembly through the rear of the casing. Mark the front of the synchronizer assembly with crayon so that it will be installed back into its original position.

The remainder of the parts of the casing can then be removed. When dismantling the output shaft assembly a press will be required to remove some of the components. The components of a later mainshaft assembly are shown in **FIG 6:11.** When dismantling the output shaft, lay the parts out in order and if necessary mark them with wax crayon to ensure that they are reassembled facing the correct direction.

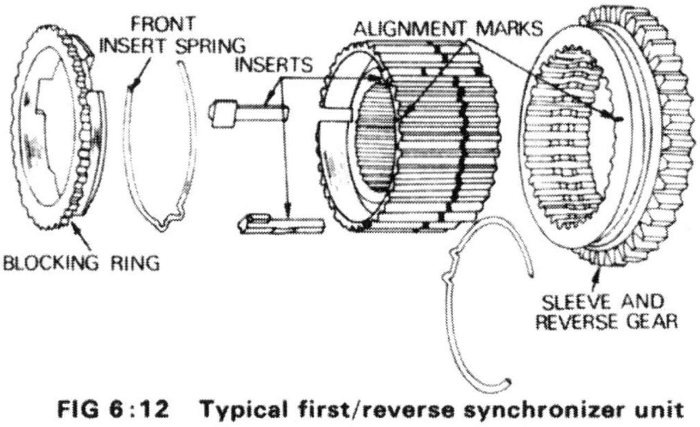

FIG 6:12 Typical first/reverse synchronizer unit

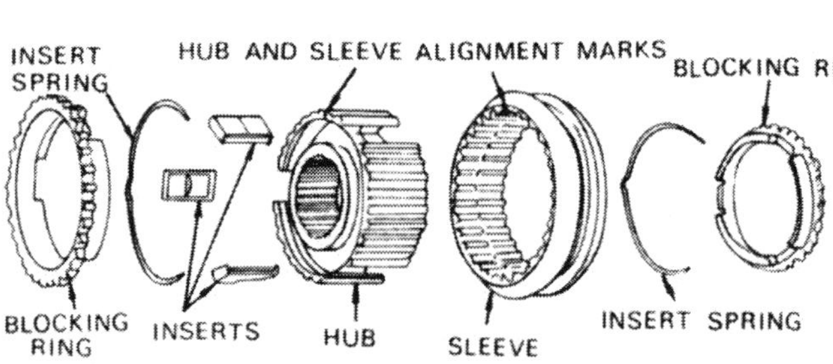

FIG 6:13 Typical second/third synchronizer unit

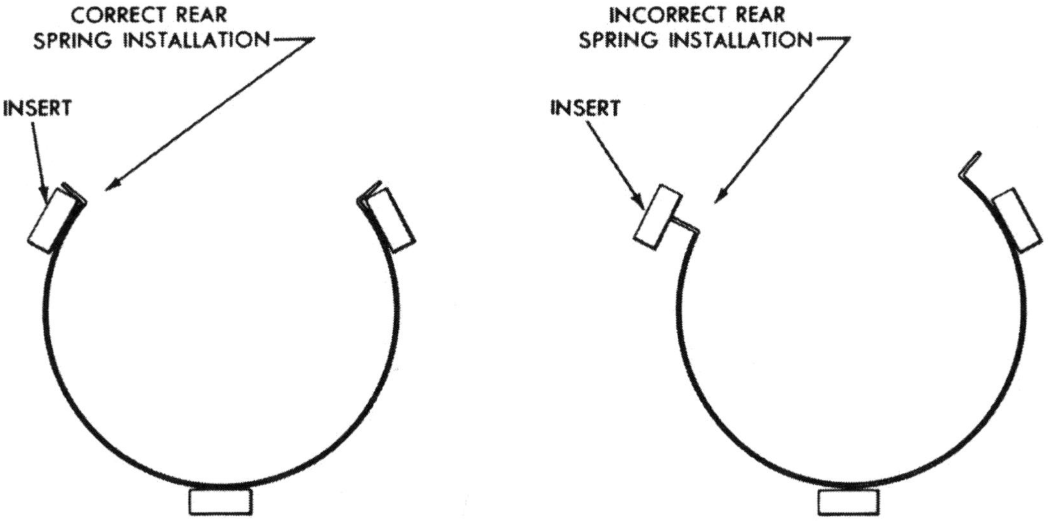

FIG 6:14 Installation of the insert spring on earlier models

Cleaning:

Wash all the parts, except the bearings and seals, in a suitable harsh solvent. Trichlorethylene is best as it removes all oil and dirt but care must be taken to wipe over bright metal parts with a piece of clean cloth dipped into a little oil, to prevent corrosion if the parts are stored for some time. Gasoline or kerosene also may be used as cleansing solvents.

Rotate the bearings in a bath of clean solvent until all dirt and oil has been washed out. The parts may be wiped dry with a piece of lint-free cloth though the bearings should be allowed to air-dry. **If compressed air is used to dry the parts, hold the bearings so that they do not spin in the air blast.**

Synchronizer units:

These should only be dismantled if they are defective. Slide the hub out of the sleeve to dismantle the unit. A typical first/rear synchronizer unit is shown in **FIG 6:12** and a second/third unit in **FIG 6:13**. On the earlier models, the insert springs may have short legs bent outwards and these must be installed as shown in **FIG 6:14**. On later models the gaps in the springs should be staggered and the springs should be installed as shown in **FIG 6:15.**

Examination:

Check all the faces on the castings for nicks or damage. Remove damage or distortion using a fine file or oilstone. Examine the castings for cracks, stripped threads or any other damage which could affect the operation of the transmission.

Examine all the teeth on the gears for chipping, wear or broken teeth. Renew any gears that are defective and examine the teeth of the gear that mates with it with extra care.

All the shafts should be true without bends or distortion and they should not show chatter or fret marks. Check bushes for wear, fretting or chatter. Inspect all splines for wear, damage or nicks.

Renew any bearings that run roughly or show any fault on the operating surfaces. It is advisable to renew bearings irrespective if they have already completed 60,000 miles.

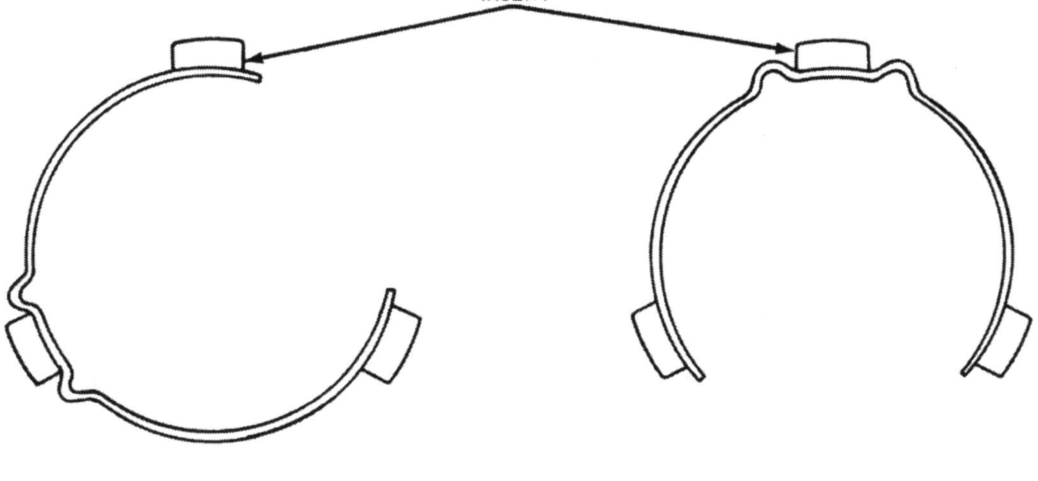

FRONT REAR

FIG 6:15 Installation of the insert spring on later models

Check the synchronizer units for wear in the index slots. Slide the hubs into the sleeves, with the etched marks aligned and check that the hub slides freely in the sleeve. Fit the blocker ring to the cone and check that there is at least .010 inch between the face of the blocker ring and the teeth on the gear. Check the conical surfaces for wear, scoring or damage. The blocker rings must be installed back into their original positions.

It is advisable to discard all seals and gaskets, installing new ones on reassembly. Check the bushing in the rear extension housing and renew it if it is defective. It should be noted that both the rear seal and bushing can be renewed with the transmssion installed in the vehicle, using similar methods as for the equivalent parts of the automatic transmission (see **Chapter 7**).

Reassembly:

Renew all defective parts. Selectively fit thrust washers to give the correct end plays. Lubricate the parts with transmission oil as they are installed and when all parts except the cover are fitted, pour transmission oil over the gear trains.

Generally assembly is the reversal of the disassembly procedure.

6:6 Trouble shooting guide

(a) Jumping out of gear

1 Shift mechanism incorrectly adjusted
2 Broken or weak detent spring
3 Worn detent grooves in shift rails
4 Worn blocking rings or synchronizer units
5 Shift fork loose on rail

(b) Noisy transmission

1 Insufficient or incorrect grade of lubricant
2 Excessive end plays on shafts and parts
3 Worn or damaged bearings
4 Worn or damaged gear teeth

(c) Hard shifting

1 Check 1, 4 and 5 in (a)
2 Stiff shift mechanism
3 Incorrect clutch adjustment
4 Defective spigot bushing in crankshaft
5 Defective clutch

(d) Lubricant leaks

1 Damaged joint faces
2 Defective gaskets
3 Defective oil seals

CHAPTER 7

AUTOMATIC TRANSMISSION

7:1 Description

A C4 automatic transmission can be fitted as an optional extra in place of the standard manually shifted transmission described in the previous chapter. A sectioned view of the unit is shown in **FIG 7:1** and the main components in **FIG 7:2.** The converter is bolted to the engine drive plate and takes the place of the mechanically operated clutch. The drive from the converter is passed through an epicyclic gearbox. The drives are selected by engaging clutches and brake bands using hydraulic pressure from the control system.

Converter:

A sectional view of the converter is shown in **FIG 7:3.** The unit is filled with automatic transmission fluid and there is a constant flow of fluid passing through the unit from and to the gearbox. The fluid flow is to ensure that the converter is always filled when the engine is running and also transfers away heat produced by slip in the converter. The impeller is attached to the casing and revolves with it. As the turbine revolves it drags the fluid around with it. Because of centrifugal force there will be a greater fluid pressure at the outlet from the impeller vanes (outside) than at the inlet and the pressure differential sets up a rotational fluid flow in the plane of the section. The combined drag and centrifugal flow sets up a flow around the unit which follows the path of an imaginary spiral spring laid around between impeller and turbine. The stator is fitted to a one-way clutch. When the turbine is stationary or rotating at a low speed compared with the impeller, the oil flow from the turbine is guided by the stator so that it impinges on the impeller at a favourable angle and assists the impeller to turn. As a result the torque on the turbine is greater than that produced by the impeller. At the stall point when the turbine is stationary the torque multiplication factor is approximately 2:1. As the turbine speeds up, the torque multiplication factor drops towards 1:1 until the factor is reached at a turbine speed of 9/10 of impeller speed. At this speed the fluid flow impinges on the back of the stator blades, but power is not lost as the one-way clutch allows the stator to revolve freely without altering the fluid flow direction.

When the car is stationary and the throttle opened, with a drive selection made, the turbine will have large torque applied to it and the torque multiplication factor will decrease until the turbine is rotating at the same speed as the impeller and the unit acts as a fluid coupling.

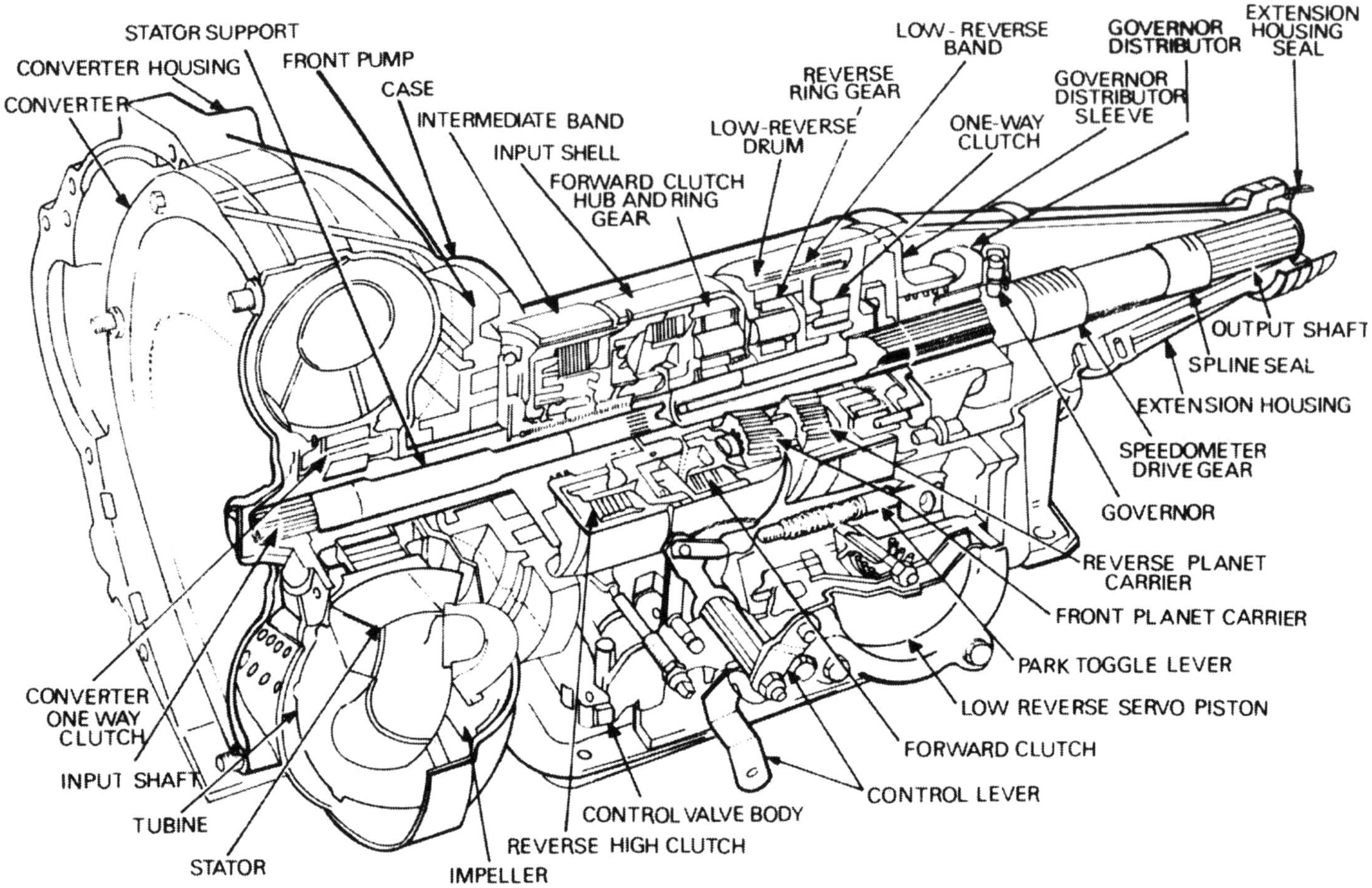

FIG 7:1 Sectional view of the transmission

There will always be a slight speed differential because of frictional losses but under optimum conditions this loss is usually in the order of 2 per cent.

The frictional losses are dissipated as heat. If the turbine is held stalled with the engine running at speed, the resultant churning of the fluid will cause rapid overheating in the unit. The turbine can only be stalled by selecting drive and holding the car stationary.

Epicyclic gearbox:

An epicyclic gear train consists of a sun gear and outer gear, with planet gears revolving on a planet carrier between the outer and sun gears. By holding any one component and driving another, drive can be taken from the third component. The ratio can be varied and the drive reversed using a single epicyclic train.

Two epicyclic gear trains are used in the gearbox to produce the three forward speeds and one reverse. The power flow through the transmission in the various selections is shown in **FIG 7:4**. The clutches are applied by hydraulic pressure from the control system. A sectional view through an intermediate servo and band is shown in **FIG 7:5**. The adjustment screw is external on the unit so that the band can be set without dismantling the transmission. When hydraulic pressure is applied to the piston, the ends of the band are squeezed together and it grips the drum to prevent it from rotating.

The gearbox should not be dismantled by the owner as special tools are essential. On reassembly all the parts must be set to correct clearances and a great deal of skill is required.

Hydraulic control system:

A schematic layout of the system is shown in **FIG 7:6**. It must be emphasised that the layout is schematic and that the fluid is led through integral passages in the castings.

The fluid is contained in the oil pan under the unit. A pump is fitted to the input shaft and draws the oil from the oil pan through an oil screen. Under normal conditions the pressure of the oil from the pump is limited to 60 lb/sq inch by the main regulator valve. When the pressure from the pump rises above this level, the regulator valve allows the surplus oil to pass through to the converter and, if the pressure is still high, the valve opens further to allow excess fluid to return directly to the oil pan. When the operation of the unit requires pressures higher than 60 lb/sq inch the pressure booster valve holds back the regulator valve.

The manual valve is connected to the selector linkage and is operated by the selector in the vehicle. This valve selects the direction of drive and also overrides the controls to lock out the higher gears as required.

The vacuum manifold moves the primary throttle valve dependent on the intake manifold suction. The governor is attached to the output shaft. **Tests on the operation of the hydraulic control system or any repair work, beyond that given in this chapter, must only be carried out by experts.**

7:2 Routine maintenance

This is confined to ensuring that the fluid level is correct, cleaning dirt from the casing and at regular intervals adjusting the brake bands.

Cleanliness:

All operations on the automatic transmission must be carried out with scrupulous cleanliness. This includes wiping the filler tube and dipstick clean before removing the dipstick.

The outside of the casing should be periodically cleaned and air vents checked for plugging. The automatic transmission operates at a high temperature and dirt on the casing will act as an insulator, causing the unit to overheat, particularly in areas of high ambient temperatures.

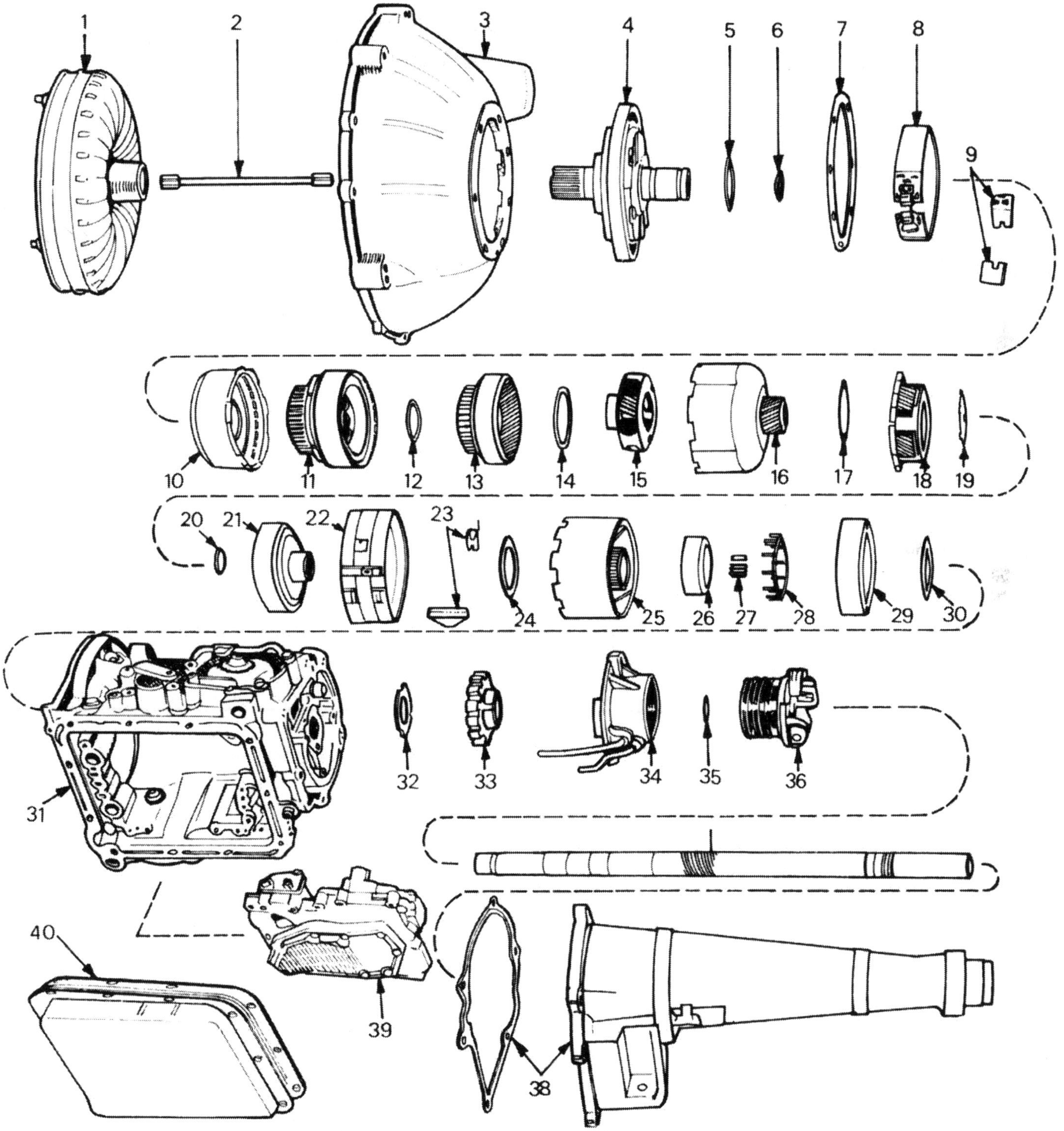

FIG 7:2 The transmission main components

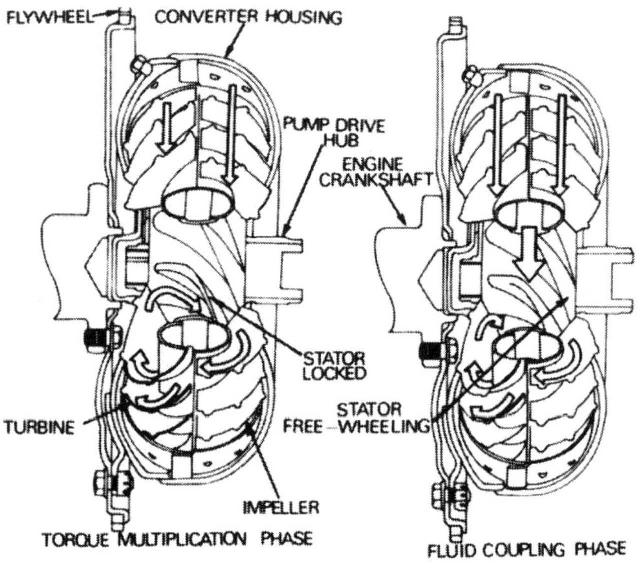

FIG 7:3 Sectional views of the torque converter

Fluid level:

Run the engine at 1200 rev/min until the automatic transmission has reached its normal operating temperature, leave the selector at the PARK position. Reduce the engine speed to normal idle and move the selector through each selection in turn before finally returning it to park. Wipe the top of the filler tube clean and withdraw dipstick. Wipe the dipstick clean on a piece of leather or lint-free cloth. Insert the dipstick back into the filler tube and press it fully home before withdrawing it again. If the level is low, top-up with automatic transmission fluid to the upper mark on the dipstick. **The level must always be taken with the transmission hot and the engine running at idle. Do not overfill the transmission. Overfilling will lead to frothing of the oil and incorrect operation of the unit.**

Linkage:

Check that the linkages operate freely and correctly and lightly lubricate pivot points.

Draining:

Routine or seasonal fluid changes are not required. **If the fluid is drained while the transmission is hot take great care, as the fluid will be hot enough to cause serious scalding.**

Remove the cover from the bottom of the converter housing and crank the engine until the two converter drain plugs can be removed. One of the drain plugs is shown in **FIG 7:7.** Allow the fluid to drain out into a suitable container before discarding it.

If the remainder of the unit requires draining, the oil pan must be removed from under it.

Oil cooler:

The oil cooler does not require routine maintenance. If there has been internal damage in the automatic transmission, the oil cooler will require flushing through.

If transmission fluid is found in the radiator header tank when checking the coolant level, there is a defect in the oil cooler. Such a defect will also allow water to enter the transmission. Water in the automatic transmission will cause a rise in the level and frothing as the water boils and it may even cause frothing fluid to overflow from the filler tube.

The oil cooler is built into the header tank and cannot be repaired, except by specialists, so if the oil cooler is leaking a new radiator assembly must be fitted.

Brake band adjustment:

These normally only require adjustment at intervals of 35,000 miles. A special tool, No. T59P-77370-B or 7345, is made for adjusting the bands and the tool is tightened until handle clicks and overruns. If the special tool is not available a torque spanner can be used instead. The method of adjusting the intermediate band is shown in **FIG 7:8** and that for the low-reverse band is shown in **FIG 7:9.** Slacken the locknut and back it off several turns so that it is well clear. Use the special tool and tighten it until the handle clicks (or a load of 10 lb ft is reached on a torque spanner). For the intermediate brake band slacken off the adjusting screw by exactly $1\frac{1}{2}$ turns and for the low-reverse band slacken back the adjusting screw by exactly 3 turns. Hold the adjusting screw and tighten the locknut to secure it in place.

7:3 Removing the transmission unit

The automatic transmission weighs considerably more than the manually operated transmission and a jack is essential for lowering and moving the unit out from under the car. A special support for the transmission should be used and it should be noted that the transmission is held down to the support using a chain.

1 Drain the transmission by removing the two drain plugs from the converter (see **FIG 7:7**). Slacken off the oil pan bolts and allow the fluid to pour out by lowering one corner of the oil pan. Install the drain plugs and tighten the oil pan attachment bolts after draining out the fluid. Disconnect the oil cooler lines from the transmission, plugging both lines and ports to prevent dirt from entering them. Remove the dipstick and oil filler tube from the case.

2 Remove the drive shaft (see **Chapter 8**) after making aligning marks so that it will be installed into its original position. Plug the extension housing on the rear of the transmission to prevent dirt from entering the unit. Disconnect the shift and kickdown rods.

3 Disconnect the speedometer drive cable, by taking out its retaining clip. Disconnect the vacuum line at the transmission vacuum unit. Disconnect the exhaust pipe at the flange on the manifolds. Free the parking brake cable from the equalizer lever. Remove the starter motor, after disconnecting the battery and the starter motor cables. Disconnect the neutral start switch wires, labelling them if necessary.

4 Remove the two bolts that secure the extension housing to the rear crossmember mounting. Remove the four bolts that secure the converter to the engine drive plate (see **FIG 7:7**). Support the engine on a jack and pad of wood under the oil pan. Support the transmission on the transmission jack and secure it with a length of chain. Remove the bolts that secure the rear mounting crossmember to the frame. Remove

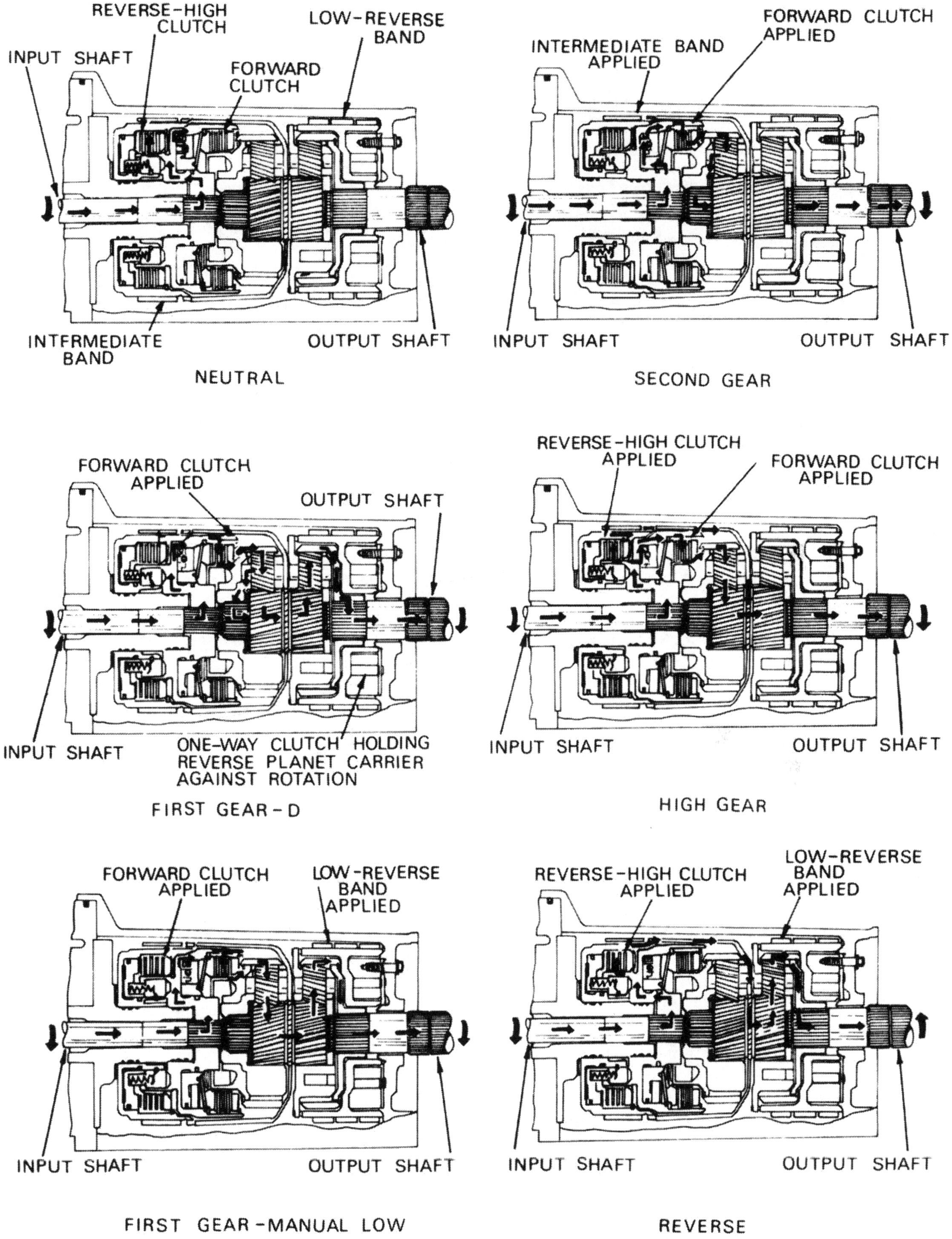

FIG 7:4 The power flow in the various selections

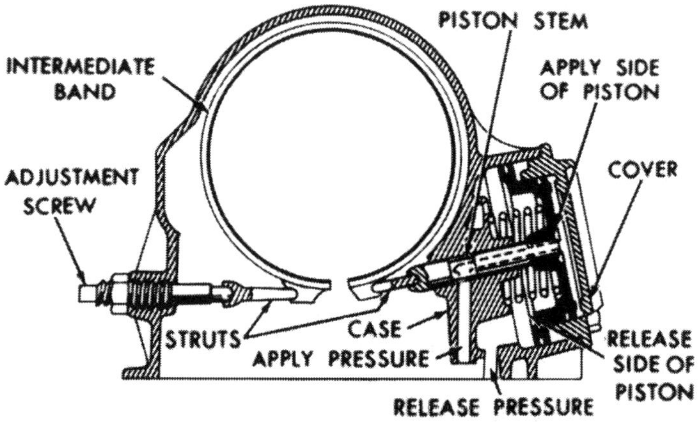

FIG 7:5 Sectional view of the intermediate brake band and servo

the five bolts that secure the converter housing to the engine. Draw the transmission slightly rearwards and then lower it down and out of the car.

Installation:

The transmission is installed in the reverse order of removal. All bolts should be torqued to the correct specifications. Check and adjust the shift and kickdown linkages. Fill the transmission to the correct level with fluid.

Oil cooler flushing:

If any dirt has entered the oil cooler or the transmission has been removed because of internal faults, the oil cooler should be flushed. With the transmission filled to the correct level with fluid, disconnect the return line from the oil cooler at the rear of the transmission. Start the engine and drain out at least two quarts of fluid from the return line. If the fluid does not flow freely from the cooler, shut off the engine and disconnect the other line to the coil cooler at the transmission. Blow through the lines and cooler using compressed air at a maximum pressure of 100 lb/sq inch in a reverse direction to the normal flow. Disconnect the lines from the oil cooler if necessary so that they can be blown through separately. If the blockage cannot be removed using compressed air, new lines or radiator assembly must be fitted. **Do not attempt to cure leaks by bypassing the cooler or plugging the lines.**

Torque converter:

Once the transmission has been taken out of the vehicle, the converter can be removed from the transmission. The converter is sealed on manufacture so if it is found to be defective it must be renewed as a unit.

Special tools are made for checking the converter. The tools are used for pressure testing the unit, as well as checking the interference fits of the stator to impeller and stator to turbine. Using a torque wrench with the special tools allows the operation of the stator one-way clutch to be checked. If it is suspected that faults in the transmission are caused by a defective converter, take the unit to a service garage holding the special tools so that the unit can be checked.

If it is suspected that dirt or sludge is in the system, drain out all the fluid from the torque converter by tilting it around with both drain plugs removed. Install the drain plugs and fill the unit with clean kerosene or suitable cleansing fluid. Preferably rotate the turbine but if this is not possible, because of lack of the correct tools, shake the converter around to swill the inside. Drain out the old kerosene and repeat the procedure as often as required to remove all dirt from inside the unit.

7:4 Stall test

This test will give a rapid guide to the condition of the automatic transmission.

Connect an accurate tachometer to the ignition circuit and fit the instrument so that it can be readily seen from the driver's seat.

Start the engine and run it until the transmission has reached its normal operating temperature. Apply the handbrake fully and hold the vehicle on the footbrake as well. Select a drive gear and floor the throttle pedal. Note the maximum engine speed and release the throttle pedal within five seconds. If the engine speed is above the normal stall speed, stop the test immediately as it is an indication of slip.

The churning of the oil in the converter with the turbine stalled will produce large amounts of heat and the full throttle period must not exceed five seconds otherwise the transmission will overheat. Check each drive selection in turn and once the maximum speed of the engine has been noted in a selection, select Neutral and allow the engine to run at 1000 rev/min for at least fifteen seconds to allow the fluid to cool. When the stall speed has been checked in all four drive selections, select Neutral and leave the engine running at 1000 rev/min to allow the transmission to cool and the temperature to stabilize. Stall speeds are given in **Technical Data.**

Low stall speed:

This can be caused by lack of power from the engine so make sure that the engine is satisfactory and correctly tuned.

The other cause of a low stall speed is a defective converter. If the stall speed is 300 to 400 rev/min low, take the car for a road test. If the car cruises normally but acceleration (particularly from standstill) is poor and if the car will only start with difficulty on gradients then the stator one-way clutch is slipping.

If acceleration and starting on gradients is satisfactory but the car drags at cruise, and maximum speeds in the gears (particularly top) are limited, and the transmission overheats easily, then the stator one-way clutch is seized. In both cases, remove the transmission so that the torque converter can be removed and checked at a service garage.

High stall speed:

A high stall speed indicates slip in the gearbox. Check each selection in turn and note the selections that give high stall speeds. Excessive speed in forward selections only indicates forward clutch slippage. Excessive speed in R selection indicates either reverse-high clutch slippage or low-reverse band slippage. Excessive speed in D1 only indicates gear train one-way clutch slippage.

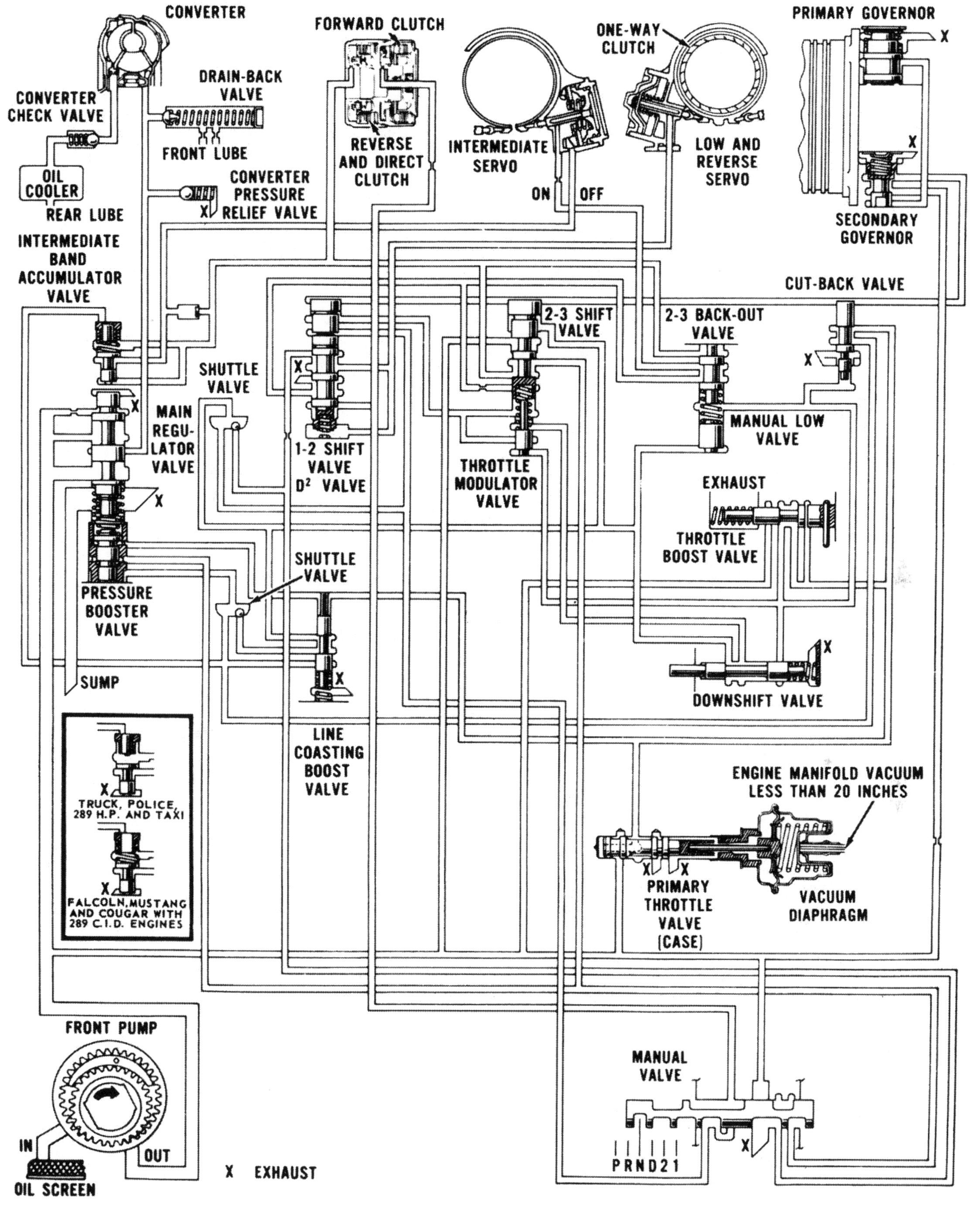

FIG 7:6 Schematic hydraulic system

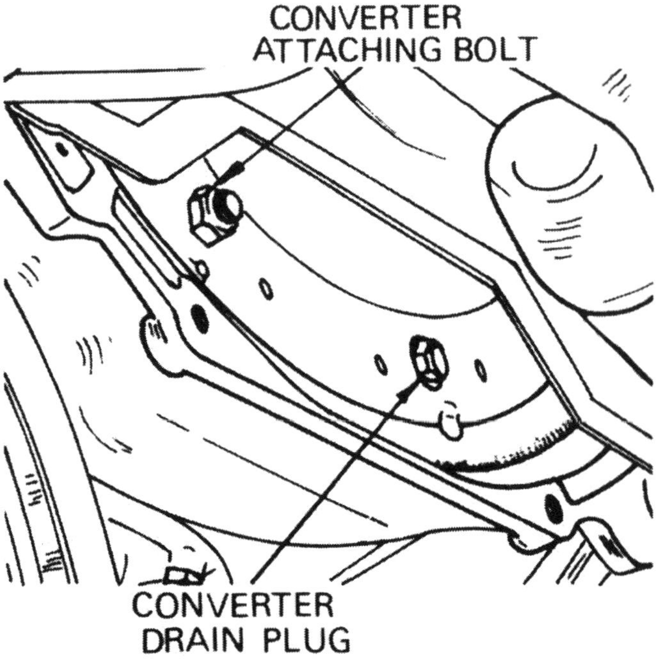

FIG 7:7 The converter drain plug and attachment

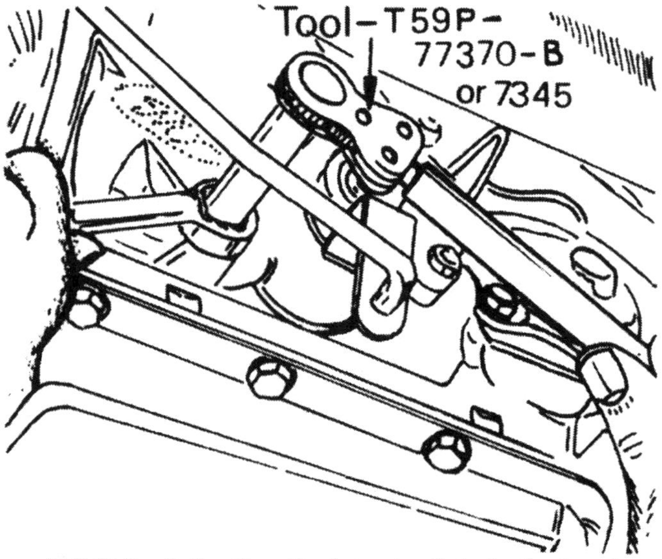

FIG 7:8 Adjusting the intermediate brake band

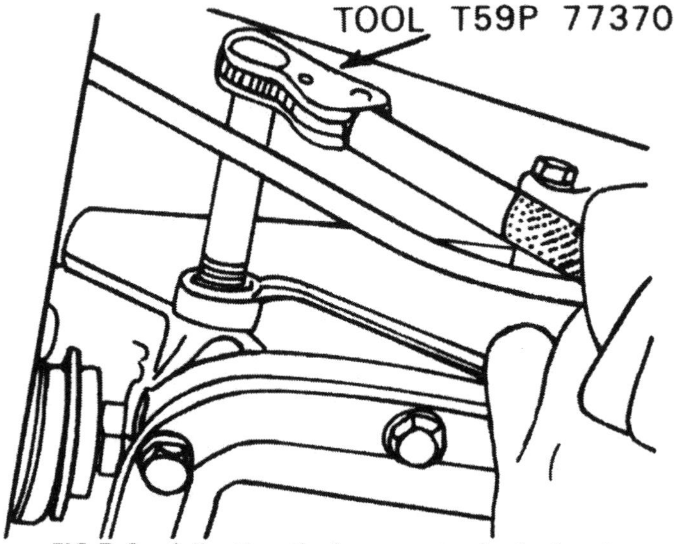

FIG 7:9 Adjusting the low-reverse brake band

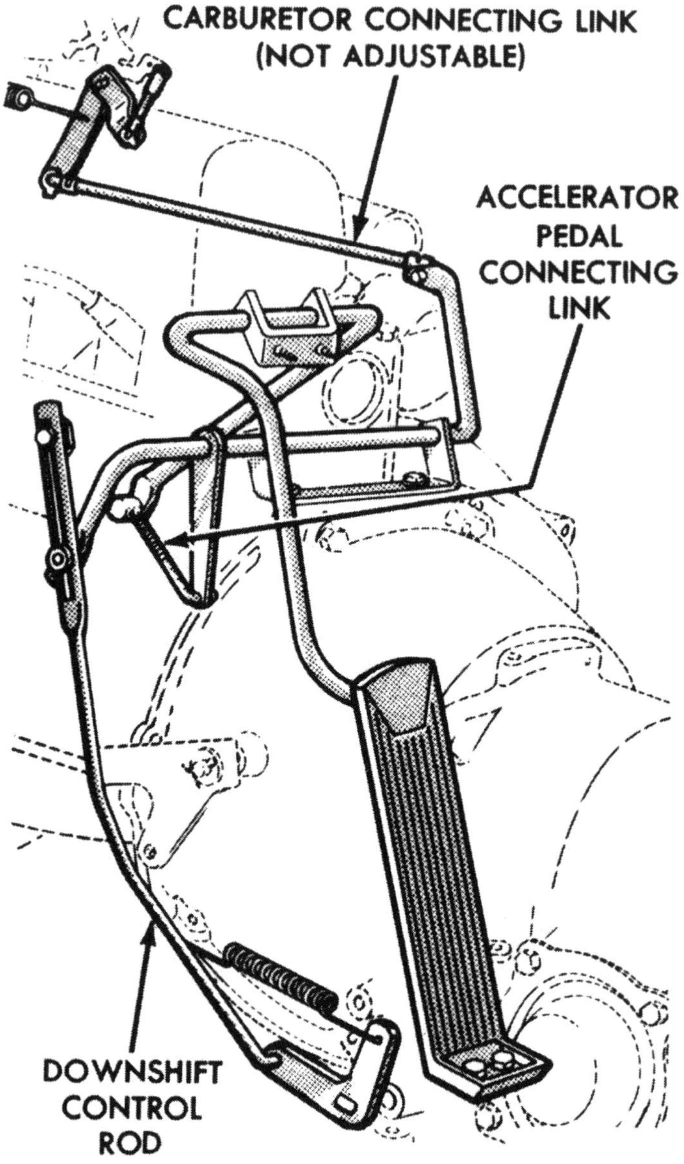

FIG 7:10 1964 model throttle linkage

7:5 Adjustments

The adjustments of the brake bands have already been dealt with in **Section 7:2**. For full and accurate checks of the transmission, test equipment is required so that the operating pressures can be noted. The vehicle should therefore be taken to a service garage for checking and diagnosis of faults that are beyond the power of the owner to adjust or correct.

Throttle linkage:

Typical throttle linkages are shown in **FIGS 7:10** and **7:11**.

Start the engine and run it until it has reached its normal operating temperature. Connect a tachometer to the ignition circuit and adjust the idling speed to the correct limits. When setting the fast-idle screw note that it must be just touching the cam at the correct mark on the cam. The large dot is for the D1 selection and the small dot for the D2 selection.

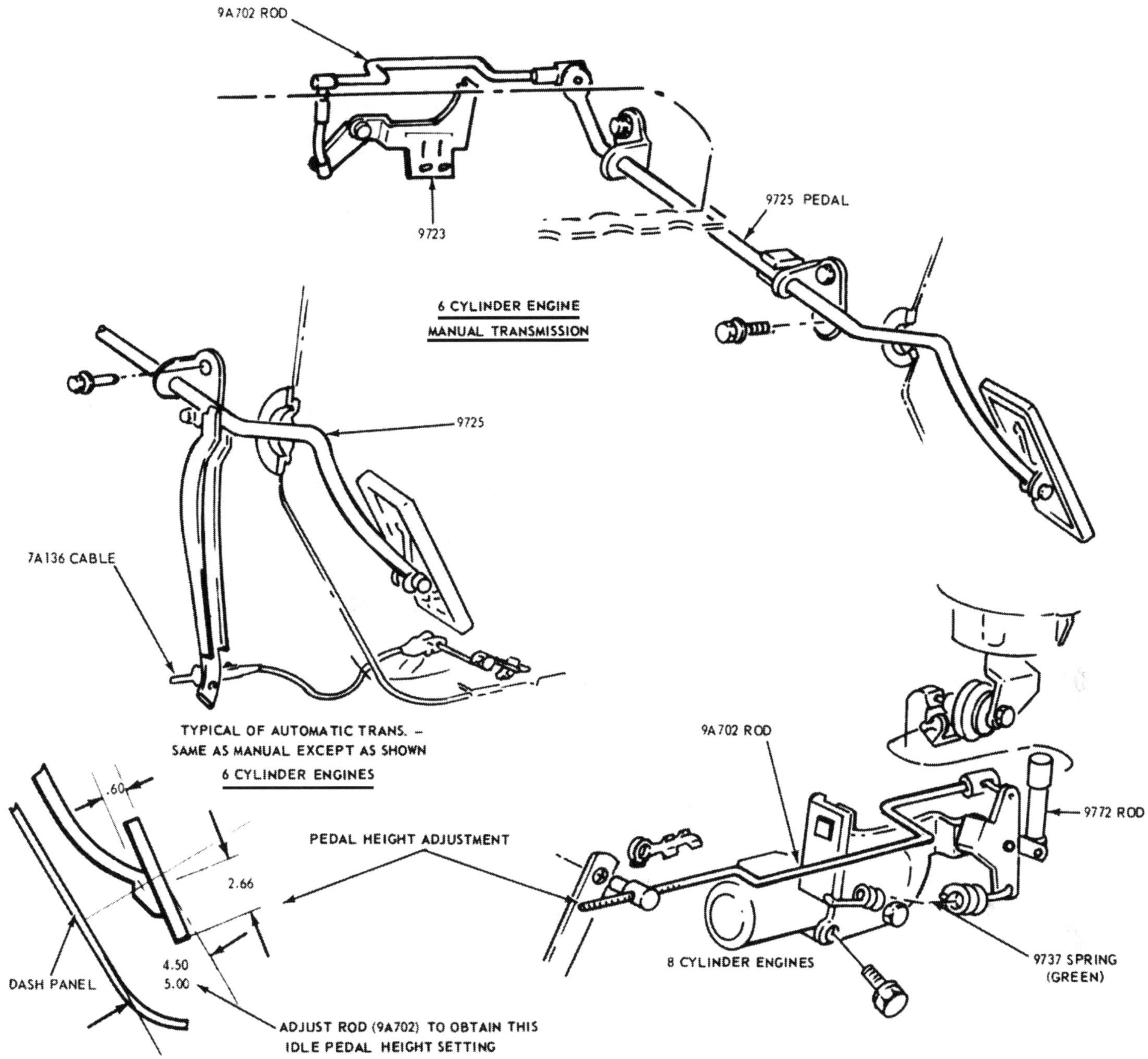

FIG 7:11 1970 model throttle linkage

Make sure that the floor covering and carpets are correctly in place, removing any matting that is not original equipment. Check the height of the accelerator pedal from the floor pan, and if it is outside limits adjust the height at the pedal link.

Turn the engine off and press the accelerator pedal right down and block it there. On models fitted with a downshift rod, adjust the rod so that the lever on the transmission is in the kickdown detent position with approximately $\frac{1}{16}$ inch of free travel on the bellcrank assembly. On models fitted with a downshift cable, disconnect the cable. Rotate the downshift lever until it is against the internal stop. Hold the lever on the transmission in this position and adjust the length of the cable until its trunnion will slide easily into the lever on the throttle linkage. Turn the adjuster one additional turn clockwise and reconnect the trunnion to the lever.

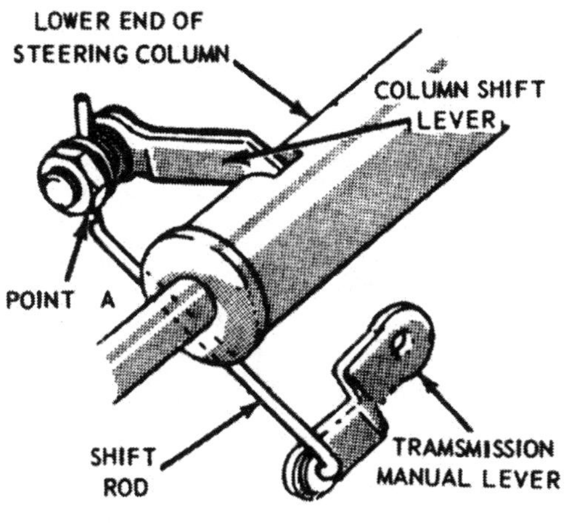

FIG 7:12 Adjusting the shift linkage

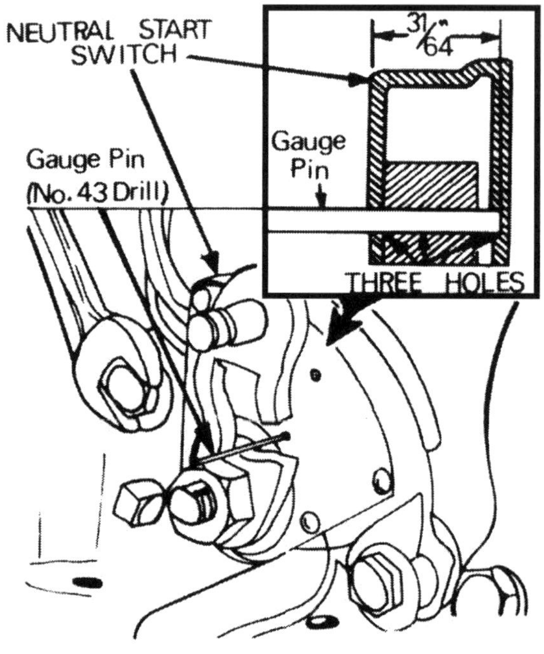

FIG 7:13 Adjusting the neutral start switch

Road test the car for idling and kickdown. A low idling speed will cause the engine to stall while a high idling speed will make the car creep when a drive selection is made.

Column shift:

The adjustment point is shown in **FIG 7:12.** Slacken the nut at the point A sufficiently to allow the shift rod to be moved along the column shift lever. Place the selector lever against the D stop in the D position. Move the transmission manual lever into the position (third from rear) and check that the selector lever has not moved. Tighten the nut A so that the shift rod is secured in position. Check the selector lever in all selections.

Neutral start switch:

This is fitted to ensure that the starter motor will only operate in the Park and Neutral selections.

The attachment of the switch to the column is shown in **FIG 7:13.** Slacken the two screws that secure the switch to the column. Set the selector level firmly against the stop of the neutral detent and rotate the switch about the column, in its elongated holes, until a No. 43 drill shank can be inserted into the gauge pin holes. Tighten the switch securing screws to hold the switch and remove the drill. Check that the starter motor only operates in the Park and Neutral selections and that it does not operate when a drive selection is made.

7:6 Governor and rear extension housing

The seal and bushing in the extension housing can be renewed without dismantling the transmission or removing it from the vehicle.

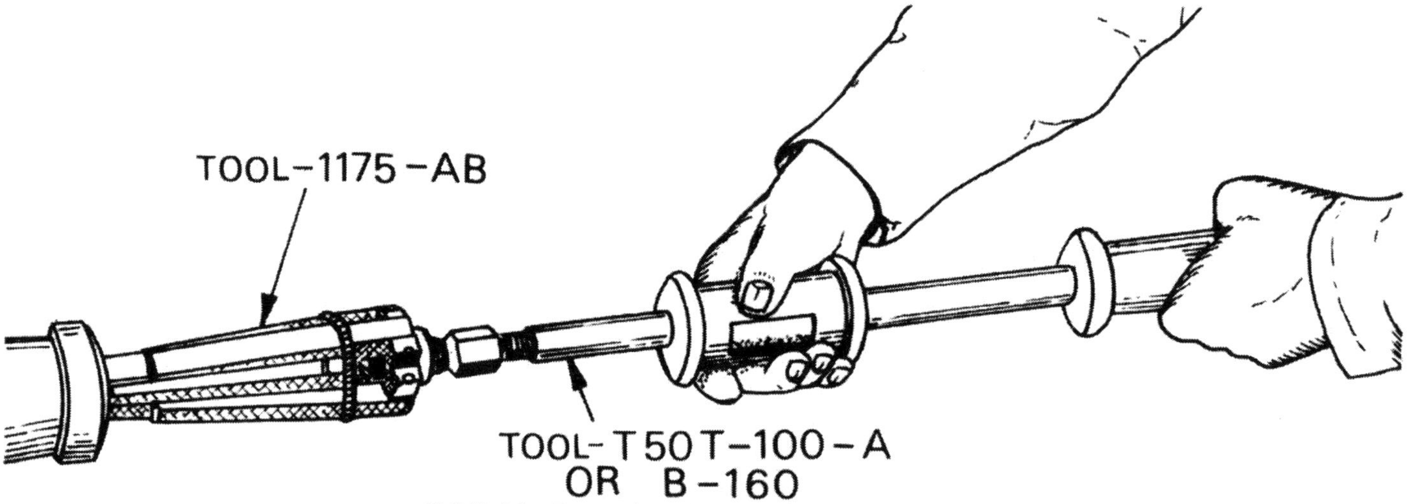

FIG 7:14 Removing the extension housing seal

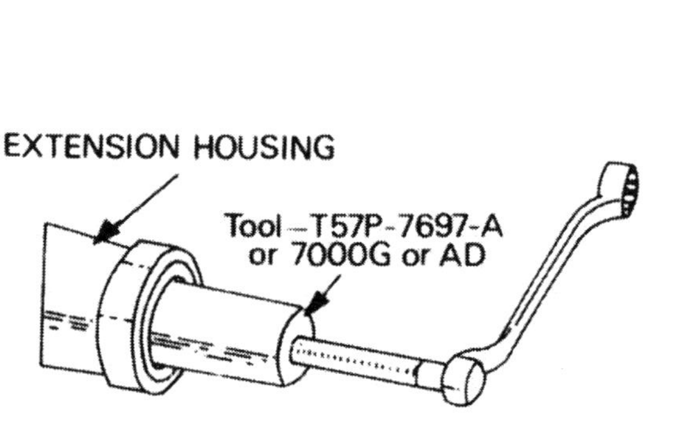

FIG 7:15 Removing the extension housing bushes

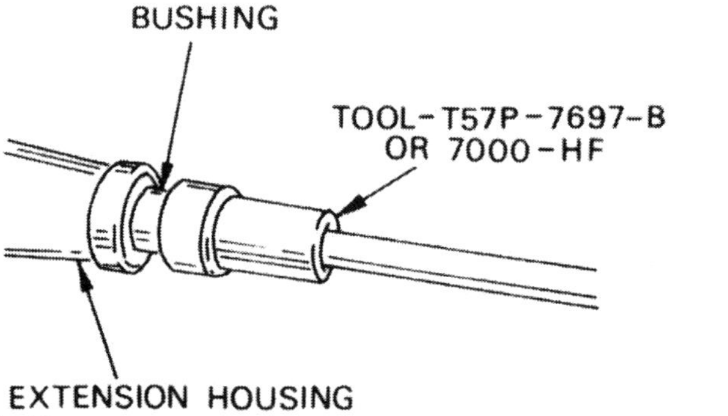

FIG 7:16 Installing the extension housing bushing

Preparation:

Raise the car on a hoist or work from a pit. Remove the drive shaft (see **Chapter 8**). Disconnect the speedometer drive cable from the extension housing if the housing is to be removed. Support the transmission on a transmission jack and free the rear mounting bolts from the extension housing.

If the extension housing is to be removed, slacken the bolts securing it so that the transmission fluid can drain out. Disconnect the muffler inlet pipes from the exhaust manifolds on the engine.

Seal and bush assembly:

Remove the extension housing seal using the tools shown in **FIG 7:14.** Carefully use the bushing remover shown in **FIG 7:15,** taking great care not to damage the spline seal, and withdraw the bushing. The new bushing is installed using the tools shown in **FIG 7:16** and the new seal using the tools shown in **FIG 7:17.** Before installing the bushing and seal, check the counterbore of the housing for nicks or burrs and use Crocus paper to remove them. Check the yoke of the drive shaft for scores or damage and renew it if it is defective. Lubricate the inside diameter of the fiber portion of the seal and the yoke splines with B8A-19589-A lubricant.

The governor:

Remove the bolts that secure the extension housing to the main casing and withdraw the extension. The attachments of the governor are shown in **FIG 7:18** and the components in **FIG 7:19.**

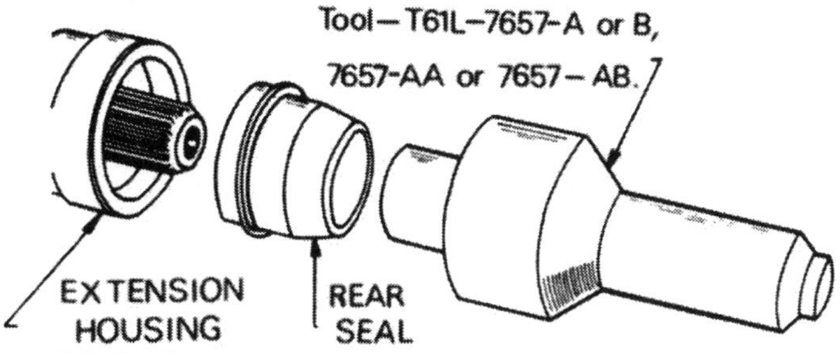

FIG 7:17 Installing the extension housing seal

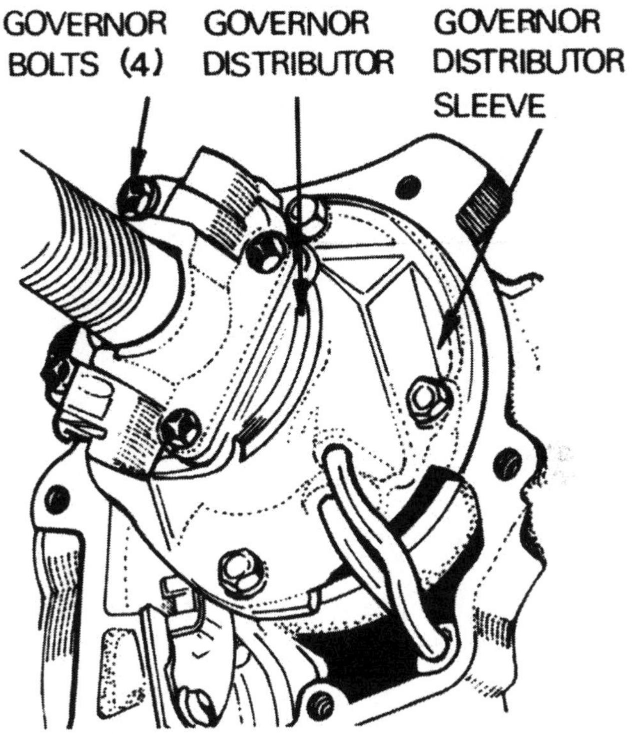

FIG 7:18 The governor attachments

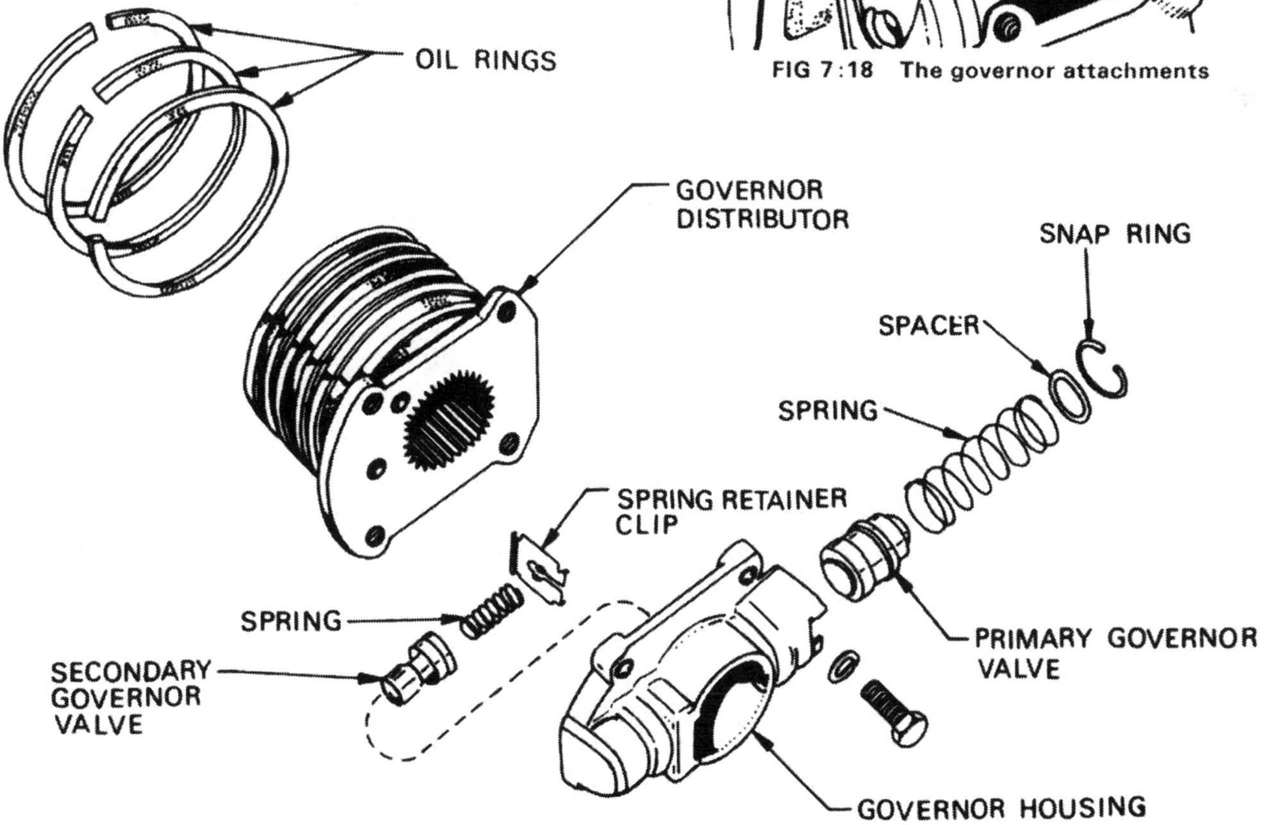

FIG 7:19 The governor components

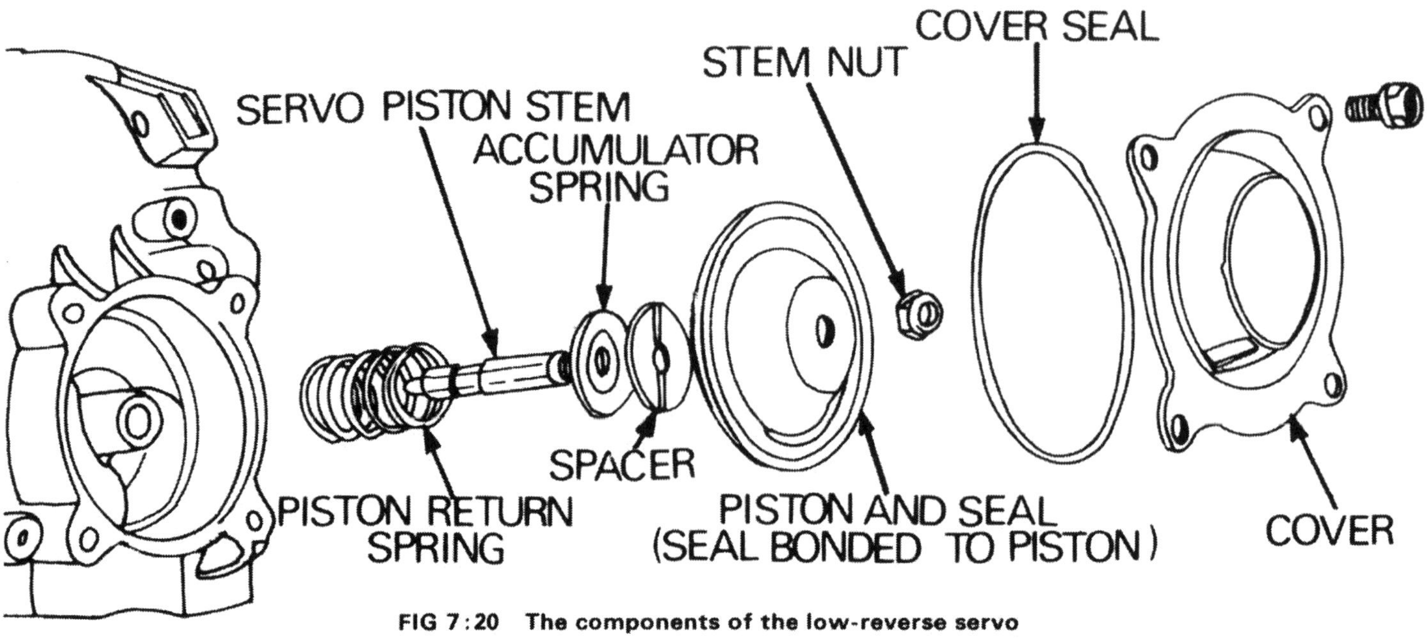

FIG 7:20 The components of the low-reverse servo

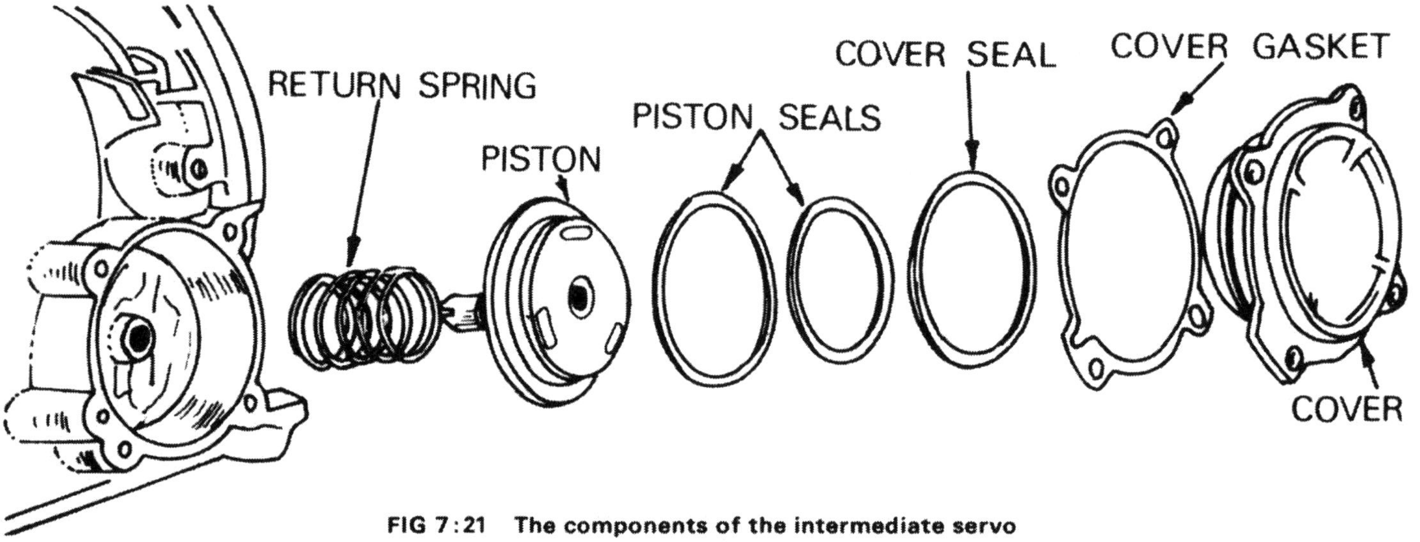

FIG 7:21 The components of the intermediate servo

Remove the governor retaining bolts and withdraw the governor assembly from the distributor.

Remove the primary valve securing snap ring and take out the primary valve parts. Remove the retaining clip so that the secondary valve parts can be removed.

Wash the parts in a suitable solvent. Check that the valves slide freely in their bores. If the valves stick, they may be lightly polished with Crocus paper but **take care not to round off the sharp edges of the valves.** Fit new parts if the old ones are scored or badly worn.

The parts are installed in the reverse order of disassembly. Lubricate them lightly with transmission fluid. When installing the secondary valve parts make sure that the retaining clip is fitted with its small concave surface downwards so that it locates the spring.

Install the extension housing using a new gasket. The remainder of the parts are then installed in the reverse order of disassembly. Fill the transmission to the correct level with fresh fluid.

7:7 Servo assemblies and vacuum unit

The components of the low-reverse servo are shown in **FIG 7:20** and those of the intermediate servo in **FIG 7:21**. The pistons and seals can be renewed but take great care to ensure scrupulous cleanliness when working on the parts. Lubricate the parts lightly with clean transmission fluid before installation.

The tool made for removing the vacuum unit and its method of use are shown in **FIG 7:22.** If the vacuum unit is suspected to be faulty, check that the vacuum line is clear and either have the vacuum unit tested at a service garage or install a new unit.

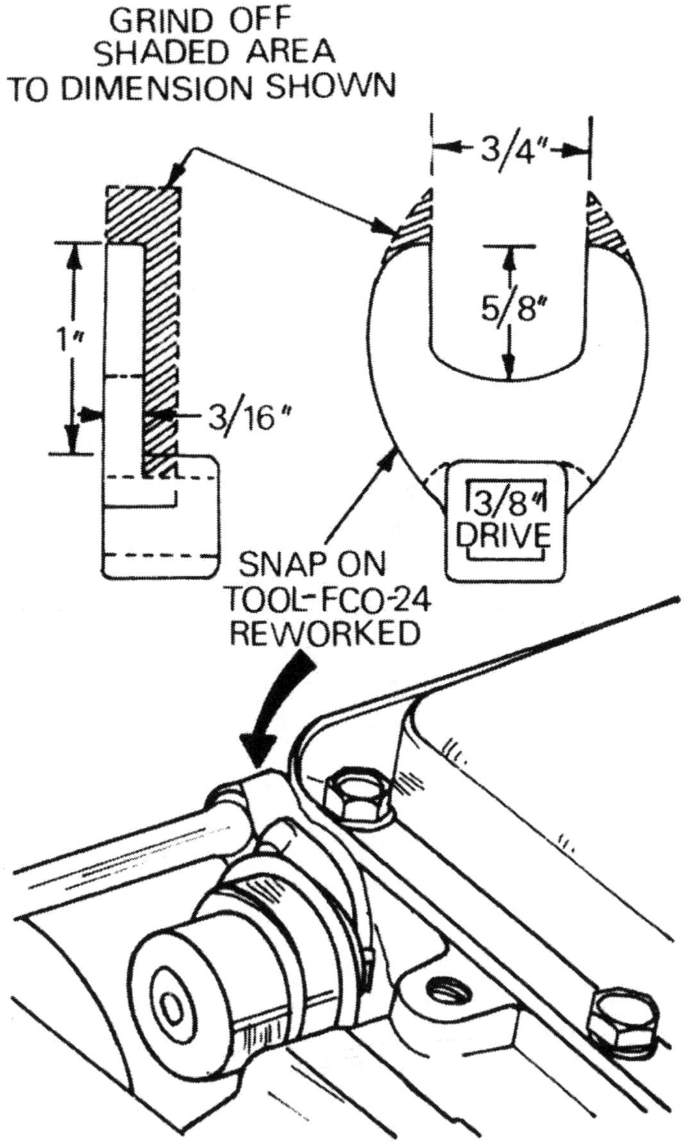

FIG 7:22 Removing or installing the vacuum unit

7:8 Trouble shooting guide

(a) Transmission overheats

1 Excessive dirt or mud on outside of unit
2 Oil cooler or lines plugged or oil cooler defective
3 Converter stator clutch seized
4 Incorrect brake band adjustment
5 Brakes binding or other drag on car causing converter to slip.

(b) Noisy operation

1 Incorrect fluid level or incorrect type of fluid
2 Incorrectly adjusted linkage
3 Oil screen to pump intake choked

(c) Incorrect shift speeds

1 Check in (b)
2 Defective vacuum unit
3 Incorrectly adjusted brake bands or defective servos

(d) Poor acceleration

1 Check 3 and 4 in (a)
2 Converter stator clutch slipping

(e) Jumps in engagement or rough engagement

1 Check 4 in (a); 2 in (b) and 2 in (c)
2 Incorrect idling speed

(f) Vehicle does not hold in Park selection

1 Check 2 in (b)

(g) Fluid frothing or being forced out of filler tube

1 Too high fluid level
2 Water in system (suspect oil cooler)
3 Air leaks on suction side of pump

This guide is not fully comprehensive and only covers faults that can be cured by the owner. For other faults take the vehicle to a service garage.

Inches	Decimals	Milli-metres	Inches to Millimetres Inches	mm	Millimetres to Inches mm	Inches
1/64	.015625	.3969	.001	.0254	.01	.00039
1/32	.03125	.7937	.002	.0508	.02	.00079
3/64	.046875	1.1906	.003	.0762	.03	.00118
1/16	.0625	1.5875	.004	.1016	.04	.00157
5/64	.078125	1.9844	.005	.1270	.05	.00197
3/32	.09375	2.3812	.006	.1524	.06	.00236
7/64	.109375	2.7781	.007	.1778	.07	.00276
1/8	.125	3.1750	.008	.2032	.08	.00315
9/64	.140625	3.5719	.009	.2286	.09	.00354
5/32	.15625	3.9687	.01	.254	.1	.00394
11/64	.171875	4.3656	.02	.508	.2	.00787
3/16	.1875	4.7625	.03	.762	.3	.01181
13/64	.203125	5·1594	.04	1.016	.4	.01575
7/32	.21875	5.5562	.05	1.270	.5	.01969
15/64	.234375	5.9531	.06	1.524	.6	.02362
1/4	.25	6.3500	.07	1.778	.7	.02756
17/64	.265625	6.7469	.08	2.032	.8	.03150
9/32	.28125	7.1437	.09	2.286	.9	.03543
19/64	.296875	7.5406	.1	2.54	1	.03937
5/16	.3125	7.9375	.2	5.08	2	.07874
21/64	.328125	8.3344	.3	7.62	3	.11811
11/32	.34375	8.7312	.4	10.16	4	.15748
23/64	.359375	9.1281	.5	12.70	5	.19685
3/8	.375	9.5250	.6	15.24	6	.23622
25/64	.390625	9.9219	.7	17.78	7	.27559
13/32	.40625	10.3187	.8	20.32	8	.31496
27/64	.421875	10.7156	.9	22.86	9	.35433
7/16	.4375	11.1125	1	25.4	10	.39370
29/64	.453125	11.5094	2	50.8	11	.43307
15/32	.46875	11.9062	3	76.2	12	.47244
31/64	.484375	12.3031	4	101.6	13	.51181
1/2	.5	12.7000	5	127.0	14	.55118
33/64	.515625	13.0969	6	152.4	15	.59055
17/32	.53125	13.4937	7	177.8	16	.62992
35/64	.546875	13.8906	8	203.2	17	.66929
9/16	.5625	14.2875	9	228.6	18	.70866
37/64	.578125	14.6844	10	254.0	19	.74803
19/32	.59375	15.0812	11	279.4	20	.78740
39/64	.609375	15.4781	12	304.8	21	.82677
5/8	.625	15.8750	13	330.2	22	.86614
41/64	.640625	16.2719	14	355.6	23	.90551
21/32	.65625	16.6687	15	381.0	24	.94488
43/64	.671875	17.0656	16	406.4	25	.98425
11/16	.6875	17.4625	17	431.8	26	1.02362
45/64	.703125	17.8594	18	457.2	27	1.06299
23/32	.71875	18.2562	19	482.6	28	1.10236
47/64	.734375	18.6531	20	508.0	29	1.14173
3/4	.75	19.0500	21	533.4	30	1.18110
49/64	.765625	19.4469	22	558.8	31	1.22047
25/32	.78125	19.8437	23	584.2	32	1.25984
51/64	.796875	20.2406	24	609.6	33	1.29921
13/16	.8125	20.6375	25	635.0	34	1.33858
53/64	.828125	21.0344	26	660.4	35	1.37795
27/32	.84375	21.4312	27	685.8	36	1.41732
55/64	.859375	21.8281	28	711.2	37	1.4567
7/8	.875	22.2250	29	736.6	38	1.4961
57/64	.890625	22.6219	30	762.0	39	1.5354
29/32	.90625	23.0187	31	787.4	40	1.5748
59/64	.921875	23.4156	32	812.8	41	1.6142
15/16	.9375	23.8125	33	838.2	42	1.6535
61/64	.953125	24.2094	34	863.6	43	1.6929
31/32	.96875	24.6062	35	889.0	44	1.7323
63/64	.984375	25.0031	36	914.4	45	1.7717

UNITS	Pints to Litres	Gallons to Litres	Litres to Pints	Litres to Gallons	Miles to Kilometres	Kilometres to Miles	Lbs. per sq. In. to Kg. per sq. Cm.	Kg. per sq. Cm. to Lbs. per sq. In.
1	.57	4.55	1.76	.22	1.61	.62	.07	14.22
2	1.14	9.09	3.52	.44	3.22	1.24	.14	28.50
3	1.70	13.64	5.28	.66	4.83	1.86	.21	42.67
4	2.27	18.18	7.04	.88	6.44	2.49	.28	56.89
5	2.84	22.73	8.80	1.10	8.05	3.11	.35	71.12
6	3.41	27.28	10.56	1.32	9.66	3.73	.42	85.34
7	3.98	31.82	12.32	1.54	11.27	4.35	.49	99.56
8	4.55	36.37	14.08	1.76	12.88	4.97	.56	113.79
9		40.91	15.84	1.98	14.48	5.59	.63	128.00
10		45.46	17.60	2.20	16.09	6.21	.70	142.23
20				4.40	32.19	12.43	1.41	284.47
30				6.60	48.28	18.64	2.11	426.70
40				8.80	64.37	24.85		
50					80.47	31.07		
60					96.56	37.28		
70					112.65	43.50		
80					128.75	49.71		
90					144.84	55.92		
100					160.93	62.14		

UNITS	Lb ft to kgm	Kgm to lb ft	UNITS	Lb ft to kgm	Kgm to lb ft
1	.138	7.233	7	.967	50.631
2	.276	14.466	8	1.106	57.864
3	.414	21.699	9	1.244	65.097
4	.553	28.932	10	1.382	72.330
5	.691	36.165	20	2.765	144.660
6	.829	43.398	30	4.147	216.990

CHAPTER 8

THE DRIVE SHAFT, REAR SUSPENSION AND REAR AXLE

8:1 The drive shaft

A typical drive shaft assembly is shown in **FIG 8:1**. The pinion flange on the rear axle forms part of the rear universal joint and two of the bearings on the spider are secured to the pinion drive flange by bolts. The slip yoke on the front of the drive shaft slides into the extension housing of the transmission and fits onto the splines of the output shaft of the transmission. Seals are fitted to prevent lubricant from leaking out of the transmission, as shown in **FIG 8:2**.

Removal:

To preserve the balance of the drive shaft the rear yoke of the drive shaft must be kept in the same alignment with the drive pinion on the rear axle. When the parts are new they are marked with yellow aligning marks but if these marks have worn off or become covered with dirt, new ones should be made.

Remove the nuts and carefully drive out the U-bolts that secure the shafts to the pinion flange. **Wrap tape around the rear universal joint to prevent the bearings from falling off from the spider.**

Lower the rear end of the shaft and withdraw the slip yoke from the transmission. **Take care not to damage or mark the slip yoke when the shaft is free,** as nicks or burrs on the yoke can cause leaks from the transmission. Light damage or burrs can be smoothed off with a stone but if the yoke is badly scored or damaged a new one will have to be fitted.

Install the drive shaft in the reverse order of removal. Lubricate the splines in the slip yoke with B8A-19589-A lubricant. Make sure that the alignment marks align after the drive shaft has been installed.

Maintenance

The spider on each universal joint is fitted with a plug. Unscrew the plug and screw a grease fitting back in its place. Inject a few strokes of Ford Universal Lube through the grease fitting and then remove the fitting and install the plug.

Vibration:

The drive shaft is balanced on manufacture by the addition of balance weights. If vibration is traced to the drive shaft, first check that the balance weights are present and have not fallen off, then make sure that mud or underseal is not clinging to the shafts and upsetting the balance. Check the universal joints for wear, The joints are checked by smartly rotating them back-

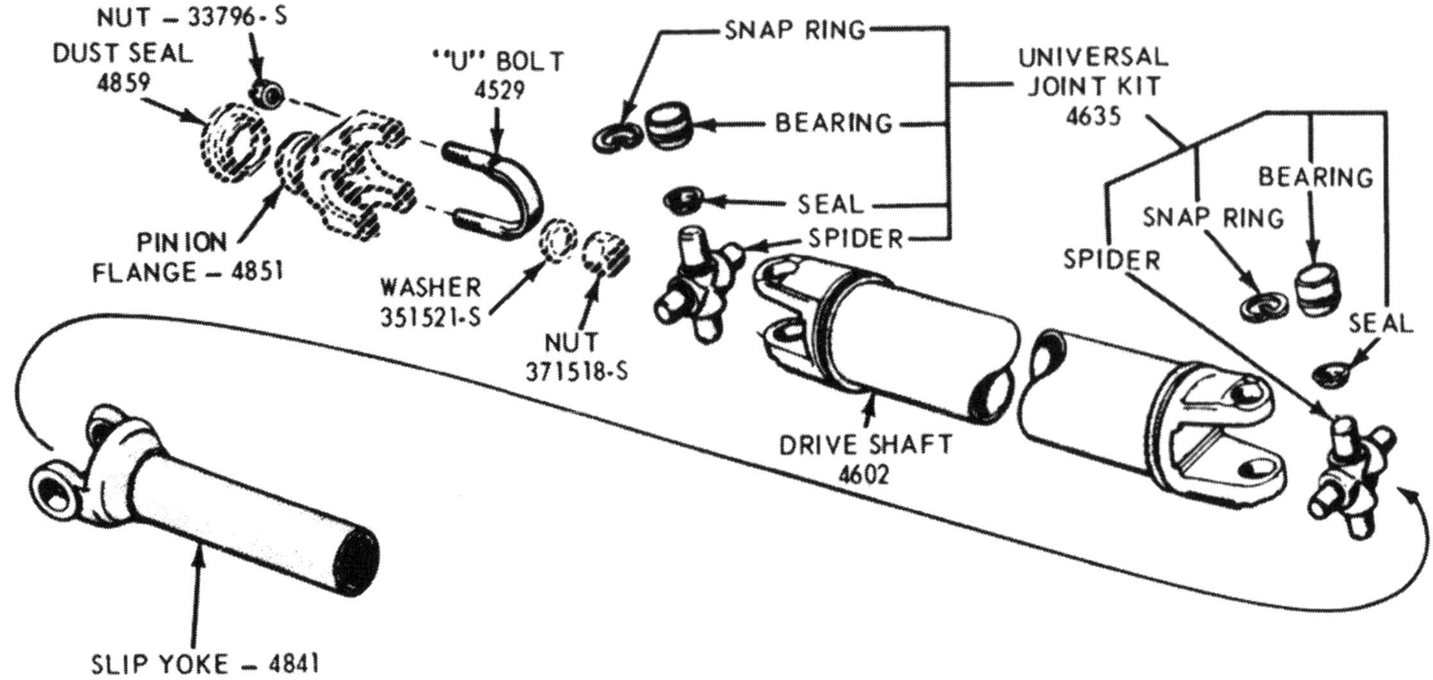

FIG 8:1 The drive shaft parts

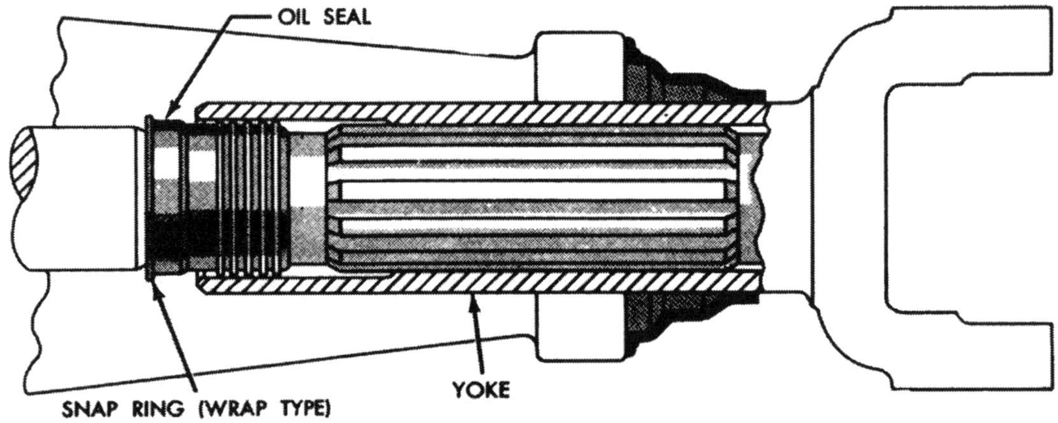

FIG 8:2 The seals and bush for the drive shaft in the transmission

wards and forwards by hand and trying to lift the shaft by hand. Wear will show up as excessive play or 'clonk' as the direction of drive is reversed.

If there is no obvious cause of vibration, try rotating the shaft 180 deg. in relation to the pinion drive flange.

If rotating the shaft through 180 deg. does not cure the vibration, jack up the rear of the car and place it **very securely** on stands. **Chock the front wheels both in front and behind.** Start the engine and select a gear. Speed the engine up until the speedometer is reading approximately 40 to 50 mile/hr. From underneath the car, bring a wax crayon slowly and carefully into light contact with the drive shaft, as shown in **FIG 8:3**. Avoid the balance weights. Stop the engine and if the mark has been made lightly enough it will only show on the heavy side of the shaft. Install two Whittek clamps to the drive shaft so that their heads are 180 deg. away from the crayon mark, as shown in **FIG 8:4**. Tighten the clamps to prevent them from moving. Start the engine and increase the speed until 65 to 70 mile/hr is shown on the speedometer. If vibration is still present, stop the engine and rotate the two clamps so that they make an angle of 45 deg. between their heads, as shown in **FIG 8:5**. If vibration is still present on test, move the heads of the clamps slightly further apart and repeat the test. **Do not run the vehicle for long periods with the wheels off the ground as there is a danger of the transmission or engine overheating.**

Servicing universal joints.

Kits of parts are available and these should be used. Individual parts of the universal joint should not be renewed separately. Before dismantling the shaft, make aligning marks on the slip yoke and shaft so that they will be reassembled and the balance of the shaft preserved.

The method of removing the bearings from a universal joint is shown in **FIG 8:6.** Remove the snap rings that secure the bearings. If the snap rings stick, clean away all rust and enamel. Mount the shaft in the padded jaws of a vice and use the tool to press out the bearing on one side as far as possible. If the bearing cannot be removed fully with the tool, grip the bearing with grips or in a vice and pull it out, using a drift as shown in **FIG 8:7** if the bearing is tight. Remove the opposite bearing in a similar manner When all four bearings have been removed, the spider can be manoeivred out of the yoke and shaft.

Before installing new parts, check the bores in the yoke arms. The bearings must be a light drive fit through them. If the bores are worn oval a new shaft or yoke assembly must be fitted.

Check that the needle rollers in the bearings are in place and held by the grease. **If a needle roller becomes displaced and pressure is applied across it, it will snap.**

Position the spider in the yoke and press a bearing into place so that it is $\frac{1}{4}$ inch below the surface. Install a new snap ring onto the bearing. The bearing is pressed back into place using the special tool as shown in **FIG 8:8.** Start a new bearing into the yoke and press it in until the original bearing contacts the snap ring. Remove the tool and secure the second bearing with a new snap ring. Install the remaining bearings using the same method.

When all the bearings are in place, check that the joint articulates freely. Binding can be relieved by tapping the ears of the yoke arms sharply, and **it is essential that the joints do move freely before installing the shaft.**

8:2 Rear suspension description

A typical rear suspension is shown in **FIG 8:9** and the components are shown in **FIG 8:10.** Semi-elliptical springs and beam rear axles are used on all the models covered by this manual. Rubber insulated bushings are fitted into the eyes at the end of the springs and insulators are fitted between the axle and spring to ensure that road shocks and noise are not transmitted to the body. The front of the spring is supported on a hanger mounting bolt but the rear end is supported on shackles to allow the spring to flex. The axle is mounted above the spring to lower the height of the suspension and is secured by U-bolts and nuts. Shock absorbers are mounted between the frame and a plate under the spring to control the movement of the suspension. The shock absorbers are inclined inwards at the top to provide lateral location for the suspension and rear axle.

Each spring is made up of several leaves. A clamp bolt secures the leaves together at their centres and the ends of the leaves are fitted with insulators between the leaves. Clamps secure the outer ends of the leaves. The springs should be kept dry and clean and do not require lubricant sprayed on them.

8:3 The shock absorbers

The owner has no means of accurately testing the shock absorbers but they should be suspected of being faulty if the car continues pitching after a disturbance or rolls excessively on curves. The average life of a shock absorber is 36,000 miles, so consideration should be given to renewal after this mileage. As a rough test, bounce the corner of the vehicle and if it continues oscillating without rapid damping action then the shock absorber is probably defective.

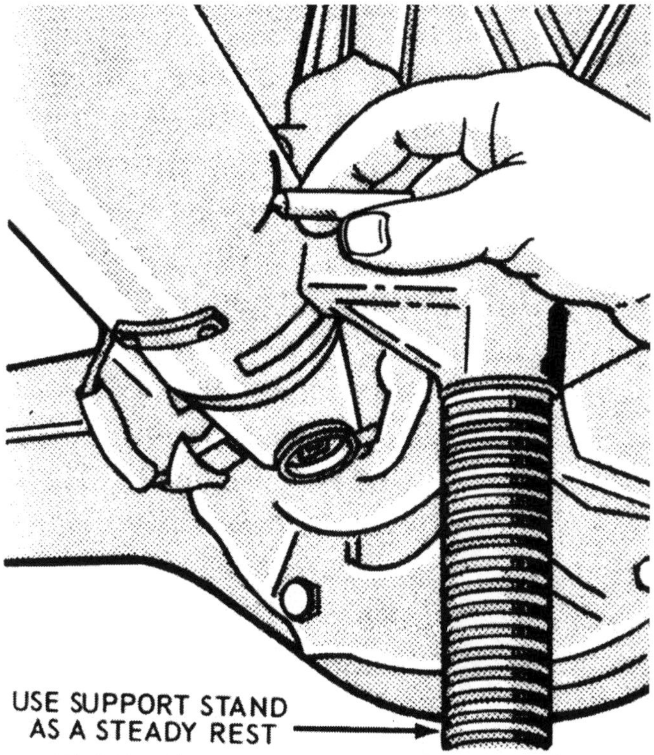

FIG 8:3 Marking the drive shaft for unbalance

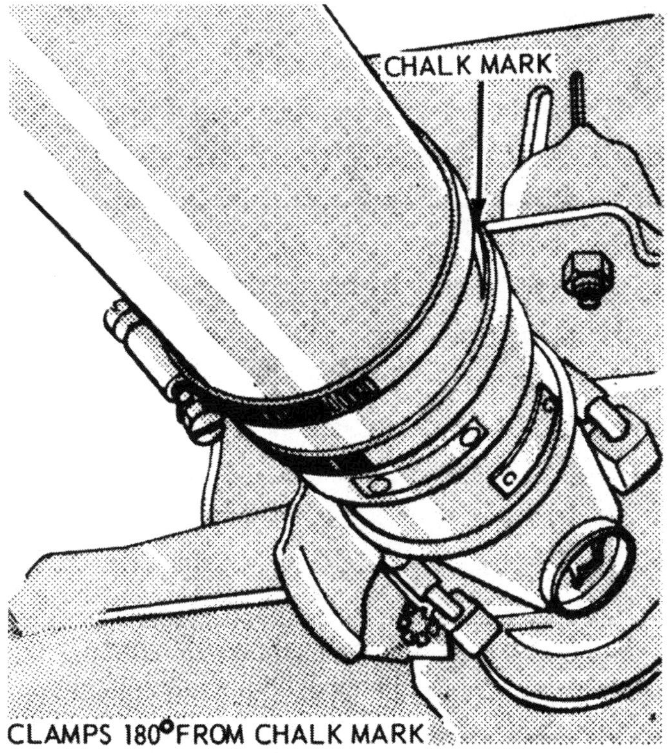

FIG 8:4 Installing the Whittek clamps

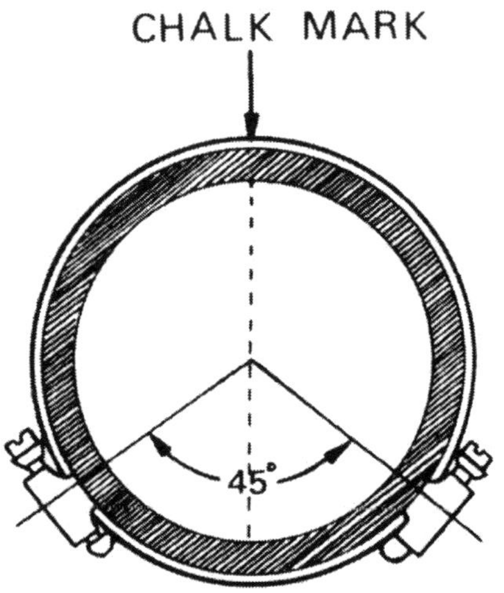

FIG 8:5 Rotating the clamps to balance the drive shaft

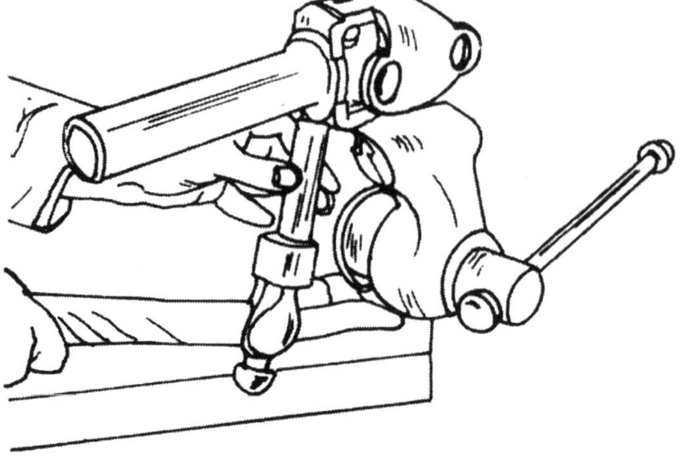

FIG 8:7 Removing a difficult universal joint bearing

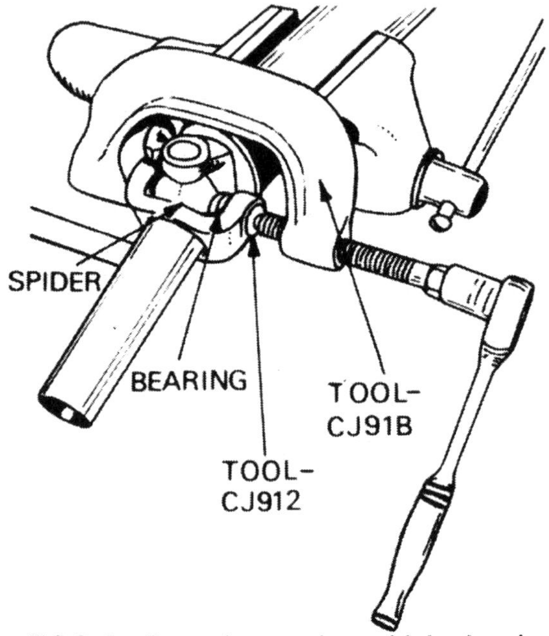

FIG 8:6 Removing a universal joint bearing

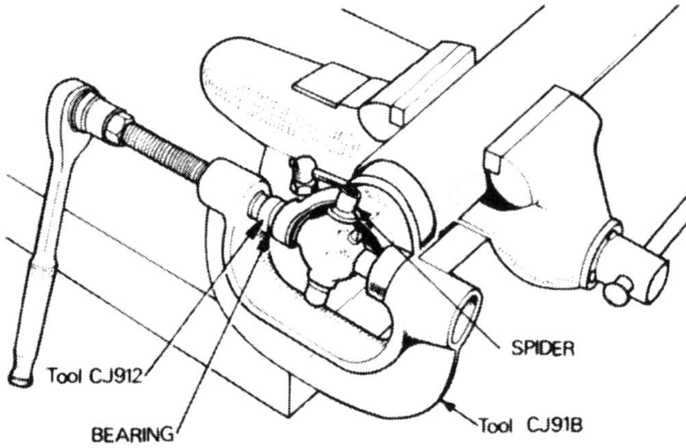

FIG 8:8 Installing a universal joint bearing

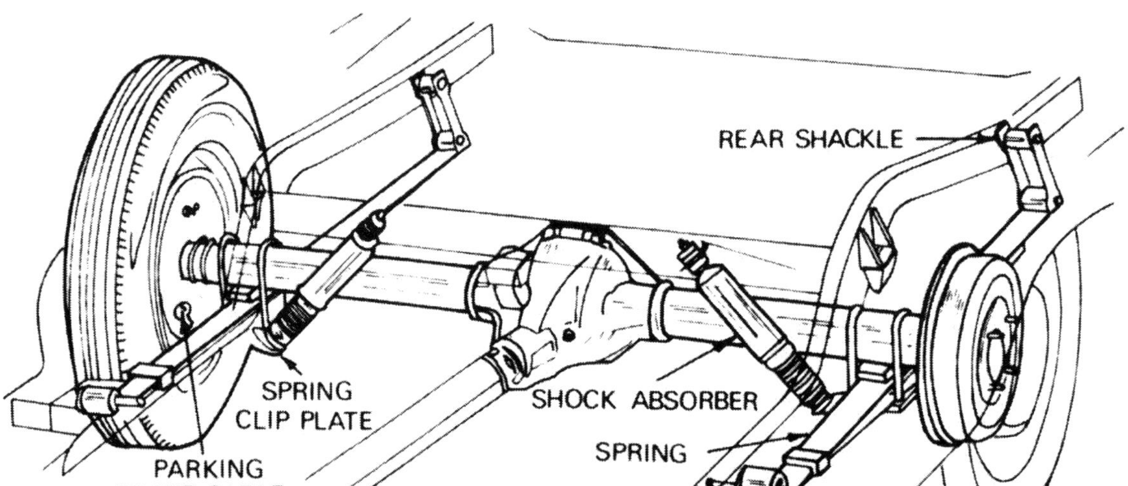

FIG 8:9 A typical rear suspension

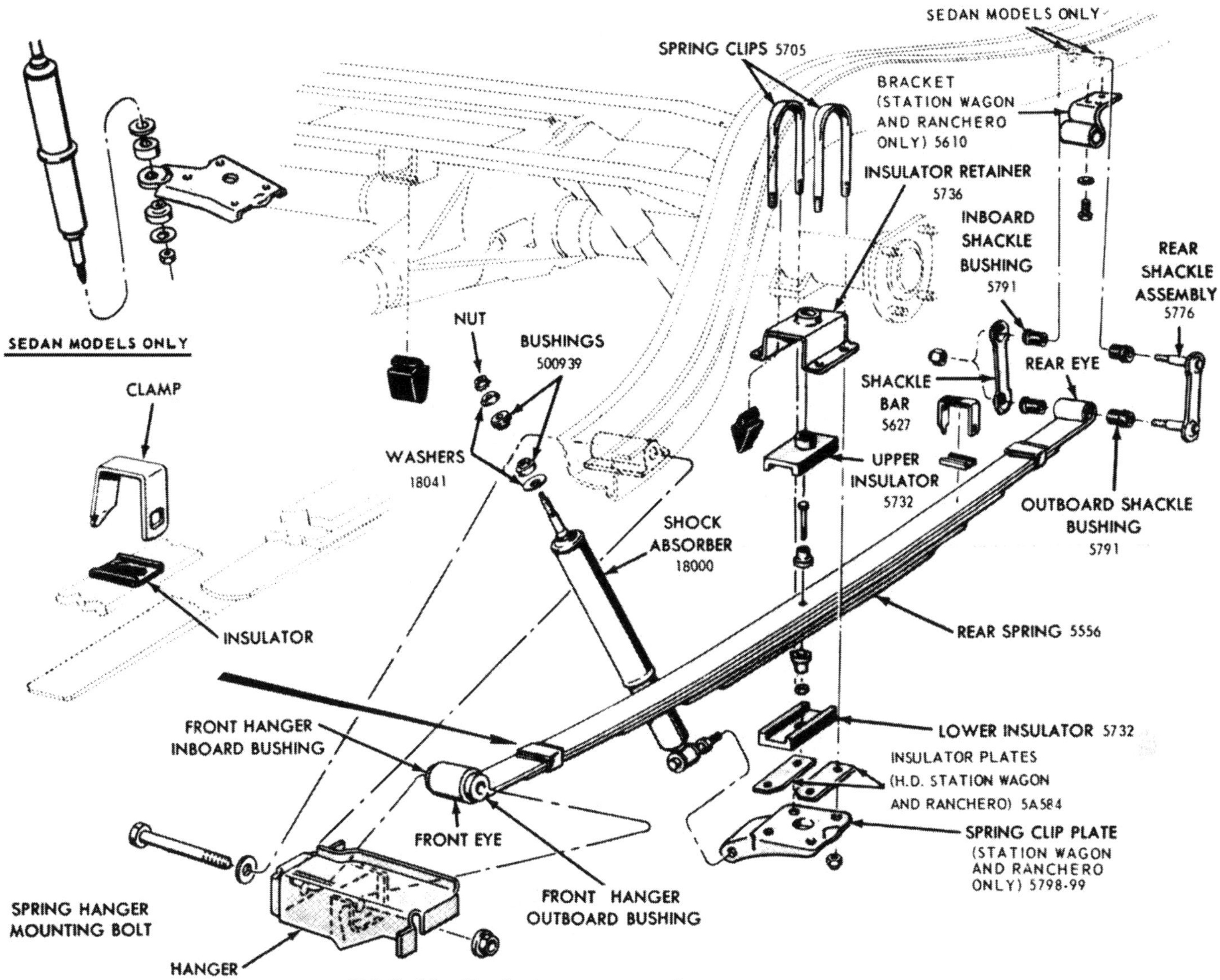

FIG 8:10 Typical rear suspension components

Once the shock absorber has been removed from the vehicle it can be checked for defects. Install a new damper if the old one has physical damage such as a dented body, bent ram, or fluid leaks. Note that it is always best to renew shock absorbers in axle pairs.

Mount the shock absorber vertically with the lower eye in the padded jaws of a vice. Pump the shock absorber, starting off with short strokes about the mid-point and gradually increasing the length of the stroke until it reaches the full travel of the shock absorber. There should be uniform and equal reistance to motion on both directions. If there are pockets of no or weak resistance, particularly when reversing stroke, the shock absorber is defective and must be renewed. Similarly the shock absorber is defective if it operates noisily or leaks. This test will ensure that any air is bled to the top of the shock absorber, if it has been in storage for some time, and it will also detect a defective shock absorber. Even if the unit passes the test this does not guarantee that it is operating at maximum efficiency.

Removal:

The task will be easier if the rear of the car is jacked up and the road wheel removed. Open the luggage compartment lid and remove the spare wheel and tire. On station wagons, remove the access cover from the opening in the seat riser. Fold back floor matting and pry out the access cover to expose the upper end of the shock absorber, as shown in **FIG 8:11.** Remove the nut, washer, and outer bushing that secure the shock absorber to the floor pan. From underneath the car, remove the nuts and bushings that secure the shock absorber to the spring plate clip. Compress the damper and remove it from the car.

Install the shock absorber in the reverse order of removal. Be sure that the spring clip plate is free from damage or burrs. Check the rubber bushings for wear or damage and renew them if they are defective.

8:4 The springs

Removal:

Raise the rear of the car and support it securely on stands beneath the underbody. It will be necessary to have jacks for supporting the rear axle.

Support the rear axle and disconnect the shock absorbers from the spring clip plates (see previous section). Push the shock absorbers up and out of the way. Lower the jacks under the axle so that it is no longer raised. Refer to **FIG 8:10.** Take off the nuts and remove

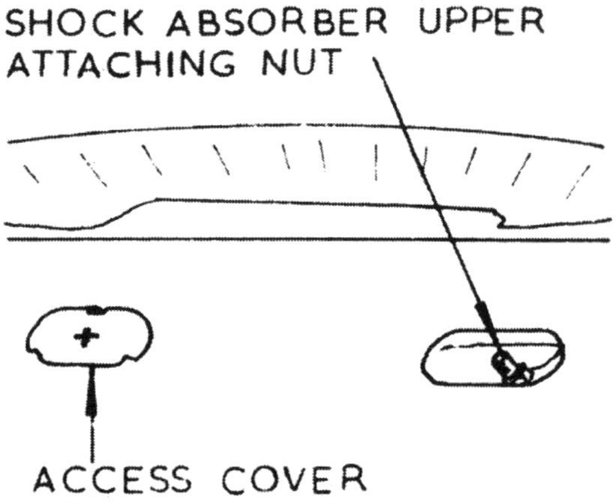

FIG 8:11 The rear shock absorber upper mounting point

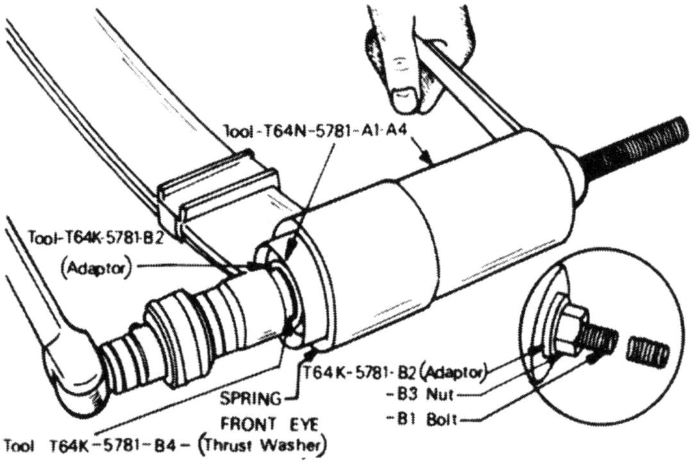

FIG 8:12 Removing the hanger bushing from the sering eye

the spring clip plate from underneath the spring, collecting the insulator and any plates fitted. Raise the rear axle on the jack sufficiently to take its weight off the spring. Remove the U-bolts, insulator retainer and upper insulator.

Remove the nuts from the rear shackle bar and take off the shackle bar so that the inboard bushings can be taken out. Withdraw the outer shackle assembly and outboard bushings. Support the spring by hand or on a suitable stand.

Remove the nut and washer from the front end. Withdraw the hanger bolt and remove the spring assembly from the vehicle.

Examination:

Check the leaves for cracks and inspect the anti-squeak inserts between the leaves. The spring can be disassembled by taking off the clamps and removing the center tie bolt. When disassembling or reassembling the spring, clamp the leaves between the jaws of a strong vice to compress them together. Slacken the vice off gradually to release the pressure once the center tie bolt has been removed.

Scrub the leaves with a wire brush to remove any dirt and corrosion before reassembling the spring. Use a long piece of rod ground to a taper at one end to align the center holes in the leaves. Align the ends of the leaves and compress them together in a vice, making sure that the anti-squeak inserts are fitted. Use a new center tie bolt and new end clamps. Be sure that the clamps are fully tightened before installing the spring.

New rear shackle bushings should be fitted if the old ones are worn, distorted or perished. Similarly the front hanger bushing should be renewed if it is defective. The special tools used for removing the old bushing are shown in **FIG 8:12** and the tools being used to install a new bushing are shown in **FIG 8:13**. If care is taken, an arbor press and suitable mandrels can be used instead of the special tools. If lubricant is required for the installation of the bushings, use water or soap and water. **Do not use grease or oil as they will perish the material of the bushings.**

Installation:

The spring is installed in the reverse order of removal. Install it first at the hanger then at the rear shackle, lower the axle back into place (with insulator and retainer in place) and secure it to the spring using the U-bolts, spring clip plate, and nuts. Leave the hanger and shackle bolts finger tight. When all the parts are installed, using new insulators if the old ones are defective, place a jack under the spring and raise the spring and axle assembly up to the normal working position, taking great care not to lift the vehicle off its stands. Torque load the nuts for the shackle and hanger with the spring held in this position, as this ensures that the bushings will have the minimum strain put on them when the weight of the vehicle is resting on the suspension.

Reconnect the shock absorber and install the road wheel. Lower the vehicle back to the ground.

8:5 The rear axle

The rear axle assembly is shown in **FIG 8:14**. A limited-slip differential can be fitted as an optional extra and a semi-sectional view of the unit is shown in **FIG 8:15**. Under cornering forces the clutch slips and allows the differential to act normally. When one wheel loses traction, such as on ice or mud, the clutch transmits torque from the wheel that has no grip to the wheel that has grip, helping to prevent wheel spin and at the same time allowing the wheel with grip to drive the vehicle.

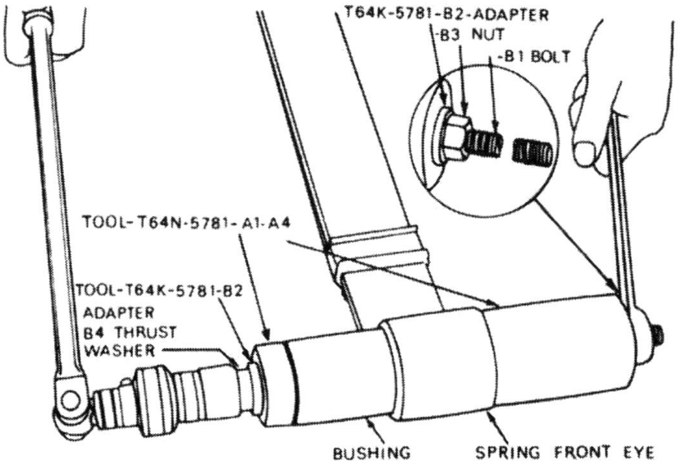

FIG 8:13 Installing the hanger bushing in the spring eye

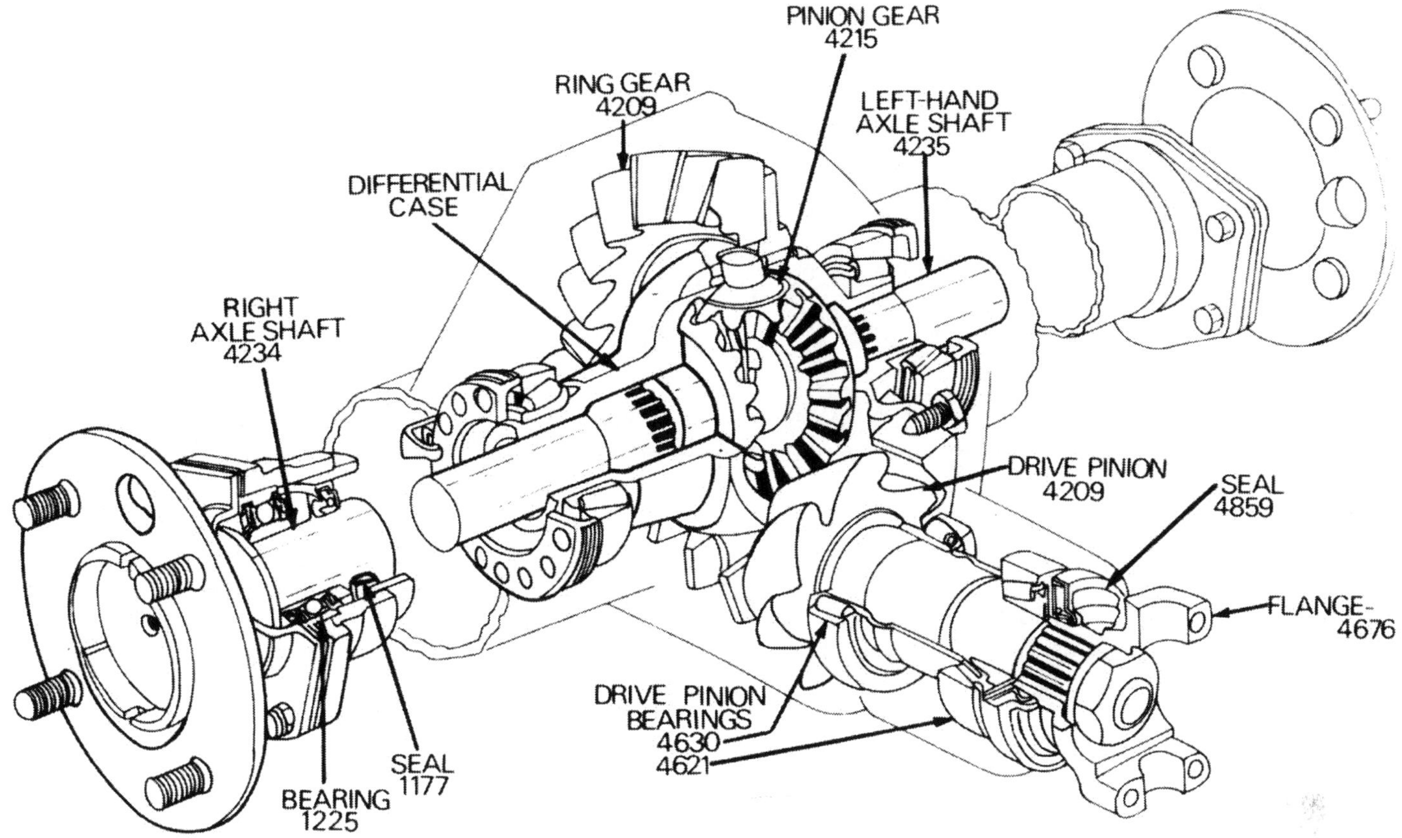

FIG 8:14 The rear axle assembly

Maintenance:

A drain plug is not fitted as seasonal lubricant changes are not required. The factory fill should remain in the housing for the life of the axle. When major repairs have to be carried out, drain the axle by slackening the rear cover bolts and allowing the lubricant to seep out.

The lubricant level should be checked at intervals of 6000 miles. Clean the area around the combined filler and level plug on the rear cover and remove the plug. Top up as required to bring the level to the bottom of the filler hole. Allow surplus lubricant to drain out before installing the filler plug. Normally Hypoid SAE.90 oil should be used (to factory specification) but in climates below —25°F use SAE.80 grade.

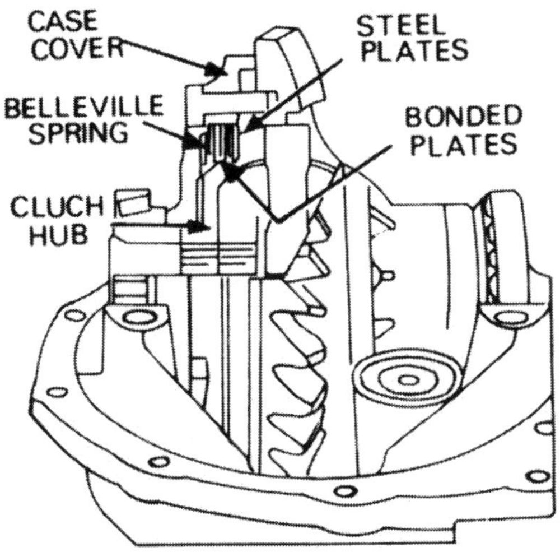

FIG 8:15 The limited slip differential

When checking the lubricant level, check the vent in the casing to be sure that it is not plugged with road dirt. A plugged vent will cause a rise in internal pressure as the unit warms up with use and lubricant will be forced past oil seals.

Hubs and axle shafts:

The components of the standard rear axle are shown in **FIG 8:16**. Synthetic material oil seals are fitted on production but leather seals only are available as replacement parts. The whole length of the axle shaft passes through the oil seal when removing or installing the shaft and great care must be taken not to damage the seal. If the seals are damaged they must be renewed, otherwise axle lubricant will leak into the brakes. **A new leather seal should be fitted in place of the original seal whenever the axle shaft is removed.** Note that if the wheel bearings are removed from the axle shafts it will make the bearings unfit for further use.

Raise the rear of the vehicle, after slackening the road wheel nuts, and place it securely onto underbody stands. Remove the road wheel. Take off the Tinnerman nuts that secure the brake drum and withdraw the drum from the assembly. Rotate the axle shaft to a convenient position for working and remove the nuts that secure the wheel bearing retainer plate, through the holes provided in the axle shaft flange. Attach the special tools shown in **FIG 8:17** to the axle flange and use the impact hammer to draw the bearing and axle shaft out of the axle. When the axle shaft has been removed, prevent the brake backplate from moving by securing it with one nut.

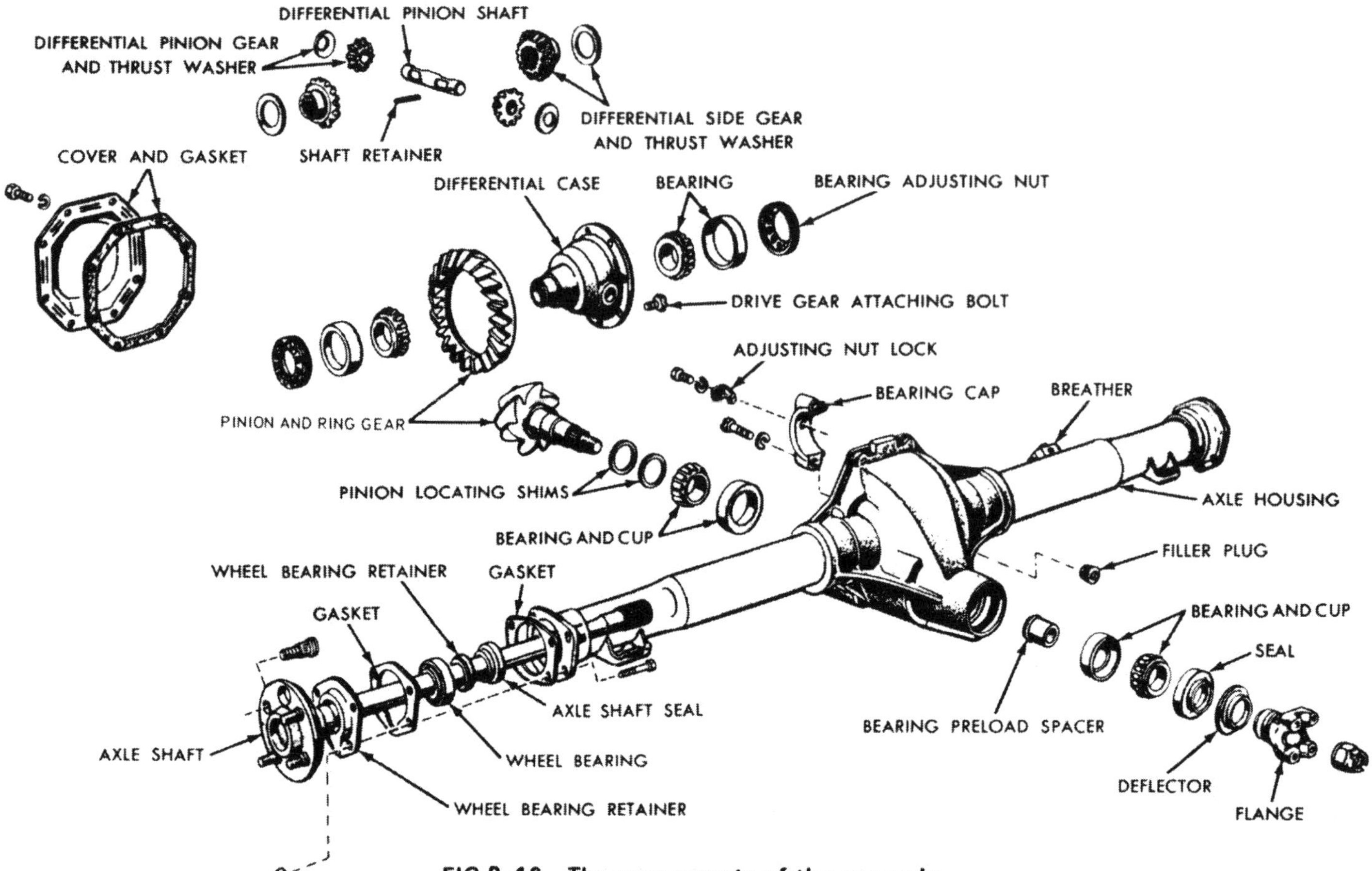

FIG 8:16 The components of the rear axle

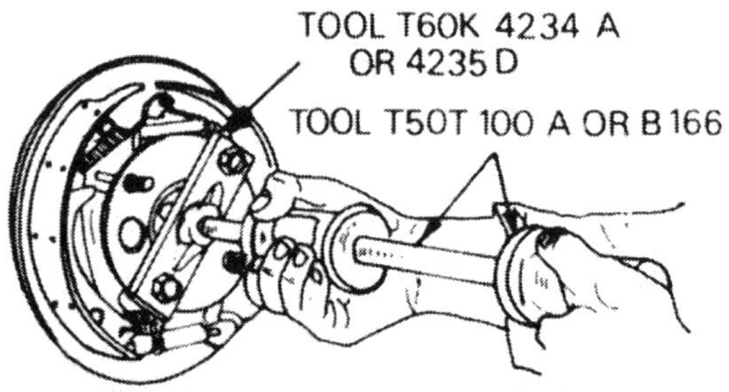

FIG 8:17 Withdrawing the axle shaft and bearing assembly

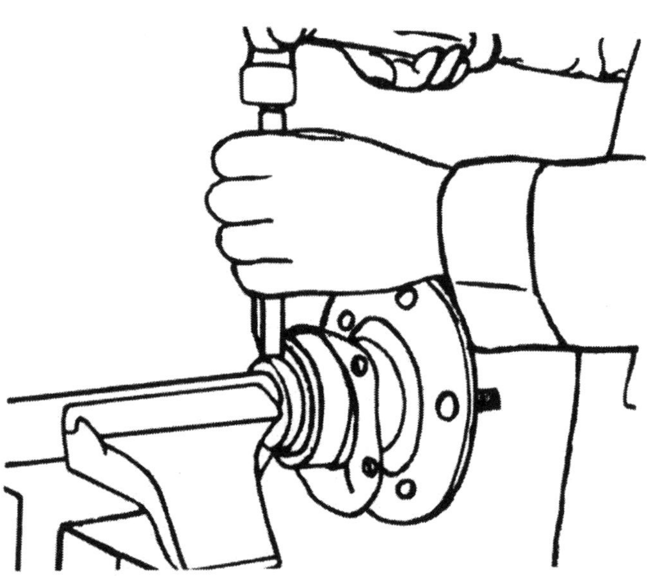

FIG 8:18 Freeing the bearing retainer ring

While checking and cleaning the parts, put the new leather oil seal to soak in light SAE.10 oil for at least $\frac{1}{2}$ hour.

If the bearing is worn or runs noisily, remove it from the shaft and fit a new bearing and retaining ring. Support the shaft on the padded jaws of a vice and use a cold chisel to nick it deeply at several points around the circumference, as shown in **FIG 8:18.** The aim is to nick deeply but not cut through, and once this has been done at several points the retaining ring will slide off easily. The methods of removing and installing the bearing on the shaft are shown in **FIG 8:19.** The bearing should be pressed on right up to the shoulder before attempting to press a new retainer ring into place.

Remove the old seal from the axle casing tube using the special tool shown in **FIG 8:20.** Install the new seal using the tool as shown in the figure. It is essential that the correct tool is used for installation of the seal, otherwise the seal may leak in service. **The seal must have had $\frac{1}{2}$ hour soak in light oil before installation.** Wipe off any oil around the circumference, smear seal with sealing compound.

Check the shaft and axle housing for irregularities, burrs, rough spots or damage. Smooth down rough spots and renew the parts if they are excessively damaged.

Place a new gasket on either side of the brake coverplate, discarding the old gaskets. Very gently and carefully slide the axle shaft assembly back into the housing, taking great care that the splines do not damage the new oil seal. Turn the shaft slightly if required and start the splines into the side gear. Push the shaft in until the bearing bottoms in the housing and

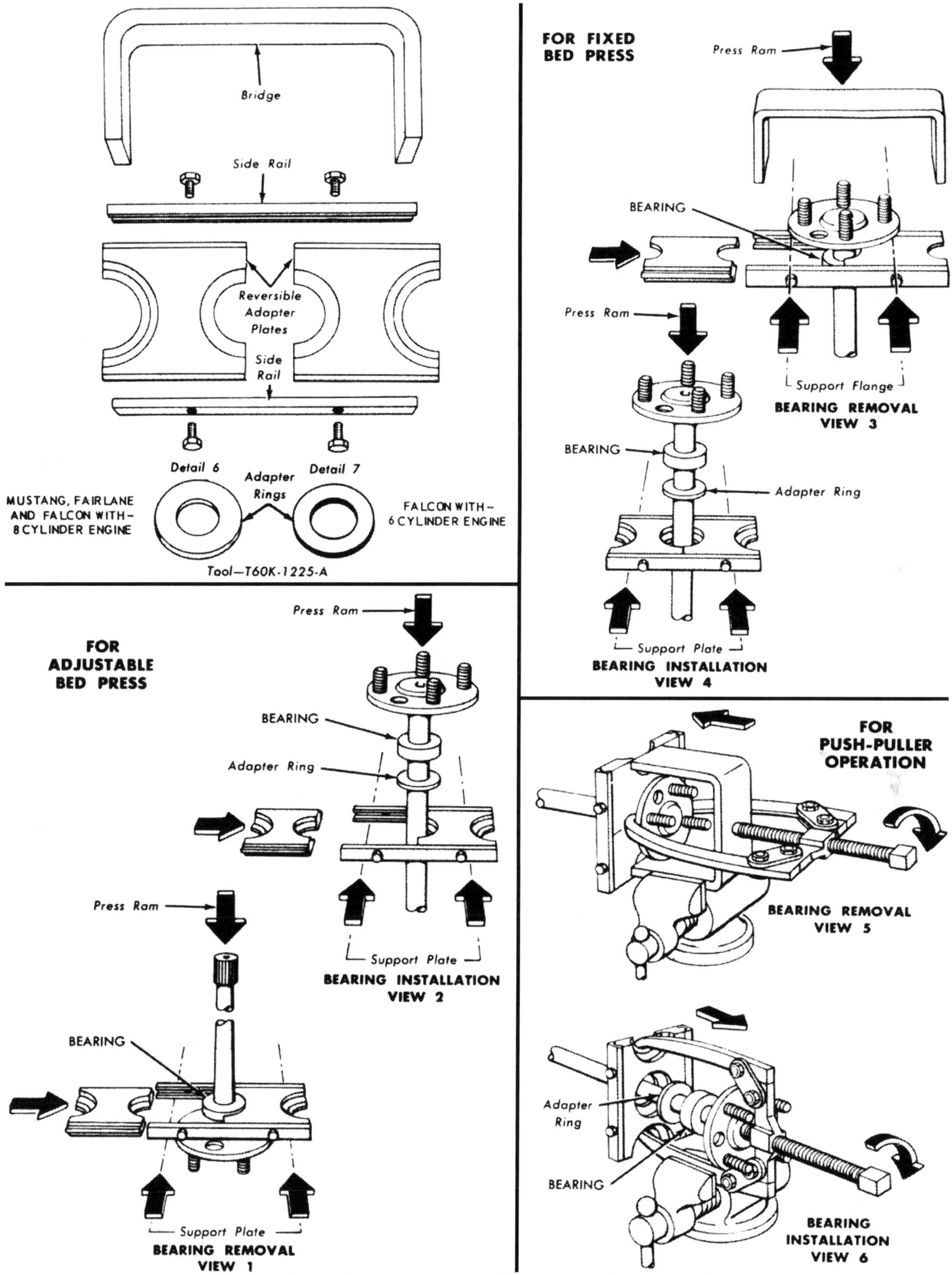

FIG 8:19 Removing and installing the bearing on the axle

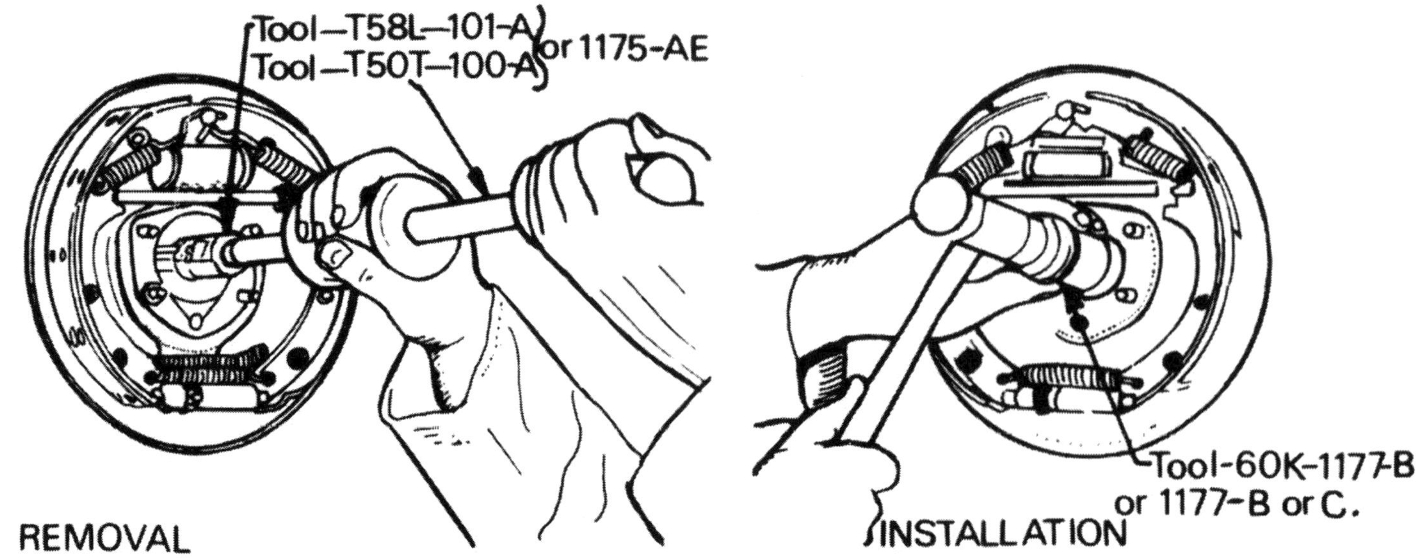

FIG 8:20 Removing and installing the axle shaft seal

install the bearing retainer onto the bolts. Secure the retainer with the nuts. Install the brake drum and road wheel and check the brake adjustment. Lower the vehicle back to the ground and check the lubricant level in the rear axle.

Pinion oil seal:

This can be renewed without having to remove the axle or differential carrier from the vehicle. Disconnect the drive shaft from the drive pinion flange, making sure that the bearings on the shaft are held in place by tape.

1 Raise the rear of the vehicle onto stands and remove both road wheel and brake drums. Attach an inch lb torque spanner to the drive flange and use the torque spanner to rotate the drive pinion for several revolutions. Note the average force required to turn the pinion.

2 Hold the flange with a tool and use a socket wrench to remove the securing nut. Remove the drive pinion flange as shown in **FIG 8:21** or with a suitable extractor.

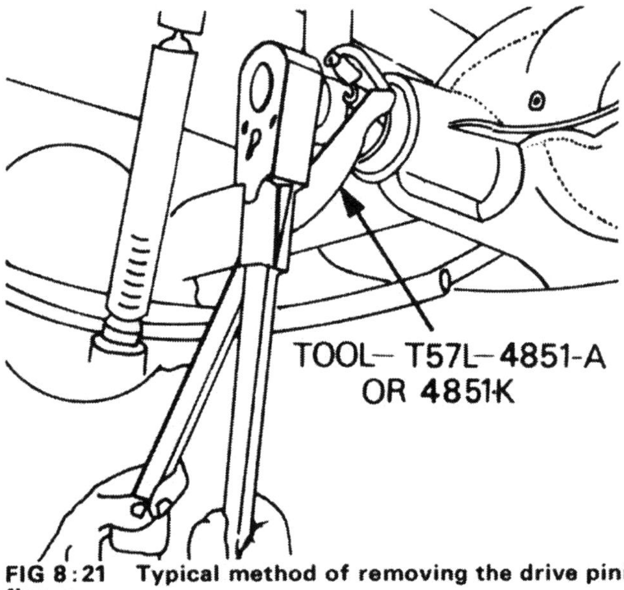

FIG 8:21 Typical method of removing the drive pinion flange

3 Remove the old seal using a similar method to the one shown in **FIG 8:20**, noting that the adaptor 1175 AB is the correct one for this seal. Lubricant will flow from the axle as the seal is removed so have a drain pan in position.

4 Check the splines on the pinion shaft for burrs or nicks. Similarly check the surface of the flange for scores or nicks. Light damage can be smoothed off using Crocus paper but if the damage is deep new parts will have to be fitted.

Wipe away oil from the seal seating area, and lightly smear the external surface of the seal with sealing compound. Do not allow sealing compound onto the lips of the seal. Install the new seal by driving it in or pulling it in using suitable tools.

5 Install the drive flange and secure it using a new washer and nut. Lightly lubricate the washer side of the nut with a little oil. Tighten the nut while holding the flange and at frequent intervals check the load required to revolve the pinion shaft, using a torque spanner. Tighten the nut until the preload is the same as before disassembly, as found in operation 1, and then tighten the nut further until the original preload is increased by 6 to 12 lb inch. **The nut must not be slackened off to reduce the preload** and if the nut is tightened too far and the preload becomes excessive the differential will have to be removed and a new spacer fitted between the pinion bearings.

6 Install the drive shaft, brake drums, and road wheels. Check the brake adjustment. Lower the vehicle back to the ground and check the lubricant level in the rear axle.

Differential:

This can be removed from the vehicle, after disconnecting the drive shaft and removing both axle shafts, leaving the axle casing fitted to the vehicle. Work on the differential should always be carried out by specialists as the meshing of the gears and the preloads on the bearings must be set with extreme accuracy.

Axle removal:

If the complete axle assembly is being removed, the parking brake cables must be disconnected from the brake carrier plates and the main hydraulic line disconnected at the T-piece on the axle. The hydraulic system will then have to be bled after installation of the axle.

The method given here obviates the necessity for disconnecting the braking system parts.

1 Raise the rear of the vehicle and support it on stands. Remove both road wheels, brake drums and then withdraw the axle shaft assemblies. Suspend the brake carrier plates, still attached to the brake lines and parking brake cables, safely out of the way.

2 Drain the unit by removing the rear cover. Disconnect the vent pipe from the axle casing. Free the hydraulic line and T-piece from the axle casing. Disconnect the shock absorber lower attachments from the spring clip plates.

3 Make alignment marks and disconnect the drive shaft from the drive pinion flange, carefully taping the loose bearings in place on the drive shaft.

4 Remove the nuts that secure the spring clip plates on the suspension. Remove the U-bolts, insulators and insulator retainers. Manoeuvre the axle assembly out from over the road springs.

The unit is installed in the reverse order of removal. The points to watch for when reconnecting and installing the parts have been dealt with in the relevant sections.

8:6 Trouble shooting guide

(a) Noisy suspension or axle

1 Lack of lubricant or incorrect lubricant used
2 Defective shock absorbers
3 Defective shock absorber mounting bushings
4 Defective axle mounting insulators
5 Defective spring mounting bushings
6 Worn drive shaft universal joints
7 Defective differential unit
8 Worn bearings

*On models fitted with limited slip differential a high pitched chattering noise on turns can be caused by lack of or incorrect lubricant. Try driving the car in five fairly tight circles clockwise and then five more circles counterclockwise. This will permit the lubricant to work into the clutch plates. Check the lubricant by draining out two tablespoonfuls of lubricant and mix it with an equal amount of white alcohol. Let the mixture stand for at least two minutes. If the sample is blue-tinted a Ford approved lubricant is used. If the sample is yellow-tinted an incorrect lubricant is being used. Drain out incorrect lubricant and refill with Ford approved lubricant. There is no need to flush the axle but the vehicle should be driven in five tight circles in both directions to distribute the lubricant onto the clutch plates.

(b) Vibration

1 Check 5 and 6 in (a)
2 Drive shaft out of balance (lost weights or dirt)
3 Drive shaft fitted 180 deg. out

(c) Lubricant leaks

1 Defective hub oil seals
2 Defective pinion oil seal
3 Defective seals on universal joint spiders
4 Plugged breather on axle
5 Too high lubricant level
6 Defective gaskets or damaged joint faces

Inches	Decimals	Millimetres	Inches to Millimetres Inches	mm	Millimetres to Inches mm	Inches
1/64	.015625	.3969	.001	.0254	.01	.00039
1/32	.03125	.7937	.002	.0508	.02	.00079
3/64	.046875	1.1906	.003	.0762	.03	.00118
1/16	.0625	1.5875	.004	.1016	.04	.00157
5/64	.078125	1.9844	.005	.1270	.05	.00197
3/32	.09375	2.3812	.006	.1524	.06	.00236
7/64	.109375	2.7781	.007	.1778	.07	.00276
1/8	.125	3.1750	.008	.2032	.08	.00315
9/64	.140625	3.5719	.009	.2286	.09	.00354
5/32	.15625	3.9687	.01	.254	.1	.00394
11/64	.171875	4.3656	.02	.508	.2	.00787
3/16	.1875	4.7625	.03	.762	.3	.01181
13/64	.203125	5·1594	.04	1.016	.4	.01575
7/32	.21875	5.5562	.05	1.270	.5	.01969
15/64	.234375	5.9531	.06	1.524	.6	.02362
1/4	.25	6.3500	.07	1.778	.7	.02756
17/64	.265625	6.7469	.08	2.032	.8	.03150
9/32	.28125	7.1437	.09	2.286	.9	.03543
19/64	.296875	7.5406	.1	2.54	1	.03937
5/16	.3125	7.9375	.2	5.08	2	.07874
21/64	.328125	8.3344	.3	7.62	3	.11811
11/32	.34375	8.7312	.4	10.16	4	.15748
23/64	.359375	9.1281	.5	12.70	5	.19685
3/8	.375	9.5250	.6	15.24	6	.23622
25/64	.390625	9.9219	.7	17.78	7	.27559
13/32	.40625	10.3187	.8	20.32	8	.31496
27/64	.421875	10.7156	.9	22.86	9	.35433
7/16	.4375	11.1125	1	25.4	10	.39370
29/64	.453125	11.5094	2	50.8	11	.43307
15/32	.46875	11.9062	3	76.2	12	.47244
31/64	.484375	12.3031	4	101.6	13	.51181
1/2	.5	12.7000	5	127.0	14	.55118
33/64	.515625	13.0969	6	152.4	15	.59055
17/32	.53125	13.4937	7	177.8	16	.62992
35/64	.546875	13.8906	8	203.2	17	.66929
9/16	.5625	14.2875	9	228.6	18	.70866
37/64	.578125	14.6844	10	254.0	19	.74803
19/32	.59375	15.0812	11	279.4	20	.78740
39/64	.609375	15.4781	12	304.8	21	.82677
5/8	.625	15.8750	13	330.2	22	.86614
41/64	.640625	16.2719	14	355.6	23	.90551
21/32	.65625	16.6687	15	381.0	24	.94488
43/64	.671875	17.0656	16	406.4	25	.98425
11/16	.6875	17.4625	17	431.8	26	1.02362
45/64	.703125	17.8594	18	457.2	27	1.06299
23/32	.71875	18.2562	19	482.6	28	1.10236
47/64	.734375	18.6531	20	508.0	29	1.14173
3/4	.75	19.0500	21	533.4	30	1.18110
49/64	.765625	19.4469	22	558.8	31	1.22047
25/32	.78125	19.8437	23	584.2	32	1.25984
51/64	.796875	20.2406	24	609.6	33	1.29921
13/16	.8125	20.6375	25	635.0	34	1.33858
53/64	.828125	21.0344	26	660.4	35	1.37795
27/32	.84375	21.4312	27	685.8	36	1.41732
55/64	.859375	21.8281	28	711.2	37	1.4567
7/8	.875	22.2250	29	736.6	38	1.4961
57/64	.890625	22.6219	30	762.0	39	1.5354
29/32	.90625	23.0187	31	787.4	40	1.5748
59/64	.921875	23.4156	32	812.8	41	1.6142
15/16	.9375	23.8125	33	838.2	42	1.6535
61/64	.953125	24.2094	34	863.6	43	1.6929
31/32	.96875	24.6062	35	889.0	44	1.7323
63/64	.984375	25.0031	36	914.4	45	1.7717

UNITS	Pints to Litres	Gallons to Litres	Litres to Pints	Litres to Gallons	Miles to Kilometres	Kilometres to Miles	Lbs. per sq. In. to Kg. per sq. Cm.	Kg. per sq. Cm. to Lbs. per sq. In.
1	.57	4.55	1.76	.22	1.61	.62	.07	14.22
2	1.14	9.09	3.52	.44	3.22	1.24	.14	28.50
3	1.70	13.64	5.28	.66	4.83	1.86	.21	42.67
4	2.27	18.18	7.04	.88	6.44	2.49	.28	56.89
5	2.84	22.73	8.80	1.10	8.05	3.11	.35	71.12
6	3.41	27.28	10.56	1.32	9.66	3.73	.42	85.34
7	3.98	31.82	12.32	1.54	11.27	4.35	.49	99.56
8	4.55	36.37	14.08	1.76	12.88	4.97	.56	113.79
9		40.91	15.84	1.98	14.48	5.59	.63	128.00
10		45.46	17.60	2.20	16.09	6.21	.70	142.23
20				4.40	32.19	12.43	1.41	284.47
30				6.60	48.28	18.64	2.11	426.70
40				8.80	64.37	24.85		
50					80.47	31.07		
60					96.56	37.28		
70					112.65	43.50		
80					128.75	49.71		
90					144.84	55.92		
100					160.93	62.14		

UNITS	Lb ft to kgm	Kgm to lb ft	UNITS	Lb ft to kgm	Kgm to lb ft
1	.138	7.233	7	.967	50.631
2	.276	14.466	8	1.106	57.864
3	.414	21.699	9	1.244	65.097
4	.553	28.932	10	1.382	72.330
5	.691	36.165	20	2.765	144.660
6	.829	43.398	30	4.147	216.990

CHAPTER 9

THE FRONT SUSPENSION AND HUBS

9:1 Description

A typical front suspension assembly is shown in **FIG 9:1**. The early models differ in the inboard attachments of the lower and upper arms, and this affects the method of setting the suspension geometry.

The upper and lower arms are attached by pivot bolts to the frame and can therefore move vertically. The outboard end of the lower arm is located by a strut to prevent it from moving in the fore and aft direction. A spindle and stub axle assembly swivels in the outboard ends of the suspension arms in ball joints. The ball joints allow the suspension to move as well as allowing the spindle to rotate about its vertical axis for the steering movement of the front wheels.

A coil spring and concentric damper are installed between the upper suspension arm and a tower on the frame. The spring takes the weight of the vehicle while the shock absorber controls the movement of the suspension and damps out oscillations.

The road wheel is attached to a hub assembly which rotates about the stub axle on two opposed taper bearings.

A stabilizer bar interconnects the two lower suspension arms. When one suspension is at a different height from the other, the stabilizer bar is twisted and exerts a restoring force between the two suspensions to bring them back to the same height.

9:2 Maintenance

The suspension ball joints should be greased at intervals of 12,000 miles. Remove the plug from the ball joint and install a grease fitting. Inject grease through the fitting until the dust cover just starts to swell. Do not inject so much grease that it exudes past the dust cover otherwise the weatherseal will be damaged. Check that the dust covers are in good condition and not starting to crack or split. If a dust cover has split, it must be renewed to prevent dirt from entering the ball joint. If dirt has entered the ball joint, try to remove it by pumping through plenty of fresh grease with the dust cover off. If the dirt cannot be flushed out, fit a new ball joint.

When the ball joints have been lubricated, remove the grease fittings and install the plugs.

At approximately 30,000 miles the wheel hub should be removed, old grease washed out and the hub installed packed with fresh grease after checking the condition of the wheel bearings. This operation is dealt with in the next section.

9:3 Wheel hubs

The components of a wheel hub fitted with drum brakes are shown in **FIG 9:2** and those fitted with disc brakes in **FIG 9:3**. Whenever the front of the vehicle is raised off the ground, briefly check the wheel bearings.

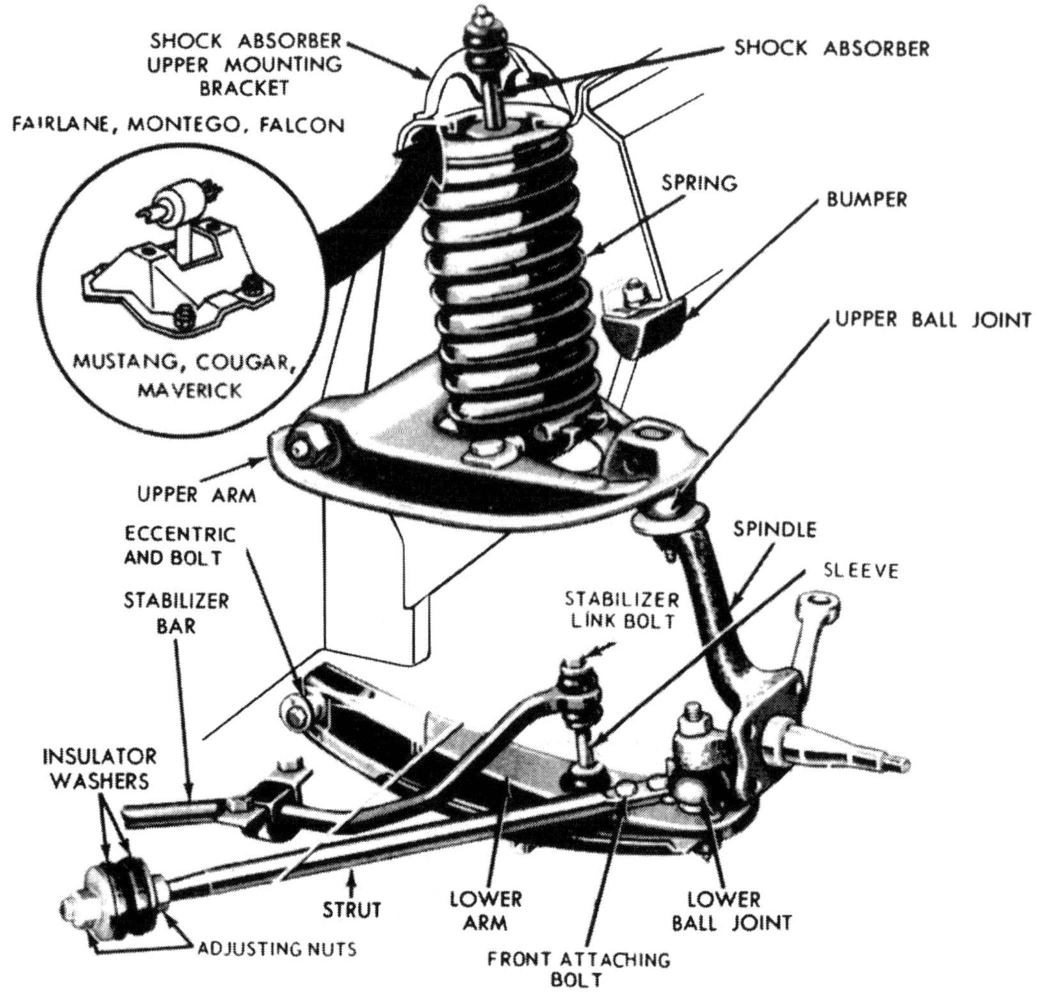

FIG 9:1 Typical front suspension assembly

Grip the tire at the nine and three o'clock positions and rock it as though trying to make it steer. If there is excessive play then the wheel bearings are likely to require adjustment. Do not confuse steering movement with play. Spin the wheel and check that it rotates freely without grinding noises. Do not confuse brake noise with that from a defective bearing. If the bearings are noisy, the wheel hub should be removed so that they can be visually examined. Packing in more grease on the outer bearing and tightening up the adjustment alone can be dangerous, as the inner bearing may be on the point of failure.

Removal:

Raise the front of the car and place it securely onto stands. On models fitted with drum brakes the hub assembly can be removed with the road wheel and tire, but on models fitted with disc brakes the road wheel must be removed in order to remove the brake caliper. Slacken the road wheel attachment nuts slightly before raising the car, otherwise difficulty will be found in slackening them once the wheel can spin freely.

Remove the grease cap from the hub and extract the cotterpin that secures the nut lock. On models fitted with disc brakes, slacken back the nut by three turns and rock the road wheel to press the friction pads and pistons back into the brake caliper. On models fitted with drum brakes, slacken the brake adjusters if necessary. Remove the road wheel. This is advisable on all models, though not necessary on models with drum brakes, as it will be found easier to handle the parts and to clean them without getting solvent or grease onto the tire.

On models fitted with disc brakes, remove the caliper from the suspension and hang it safely out of the way so that the flexible hose is not strained (see **Chapter 11**). Remove the locknut, nut and special washer. Pull the hub off from the spindle, carefully collecting the inner race of the outer bearing as it comes free.

Pry out the grease retainer so that the inner race assembly of the inner bearing can be removed from the hub. The bearing cups can be left installed in the hub, unless the bearings are defective. If the bearing cups must be removed, drive them out evenly and squarely with a suitable drift.

Cleaning and inspection:

Wipe out the surplus old grease using rags. Wash the hub assembly in a suitable solvent (such as gasoline or kerosene) to remove the remainder of the old grease. Rotate the inner races of the bearings under clean solvent until all old lubricant and dirt has been washed out. Dry the parts with an air-line. **When drying the bearings, do not allow them to rotate under the air pressure or they will become chipped and damaged.**

Brush away all dust and dirt from the brake carrier plate, or disc brake dust shield. Wipe down the spindle with a piece of cloth wetted with solvent.

Lubricate the bearings with thin oil and insert each one in turn into its bearing cup. Press the race firmly into the cup and while applying pressure oscillate the bearing backwards and forwards. If the bearing feels at all rough clean it again, as specks of dirt may be the cause. If cleaning cannot cure the roughness, drive out both bearing cups and drive new ones back into place, using the new inner races on installation of the hub. Similarly both bearings complete must be renewed if either one of them shows any signs of wear, pitting, cracking or fretting.

Check the spindle for cracks or corrosion. Give the spindle a wipe over with a clean oily or greasy cloth to prevent corrosion.

Installation and lubrication:

The points to apply grease are shown in **FIG 9:4**. Pack the inside of the hub to the depth shown with fresh grease. Work ample grease into the inner races of both bearings and apply a light smear of grease to the bearing cup. Install the inner race of the inner bearing and hold it in place by installing a new grease retainer. The new grease retainer should have been soaked in light oil for 30 minutes before installation.

Slide the hub assembly back onto the spindle, taking great care to keep it central and avoid damaging the grease retainer.

Press the inner race of the outer bearing back into position. Install the special washer and nut followed by the locknut left loosely in place.

Once the adjustment of the bearings has been completed, install the brake caliper, hub grease cap, and road wheel. Check that the road wheel spins freely before lowering the vehicle back to the ground. On models fitted with disc brakes, pump the brake pedal hard several times to take up the adjustment in the brakes before using the vehicle.

Before installing the hub assembly, use a harsh solvent (such as carbon tetrachloride) to clean the operating face of the disc or drum to remove all lubricant or finger marks.

Adjustment:

The method of setting the correct adjustment is shown in **FIG 9:5**. Spin the hub and tighten the nut to a torque of 17 to 25 lb ft. If a torque spanner is not available, tighten the nut until there is a definite but slight resistance to spinning the hub, showing that the bearings are under slight preload and have been pulled fully back into place, and slacken the nut again. Tighten the nut using moderate pressure until all end play is taken up. Install the locknut so that the castellations are aligned with a cotterpin hole. Back off the nut and locknut together by one castellation and then lock them using a new cotterpin. Note that the ends of the cotterpin must be opened circumferentially as shown, otherwise the ends may damage the static pick-up point in the grease cap.

9:4 The shock absorbers

Testing of the shock absorbers for defects is the same as for the shock absorbers on the rear suspension (see **Chapter 8, Section 8:3**).

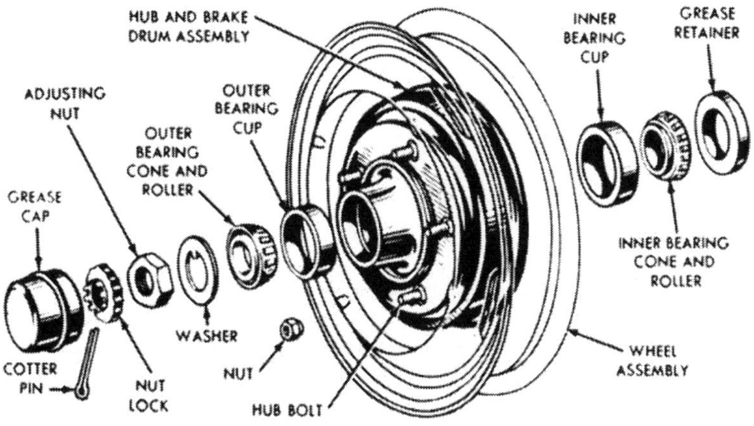

FIG 9:2 The hub assembly fitted with drum brakes

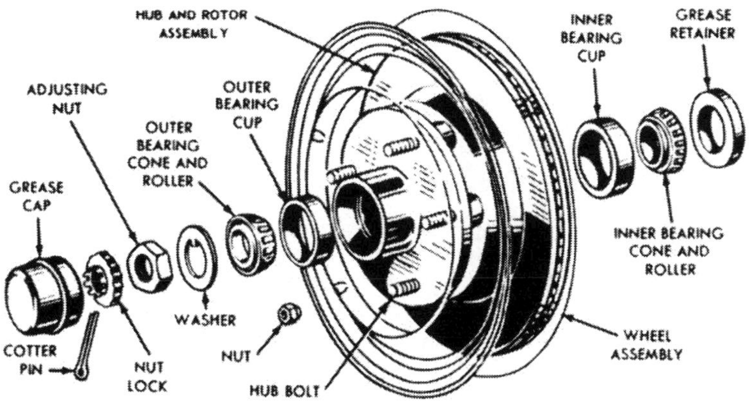

FIG 9:3 The hub assembly fitted with disc brakes

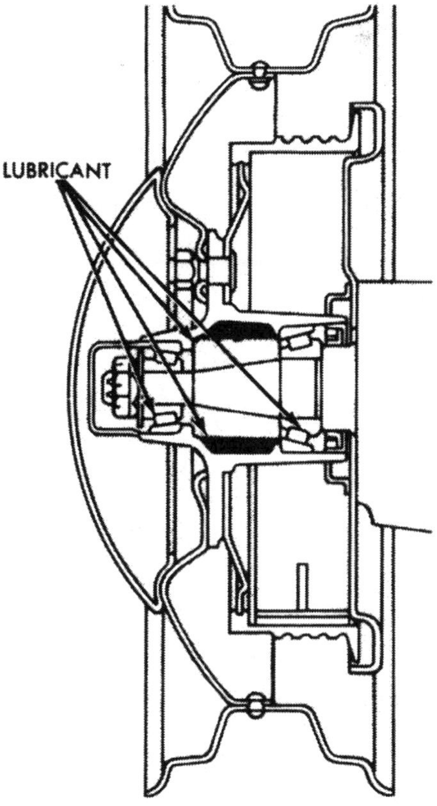

FIG 9:4 Front wheel hub lubrication

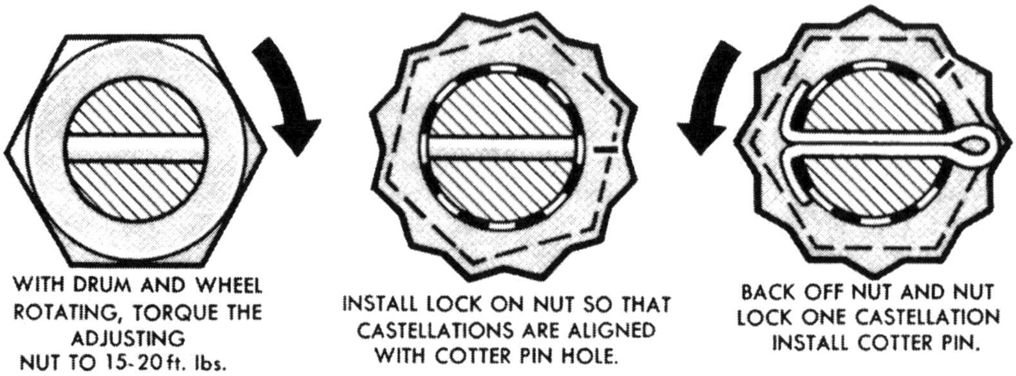

FIG 9:5 Front wheel bearing adjustment

The components of the later suspension are shown in **FIG 9:6.** On some of the earlier models, the shock absorber may have the eye at the bottom and be secured to the upper attachment bracket by nuts and bushings. A typical damper and its attachments are shown in **FIG 9:7.**

Removal:

Raise the front of the car onto stands and support the suspension with a jack under the lower arm. Keep the suspension low enough in relation to the car to be able work on the lower attachments.

Remove the nuts that secure the lower attachment and remove the nuts that secure the upper mounting bracket to the frame. Lift the shock absorber and its bracket out of the vehicle and separate the shock absorber from the bracket.

Install the shock absorber in the reverse order of removal, making sure that all bushings are in good condition.

9:5 The stabilizer bar

The attachments of the stabilizer bar are shown in **FIG 9:6** and the figure is typical for all model stabilizer bars.

Raise the front of the vehicle and place it securely on stands. Access will be easier if the road wheels are removed. Support each suspension with a jack or small stand under the lower arm, ensuring that both suspensions are set to the same height.

Take off the nuts and withdraw the stabilizer link bolts from each lower arm, carefully collecting the spacer, bushings and washers as they come free.

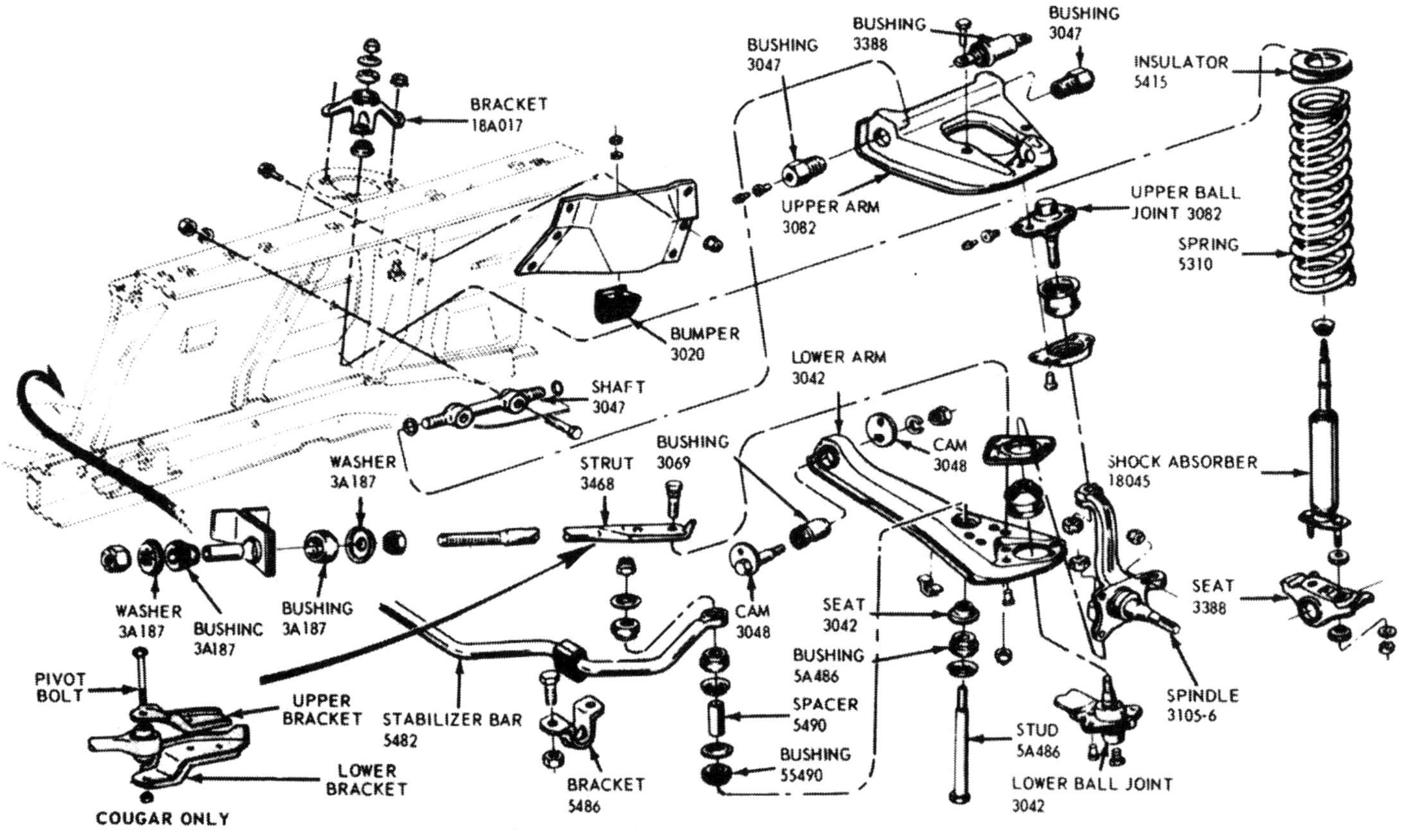

FIG 9:6 Typical later suspension components

FIG 9:7 Typical front shock absorber

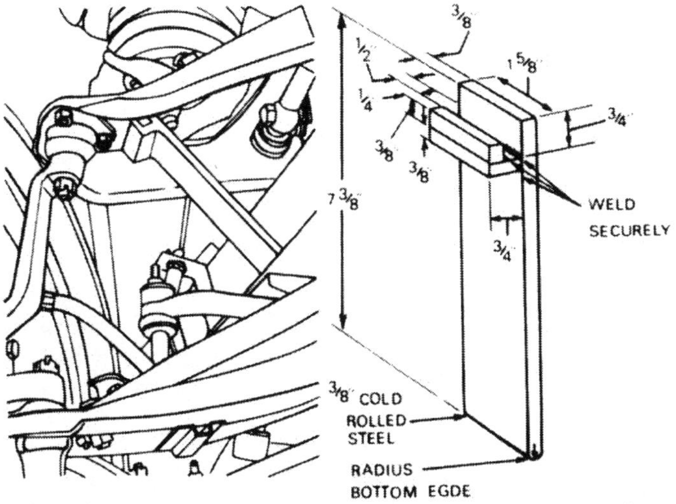

FIG 9:8 The upper arm support for 1964 and 1965 models

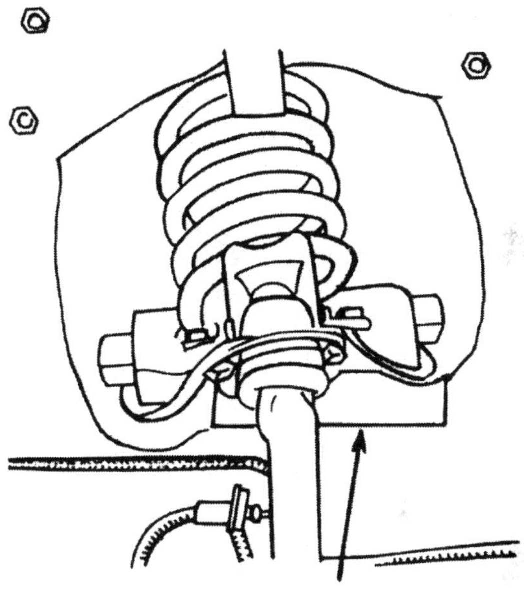

FIG 9:9 The upper arm support for all later models

Remove the bolts that secure the stabilizer brackets to the frame and withdraw the stabilizer from under the vehicle.

Before installing the stabilizer bar, check all the bushings for wear, perishing or deformation. Renew all defective bushings.

Lubricate the bushings with RUGLYDE or comparable lubricant and install the parts in the reverse order of disassembly. Tighten the bolts and nuts to the correct torque loads before fitting the road wheels and lowering the vehicle back to the ground.

Strut rod:

Typical strut attachments are shown in **FIG 9:6.** The strut can be removed by removing the nuts and bolts that secure it to the lower arm and then taking off the nut that secures it to the frame bracket. Before installing the strut, be sure that the bushings are in good condition and that the washers are not distorted or damaged.

9:6 Suspension ball joints

Typical suspension components are shown in **FIG 9:6.**

Upper ball joint inspection:

Raise the front of the vehicle and support it on underbody stands so that both suspensions hang freely at the end of their range of movement. This ensures that the upper ball joints are unloaded. Be sure that the wheel bearings are correctly adjusted (see **Section 9:3**).

Mount a dial indicator on the upper arm and position it so that its plunger rests vertically on the underside of the spindle at the upper ball joint. Lift and lower the spindle and suspension assembly, noting the play on the dial indicator. The play should not exceed .200 inch. Mount the dial indicator on the upper arm so that the plunger now rests vertically against the inside wheel rim. Grasp the top of the tire and rock it in and out so that the radial play on the ball joint can be measured on the dial indicator. The radial play must not exceed .250 inch. If the play is excessive in either direction, a new ball joint must be fitted.

Lower ball joint inspection:

With the front of the vehicle raised off the ground on underbody stands, have an assistant rock the lower edge of the tire in and out. Any movement between the lower end of the spindle and the lower arm indicates that the ball joint is worn and must be renewed. While carrying out this check there will be movement on the upper ball joint, but this movement may be safely ignored unless it is found to exceed the limits given.

FIG 9:10 Freeing a ball joint. Apply pressure and then lightly tap on the spindle

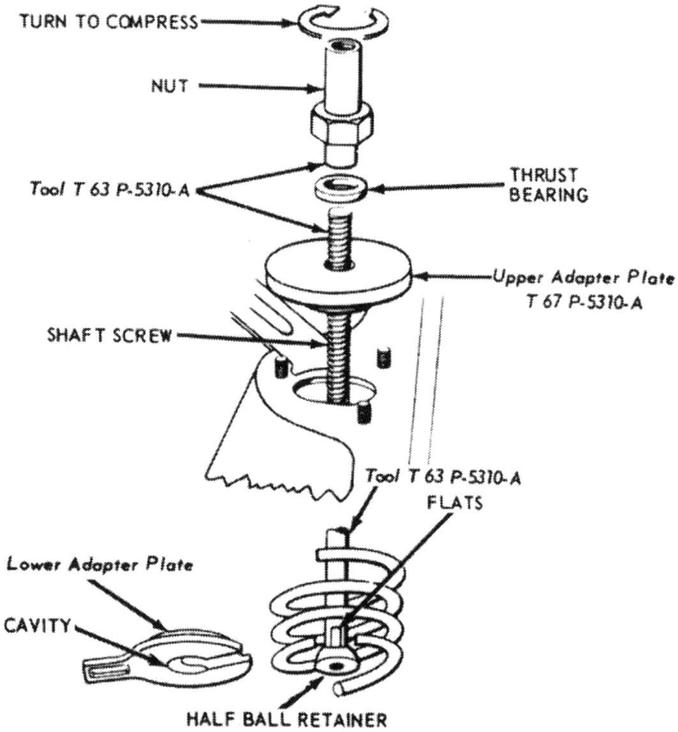

FIG 9:11 Typical spring compressor for removing front road spring

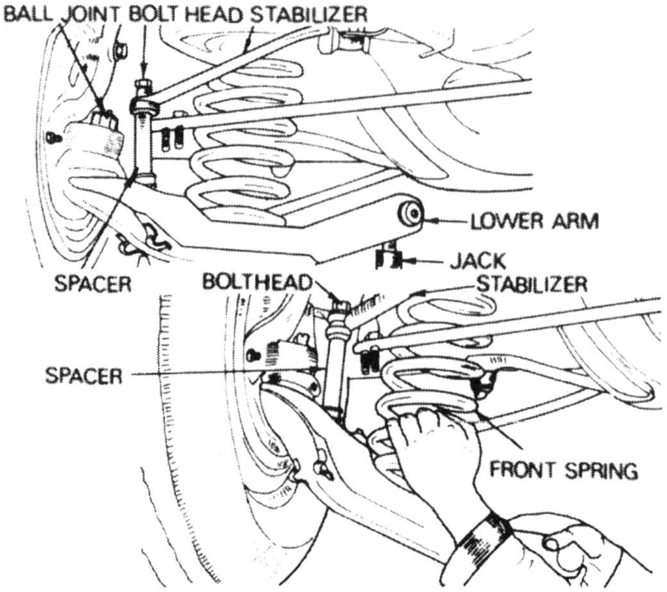

FIG 9:12 Removing or installing the front road spring

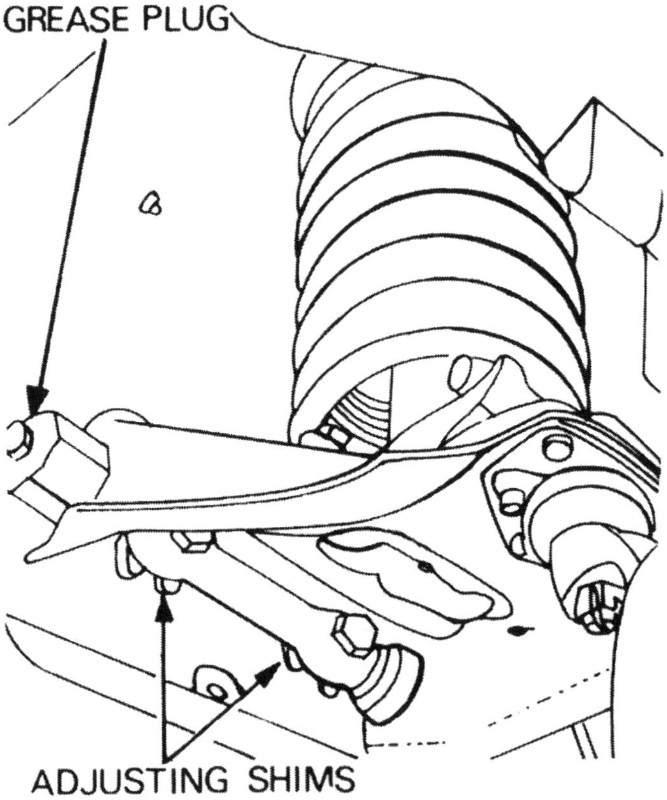

FIG 9:13 Setting the castor and camber on 1964 and 1965 models

Renewing upper ball joint:

On the 1964 and 1965 models, have fabricated a support to the dimensions shown in **FIG 9:8**, and mount it as shown in the figure. On later models a block of

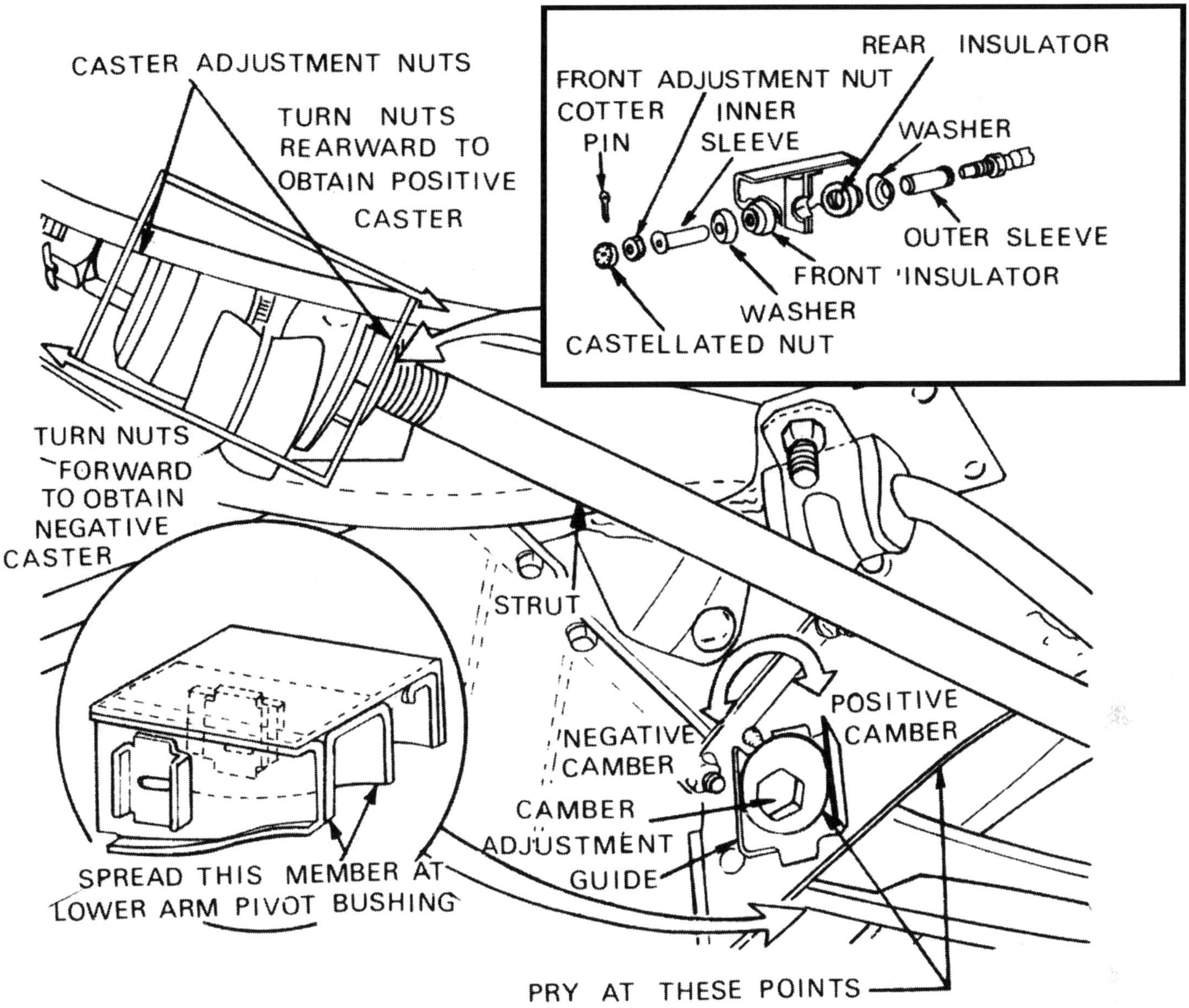

FIG 9:14 Setting the castor and camber on later models

wood fitted between upper arm and frame, as shown in **FIG 9:9**, will take the pressure of the road spring. Raise the front of the vehicle and make sure that the support fits securely and safely into place. Stand the vehicle on underbody supports.

Remove the cotterpin from the ball joint nut and slacken the nut by several turns. Install the tool as shown in **FIG 9:10**. The ball joint can be freed from the spindle by pressing out the threaded pin with the tool. **Under no circumstances hit on the end of the threaded portion, as even with a slave nut fitted to protect the threads internal damage will be caused to the ball joint.** If the special tool is not available, pull firmly down on the spindle, lay a block of metal on one side of the spindle eye and hit on the opposite side of the eye with a copper mallet. The tapers will quickly free.

Partially drill through the heads of the rivets that secure the old ball joint, ensuring that the upper arm itself is not marked. Knock off the heads with a cold chisel and punch out the stems so that the ball joint is freed from the upper arm.

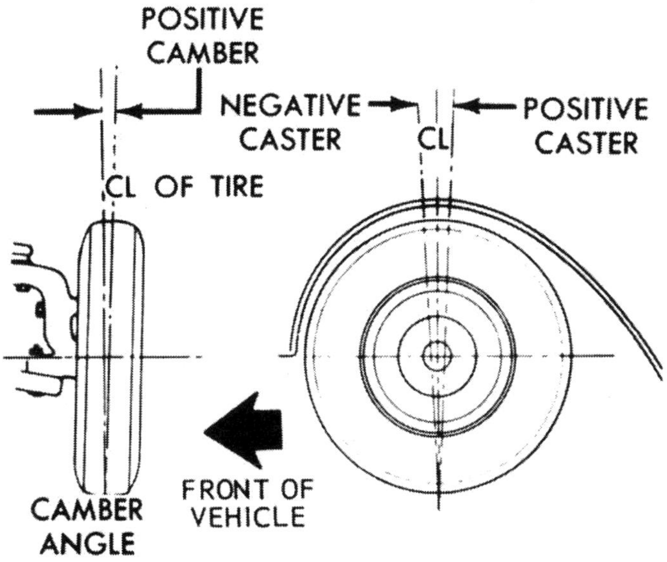

FIG 9:15 Castor and camber angles illustrated

Check the end of the arm for cracks or damage. Clean the end of the arm and remove any burrs from the edges of the mounting holes.

Secure the new ball joint in place with the special nuts and bolts supplied. **Do not use ordinary bolts and do not attempt to rivet the new ball joint into place.** Install the spindle onto the ball joint pin and tighten the nut to the specified torque. Lock the nut with a new cotterpin, tightening the nut if required to align the cotterpin hole and castellations.

Lubricate the ball joint until the dust cover just starts to swell. Install the road wheel and tire, lower the vehicle back to the ground and remove the block or support that held the upper arm.

Renewal of lower arm:

This is best left to a service garage, not because the work is beyond the owner's capabilities, but because the positioning of the shims and bushes on the lower arm inboard pivot controls the camber of the suspension and the position of the strut in its frame bracket controls the castor of the suspension. If the owner does carry out the work, the car must be taken to a service garage for checking and adjusting the suspension geometry.

If the arm is to be renewed, raise the car with the supports holding the upper arm, as described for renewing upper ball joint. Disconnect the stabilizer bar link. Remove the nuts and bolts that secure the strut to the lower arm. Note that the strut bushings must be renewed if they are worn or perished and the strut cup washers must be renewed if they are distorted or damaged.

Disconnect the lower ball joint, using the tool shown in **FIG 9:10.** It may be necessary to remove the cotterpin from the upper ball joint nut in order to position the tool satisfactorily on the upper ball joint. Remove the pivot bolt and take off the lower arm.

Install the arm in the reverse order of disassembly.

9:7 The springs

Removal:

1 Remove the air cleaner from the engine to gain access room. Raise the front of the car and place it securely onto underbody stands. Remove the road wheel.

2 Remove the shock absorber as described in **Section 9:4.** Install the parts of the special tool shown in **FIG 9:11.** The actual part numbers of the tool will vary, so check with a dealer for the correct parts. The lower plate is installed under the third coil from the bottom and locked into the half ball retainer. The upper adaptor plate is held by the three studs of the damper upper attachment. Use a suitable socket and ratchet drive to compress the spring until it is clear of the upper arm.

3 Disconnect the upper ball joint from the spindle. Remove the nuts that secure the upper arm pivot to the frame and take out the upper arm assembly, carefully collecting the shims fitted.

4 Slowly and carefully release the spring compressing tool and when the pressure is released, remove the tool parts and remove the spring as shown in **FIG 9:12.**

Check the spring for cracks, damage or weakening and renew it if it is defective.

The parts are installed in the reverse order of disassembly. Insert the spring into position and compress it with the special tools. Install the upper arm. When installing the upper arm use the specified type keystone lockwashers and fit the shims back into their original positions.

9:8 Suspension geometry

On the 1964 and 1965 models, shims are fitted between the lower arm pivot and frame to control both the castor and camber angles of the suspension, as shown in **FIG 9:13.** On all the later models the castor is set by the positioning of the support strut in the frame bracket and the camber is set by the position of the eccentric on the lower inboard pivot, as shown in **FIG 9:14.** The suspension angles are illustrated in **FIG 9:15.**

Incorrect suspension geometry will cause poor handling and excessive tire wear. Accurate gauges and instruments are required to check the angles and for this reason the vehicle should be taken to a service garage if it is suspected that the geometry is incorrect. The cost of a check will be less than the cost of new tires, so always have the geometry checked after major repairs or 'curbing' a wheel.

9:9 Trouble shooting guide

(a) Wheel wobble or tramp

1 Unbalanced front wheels and tires
2 Broken or weak road springs
3 Defective suspension arm inboard pivots
4 Worn or loose hub bearings
5 Worn or defective suspension ball joints
6 Loose road wheel attachments
7 Uneven tire wear

(b) Vehicle bottoms on suspension

1 Check 2 in (a)
2 Bumpers missing or worn
3 Defective shock absorbers

(c) Heavy steering

1 Neglected lubrication
2 Incorrect suspension geometry

(d) Excessive tire wear

1 Check 2, 3, 4 and 5 in (a); 2 in (b)

(e) Rattles

1 Check 2, 3 and 5 in (a) and 3 in (b)
2 Stabilizer bar broken or bushings worn

(f) Excessive rolling

1 Check 2 in (a); 3 in (b) and 2 in (e)

CHAPTER 10

THE STEERING SYSTEM

10:1 Description

A recirculating ball steering unit is fitted to all models covered by this manual. A sectional view of a typical unit is shown in **FIG 10:1**. The steering shaft is connected to the steering wheel, through the steering column, and therefore revolves as the wheel is turned. The ball nut moves up or down the spiral of the steering shaft, with the balls reducing friction, and as it moves it rotates the sector shaft by the sector gear teeth.

The sector shaft is linked to the steering linkage, shown in **FIG 10:2**, by the Pitman arm. An idler arm is mounted on the opposite side of the frame to the steering unit and a rod interconnects the Pitman arm to the idler arm so that the tie rod will always be at right angles to the fore and aft axis of the vehicle. Spindle connecting rods transmit the motion of the tie rod to the spindles on the suspension so that the front wheels move for steering.

Power steering can be fitted as an optional extra. The parts are fitted as an addition to the normal system, as shown in **FIG 10:3**.

The pump is integral with the reservoir and mounted inside it. The pump and reservoir assembly is mounted on the engine and the pump is driven by a belt from the engine. The control valve is operated when the steering is turned and it returns fluid from the appropriate side of the power cylinder to assist in turning the steering. The valve is designed to ensure that the fluid circulates through the pump and valve when the steering is in the straight-ahead position. The pump is of the slipper type.

Eight slippers are fitted and as the pump revolves they move in and out in the track, altering the volume of the space between them. The slippers are so arranged that as they part the fluid is drawn in through the inlet and as the track forces them together the fluid is pumped out through the outlet port. A sectional view of the pump and reservoir assembly is shown in **FIG 10:4**.

10:2 Maintenance

On models equipped with manual steering, check that the steering operates fully and freely without excessive play or stiffness. Check the dust covers on the spindle connecting rod ends to make sure that they are in good condition and not allowing dirt to enter the ball joint.

On models fitted with power steering, check the fluid level in the reservoir, check the drive belt tension and lubricate.

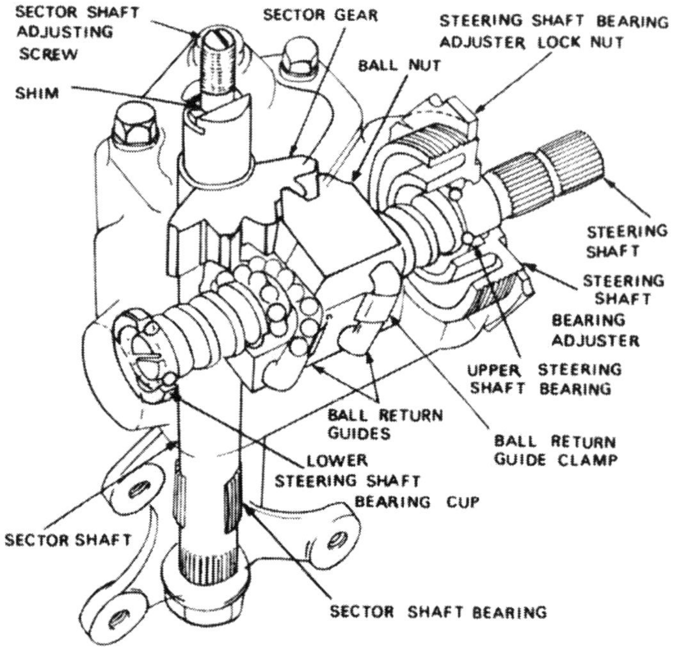

FIG 10:1 Typical steering gear unit

Start the engine and operate the steering from full lock to full lock several times and then shut off the engine. On later models the engine should be run until the power steering has reached its normal operating temperature. Withdraw the dipstick and check the level. If the level is low, top up with automatic transmission fluid to the correct mark. If the level has been allowed to fall so low that air has been drawn into the pump and the fluid appears milky or frothing, turn the steering from full lock to lock several times while topping up.

The method of setting the drive belt tension is given in **Chapter 4, Section 4:4. Do not lever against the reservoir when setting the belt tension** otherwise the reservoir will distort and cause fluid leaks. A boss is cast into the face of the pump and a wrench should be applied to this boss when moving the pump into the position for the correct tension.

When lubricating the actuator ball stud, wipe the grease fitting clean before applying a grease gun. Inject grease until the dust cover just starts to swell or expand.

Care must be taken on all models, whether power steering is fitted or not, when turning the steering from full lock to full lock with the front wheels clear of the ground. **Do not turn the steering vigorously as the steering unit will be damaged when it comes to a sudden halt at the end of its travel.** For the same reason, never apply heavy pressure to the steering wheel when the steering is on full lock when power steering is fitted.

10:3 The steering linkage

A typical steering linkage is shown in **FIG 10:2**. The spindle connecting rods are made up of the ball joint ends and an internally threaded sleeve. One end of the sleeve has a lefthand thread while the other end has a righthand thread. Clamps lock the sleeve to the ends to prevent the adjustment from altering in service. The effective length of the spindle connecting rods is altered to set the correct front wheel alignment. To set the steering wheel spokes see **Section 10:8**.

The ball joints on the ends of the spindle connecting rods are non-adjustable and non-repairable and the complete ends must be renewed if there is excessive

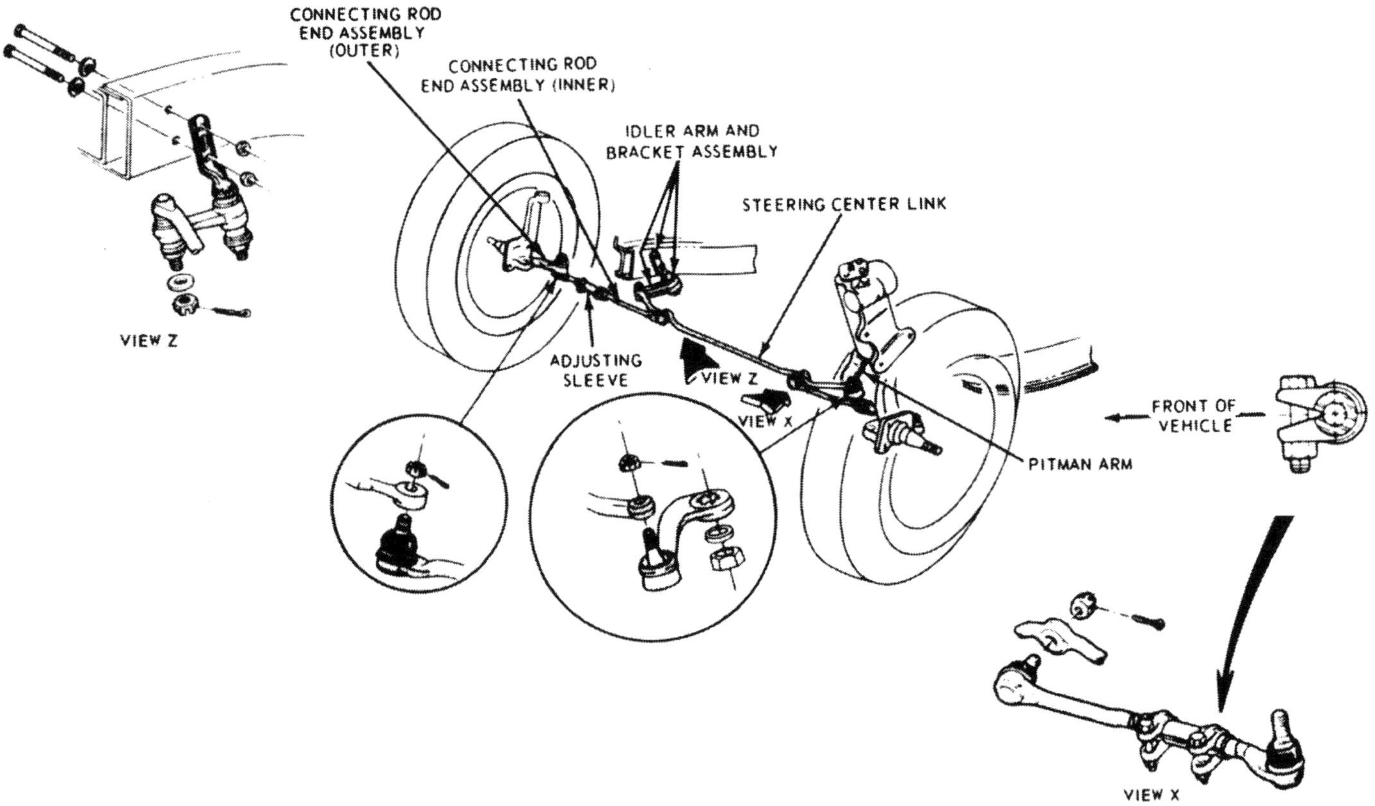

FIG 10:2 Typical steering linkage for manual steering

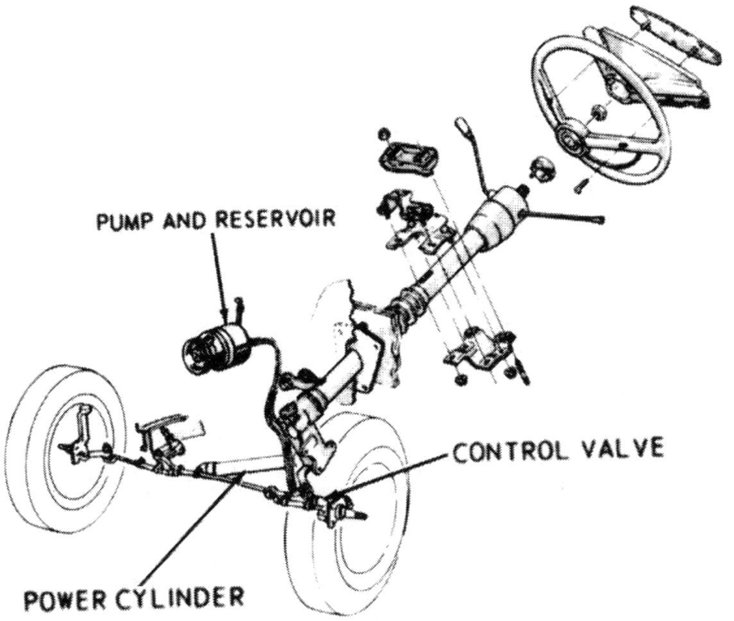

FIG 10 : 3 **Power steering installed**

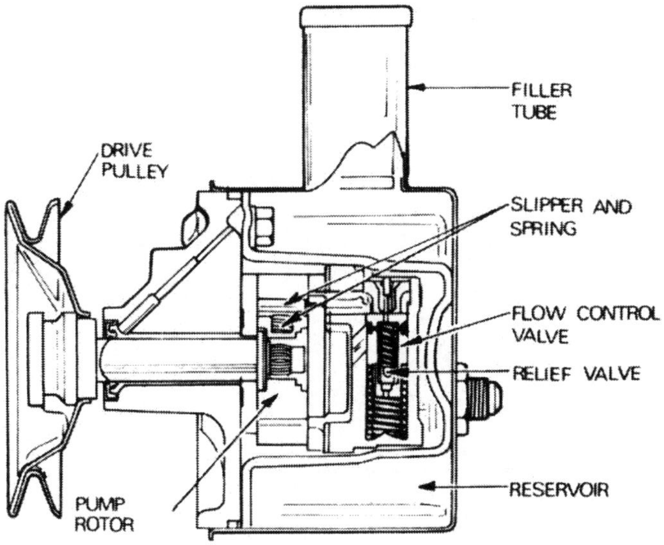

FIG 10 : 4 **Sectional view of steering pump and reservoir**

play in a ball joint. Check for play by having an assistant gently turn the steering backwards and forwards, with the weight of the vehicle on the road wheels, and watch the linkage as it moves.

Freeing ball joints:

The safest and best method of disconnecting a ball joint from its attachment is to use the special tool, as shown in **FIG 10 : 5.** Extract the cotterpin and remove the nut that secures the pin of the ball joint. Fit the tool as shown and by tightening the bolt of the tool the pin will be pressed out.

If the special tool is not available the ball joint can still be freed. **Never hammer on the end of the threaded portion of the pin, even with a slave nut fitted.** Undo the nut that secures the ball joint to the last few threads. Lay a block of metal on one side of the tapered eye and pull the end with the ball joint firmly away from the other member. Hit firmly with a copper mallet on the side of the tapered eye opposite to the block of metal and the tapers will quickly free.

The ball joint is reconnected by pressing the pin into the tapered eye, by hand, and tightening the castellated nut to the correct torque before locking it with a new cotterpin.

If a new spindle connecting rod end is being fitted, the alignment of the front wheels and the setting of the steering wheel will not be unduly altered if the number of threads required to unscrew the old end are counted and the new end screwed back by the same number of turns. However, it is advisable to have the front wheel alignment checked by a service garage.

Note that the sleeve itself must be renewed if it is damaged or bent. Do not attempt to straighten a bent sleeve.

Pitman arm:

The ball joint on the Pitman arm is disconnected as described previously. On the models where the ball joint is integral with the Pitman arm a new arm must be fitted if the ball joint is defective or worn.

When removing the Pitman arm, take off the nut and washer that secure the arm to the sector shaft of the steering unit. A suitable extractor, as shown in **FIG 10 : 6,** must be used. Hammering on the sector shaft or Pitman arm is liable to cause internal damage to the steering shaft.

Before the Pitman arm is fully removed from the sector shaft, make light alignment marks on the arm and shaft so that the arm will be installed into its original position.

Install the Pitman arm in the reverse order of disassembly.

Idler arm:

On earlier models the idler arm is fitted with a renewable bush but on later models the complete idler arm must be renewed if there is excessive play.

Free the idler arm from the central rod. On the earlier models, remove the nut and cotterpin that secure the arm to the bracket on the frame. The idler arm bushing can then be renewed using the special tool as shown in **FIG 10 : 7.**

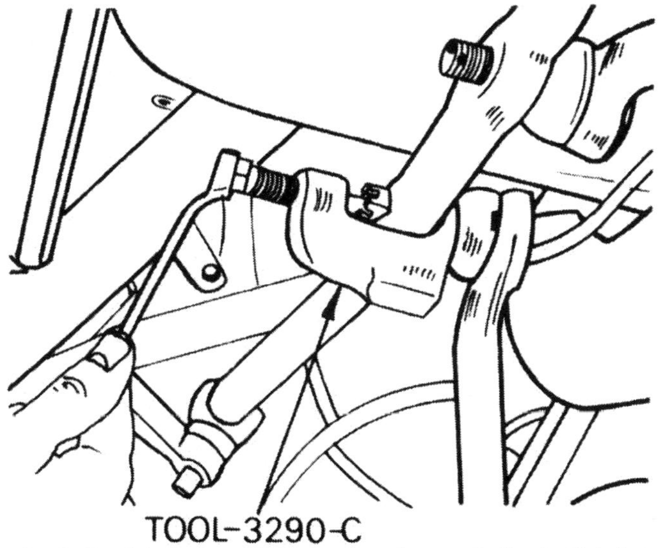

FIG 10 : 5 **Disconnecting connecting spindle ball joint**

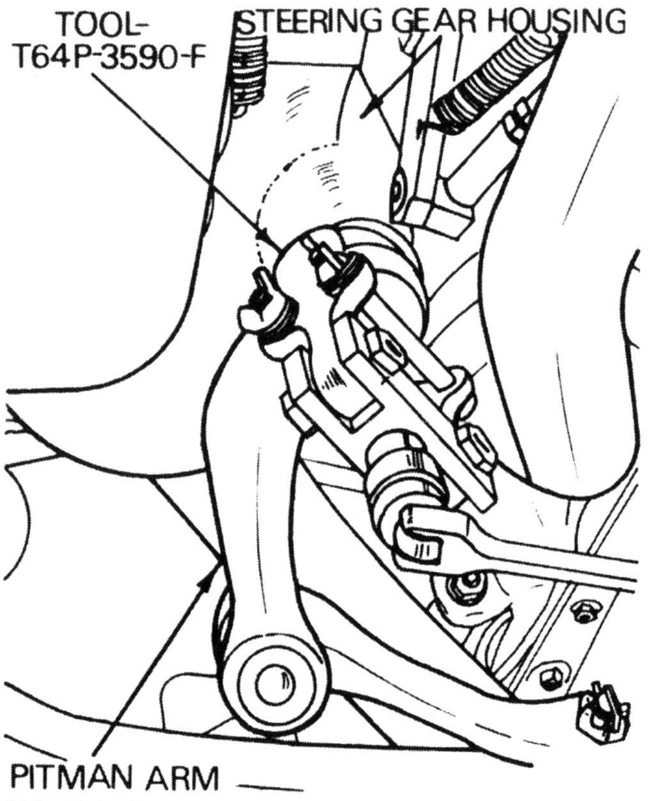

FIG 10:6 Removing the Pitman arm from the steering gear

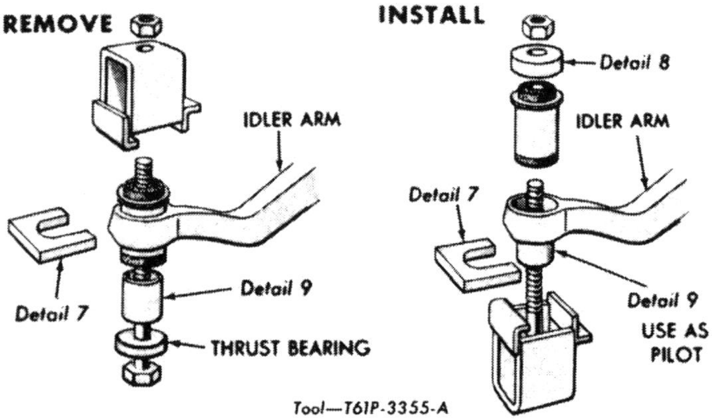

FIG 10:7 Renewing the bush on an early idler arm assembly

On the later models the idler arm complete with bracket is removed from the frame and a new assembly installed.

10:4 Steering gear adjustments

There are only two adjustments that require to be made on the steering gear unit and they are both carried out, in the correct sequence, with the unit fitted to the vehicle.

Lubrication:

Before carrying out adjustments check the lubricant level in the unit. The filler plug and adjustment points are shown in **FIG 10:8**. Center the steering and remove the filler plug. Remove the lower cover-to-housing attachment bolt. Use a piece of clean rod or similar tool to push all loose lubricant in the filler hole back into the unit so that the hole is clear. Slowly turn the steering to full left lock and lubricant should rise within the lower cover bolt hole. Slowly turn the steering to the full opposite lock and the lubricant should rise in the filler hole. If the lubricant does not rise in the holes correctly, further lubricant should be added through the filler hole until the level is correct.

Steering column:

Remove the crash pad and sufficient hardware to expose the steering shaft nut that secures the steering wheel. Check the alignment of the shaft in the bearing. If the shaft is off center in the bearing, fit a spring scale (capacity up to 30 or 50 lb) to the end of the shaft. Center the shaft by pulling on the spring scale and note the reading on the scale. If the pull required is less than 20 lb no action need be taken. If the pull required exceeds 20 lb then the steering column must be realigned. A steering column out of alignment will cause the steering to be heavy.

Steering shaft in steering gear:

Disconnect the Pitman arm from the sector shaft. Slacken the locknut on the sector adjusting screw (see **FIG 10:8**), and turn the adjusting screw counterclockwise by at least two full turns.

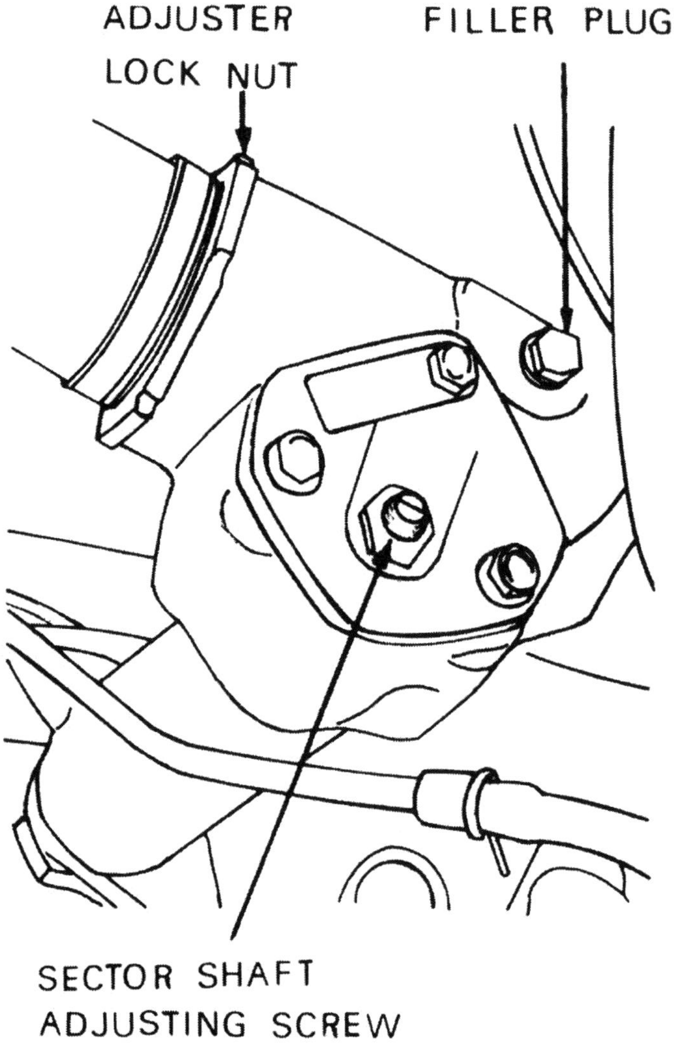

FIG 10:8 Lubrication and adjustment points on the steering gear

Attach a torque wrench (measuring in inch lb) to the steering wheel nut. Use the torque spanner to turn the steering approximately $1\frac{1}{2}$ turns either side of the center position. If the torque required is not within limits (see **Technical Data**), slacken the adjuster locknut and tighten or back off the bearing adjuster to bring the torque within limits. When the bearing preload is correctly set, hold the adjuster and tighten the adjuster locknut. If the bearings feel rough or noisy, the steering unit must be removed from the vehicle for dismantling and further examination.

Sector shaft adjustment:

When the bearings have been correctly set, set the sector shaft adjusting screw. Turn the steering gently from full lock to full lock, counting tne turns required. Do not slam the steering against the stops as internal damage will be caused. Turn the steering back by half the number of turns required from lock to lock so that it is now in the central position.

Use the torque spanner on the steering wheel nut to turn the steering, and turn the sector shaft adjusting screw clockwise until the torque is within the correct limits as the steering passes through the center point. There should be no perceptible backlash with the steering at 30 deg. either side of center. Hold the adjuster screw in position and tighten the locknut.

Install the hardware and crashpad to the steering wheel. Install the Pitman arm back onto the sector shaft.

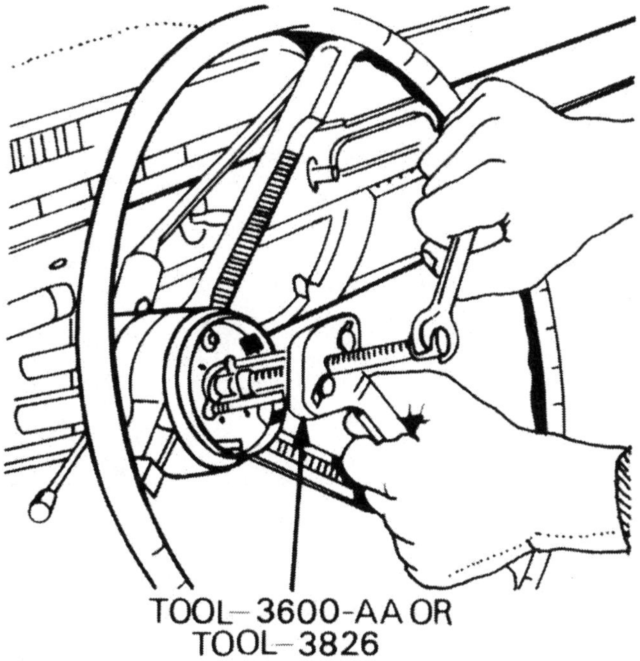

FIG 10:9 Typical method of removing the steering wheel

10:5 The steering column

On later models a safety collapsible steering column is fitted. Under heavy impact the column collapses approximately 6 inches in a controlled manner to prevent injury to the driver. On this safety column care must be

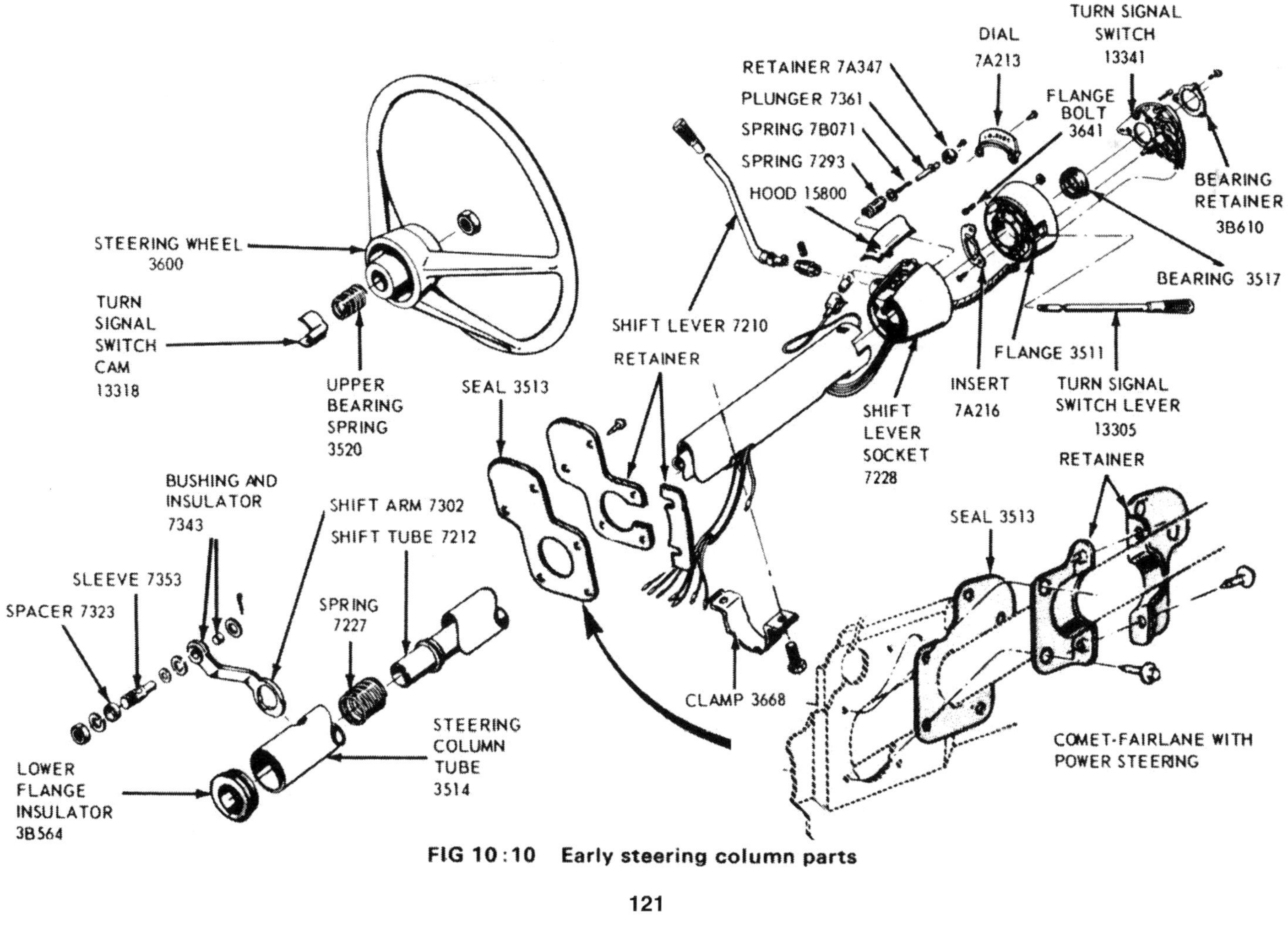

FIG 10:10 Early steering column parts

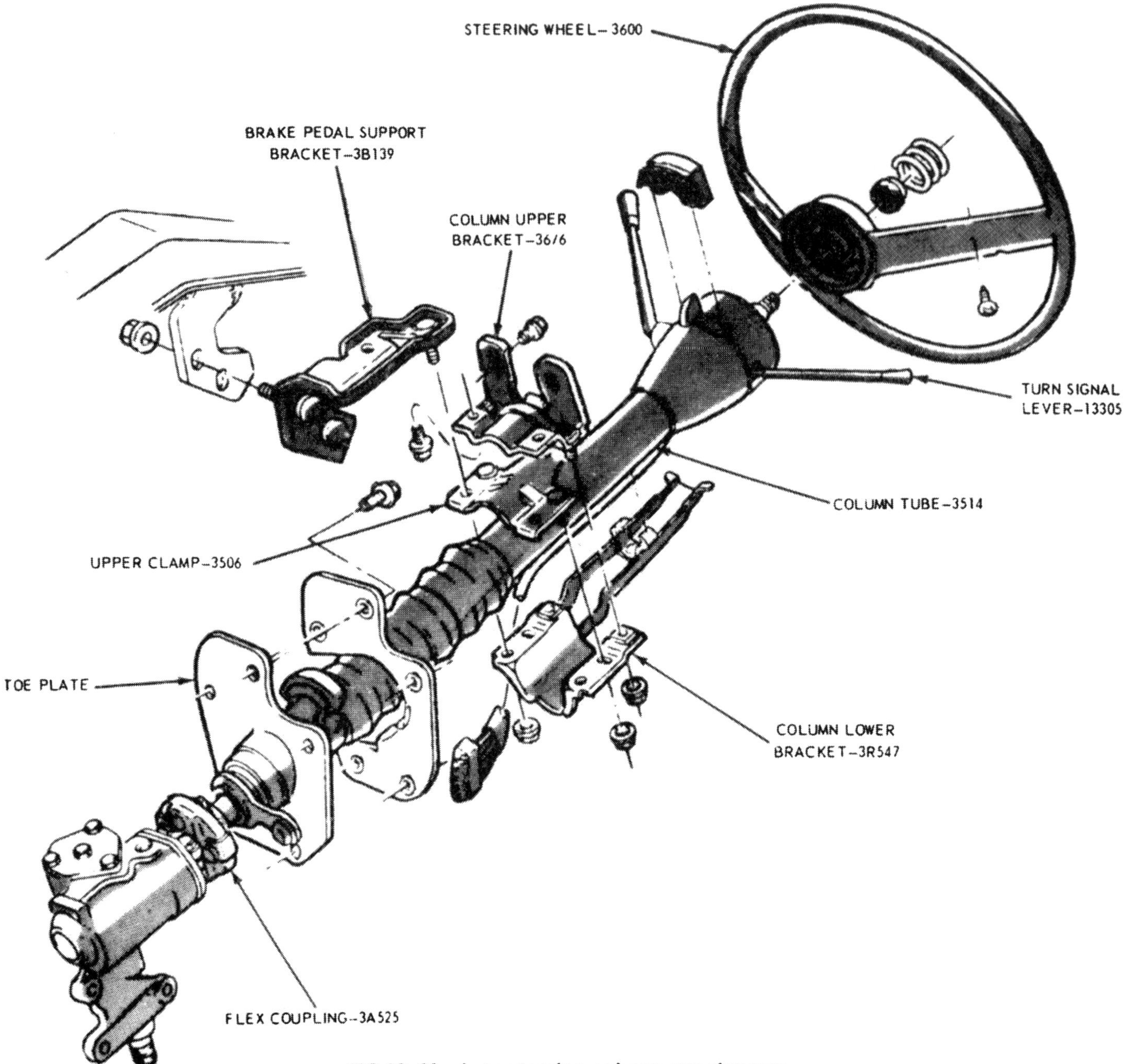

FIG 10:11 Late steering column attachments

taken when removing or installing either the steering wheel or column itself not to damage the collapsible portion inside the bellows. This is particularly important when removing the steering wheel, after its attachment nut has been taken off, and an extractor (as shown in **FIG 10:9**) should always be used to free the wheel from its splines.

Removal:

Before attempting to remove any part of the steering column or disassemble the parts, disconnect the battery.

On the very early models, the steering column and steering gear are removed as a unit, from the engine compartment. Disconnect the steering column attachments, including the wires, transmission shift linkage, brackets and Pitman arm. Remove the steering wheel and wrap the end of the steering column with rags to prevent it from damaging the fabric of the seats. It will be necessary at least to loosen the air cleaner so that access room can be made, as well as removing the exhaust manifold on the lefthand side. On models fitted with power brakes it will also be necessary to remove the power brake unit.

On earlier models the steering column is removed from inside the car (typical attachments are shown in **FIG 10:10**) and the steering unit removed separately.

On all models slotted holes are provided to ensure that the steering column is correctly aligned. When the steering column is in position, check the position of the inner column in the upper bearing and make small adjustments to centralize it. If the slotted holes do not allow sufficient movement for adjustment, shims will have

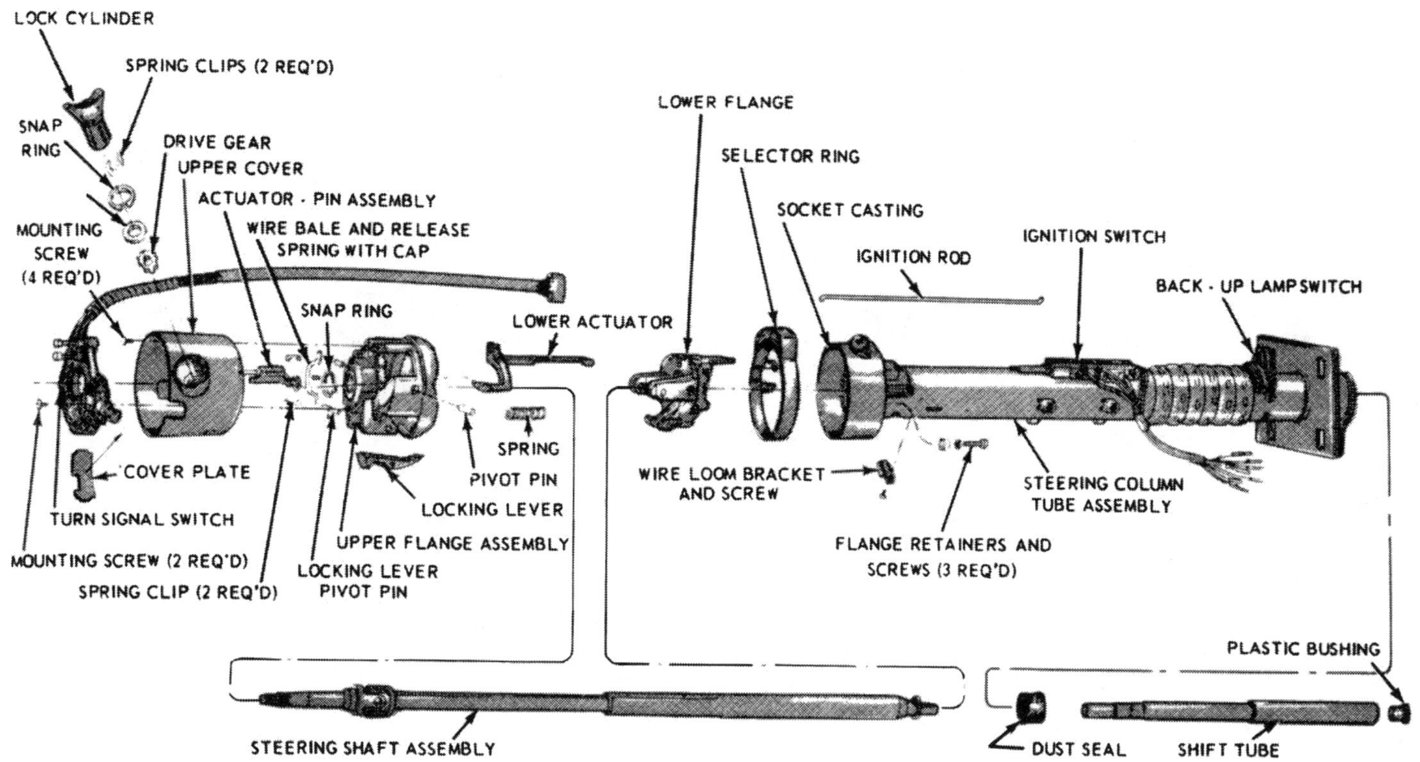

FIG 10:12 Late steering column parts

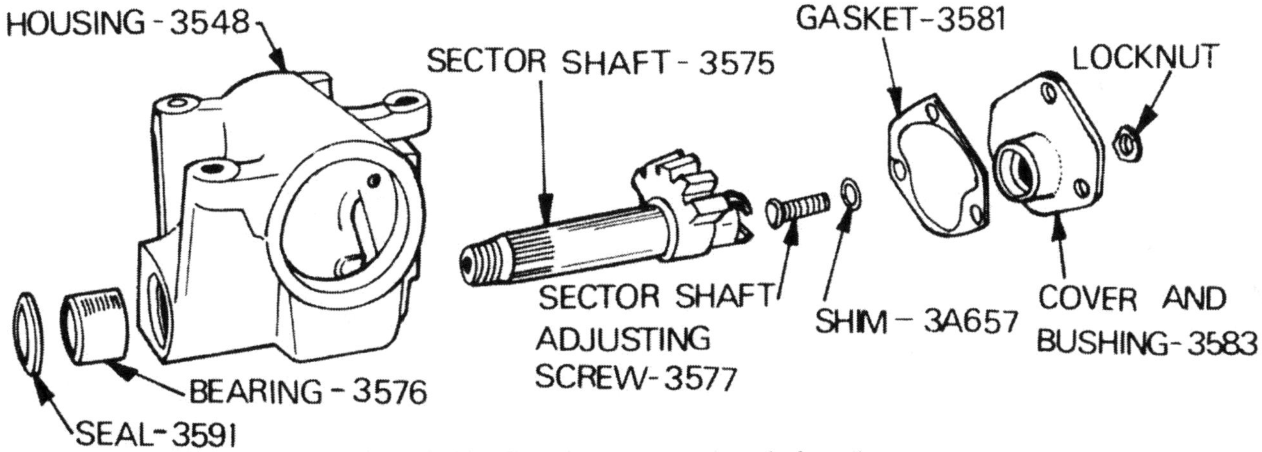

FIG 10:13 Steering gear sector shaft and parts

to be fitted. The installation of the steering column on a late model is shown in **FIG 10:11**, and its components in **FIG 10:12**.

10:6 Steering gear servicing

The components of the later steering gear unit are shown in **FIG 10:13** and **FIG 10:14**.

Strict cleanliness is essential when reassembling the parts and all the parts should be washed in a suitable solvent and lubricated with clean lubricant before installation and assembly.

Disassembly:

1 Turn the sector shaft gently from stop to stop, counting the turns, and then rotate it back by half the number of turns to set it in the midway position. Slacken the locknut on the sector shaft adjuster and remove the three bolts that secure the cover. Keep turning the adjuster screw clockwise until the cover comes free. Remove the adjusting screw from the sector shaft, but keep the shim with it. Withdraw the sector shaft from the housing.

2 Slacken the locknut for the ballbearing adjuster and unscrew the adjuster nut. Remove the steering shaft upper bearing assembly. Carefully withdraw the steering shaft and ball nut assembly from the housing. **Do not allow the ball nut to run down the spiral as the ball guides will be damaged when the ball nut reaches the end of its travel.**

3 Clean the ball nut and steering shaft assembly by washing the parts in solvent. Rotate the steering shaft in the ball nut but do not force it to the end of its travel. Dry the parts with compressed air and

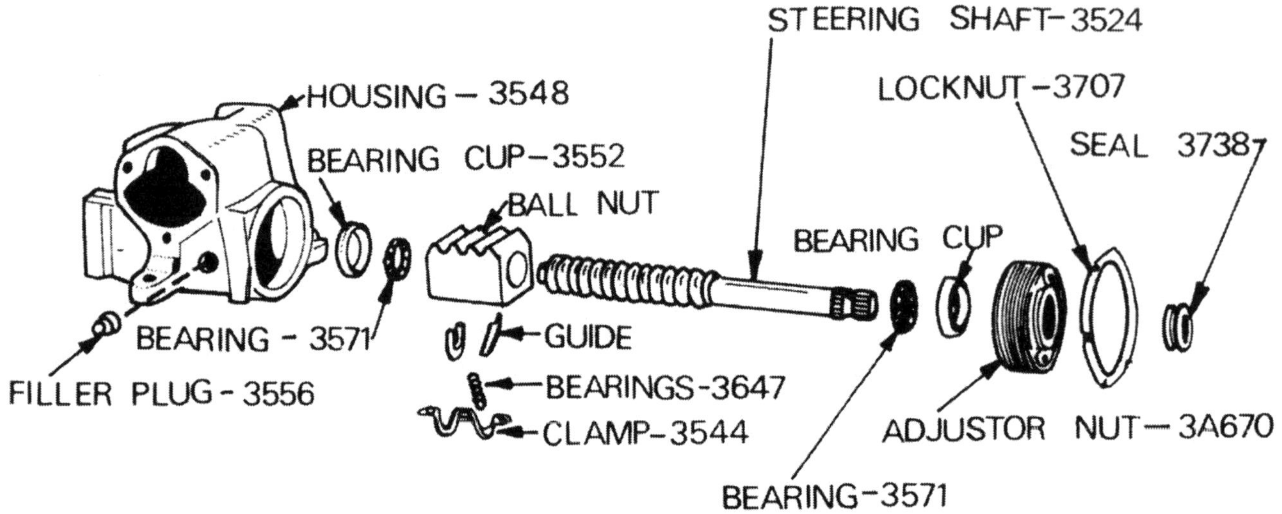

FIG 10:14 Steering gear steering shaft and parts

lubricate them with light oil. Rotate the shaft in the nut and check for stiffness or binding. If the shaft rotates freely then there is no need to dismantle the parts. Only if there is stiffness or binding should the parts be dismantled.

4 If the sector shaft bearing and oil seal show damage or wear, they should be pressed out of the housing. If the bearing has rotated in the housing, a new housing must be fitted. If the bearing and oil seal are satisfactory, leave them in place but avoid washing the oil seal with solvent.

Ball nut assembly:

If there is stiffness in the assembly, dismantle the parts and check for wear or damage. Hold the nut so that the ball guides are uppermost and keep the assembly over a container with a piece of cloth laid in the bottom. The cloth is to prevent the balls from bouncing as they fall and possibly becoming lost. Remove the clamp and withdraw the two ball guides, taking care not to lose the balls in the guides. Turn the nut over and rotate the steering shaft backwards and forwards until all the balls are out of the assembly. The number of balls varies but it is always an even number.

Fit the nut back onto the shaft, as shown in **FIG 10:15.** Install the ball guides, tapping them back into place with a wooden handle if required. Divide the balls into two equal amounts, noting that each half of the set is fitted to one ball guide and channel. Hold the nut with the ball guides uppermost and feed the balls back

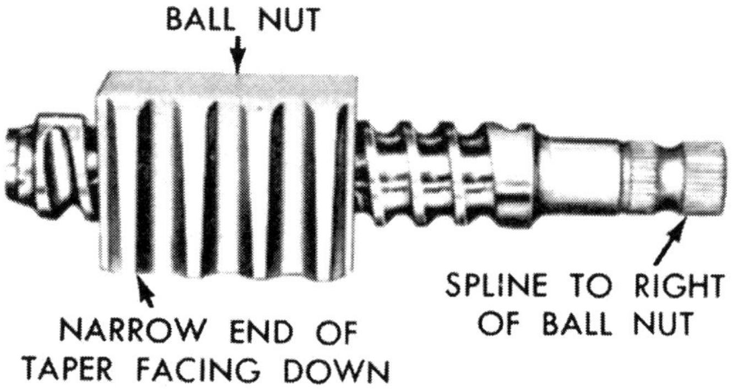

FIG 10:15 Positioning the ball nut on the shaft

through the hole in the top of each ball guide. Rotate the steering shaft backwards and forwards to distribute the balls in the circuit. When all the balls are fitted, secure the guides in place with the clamp, which also covers the holes in the ball guides. Lubricate the nut with the correct lubricant (C3AZ-19578-A).

Assembly:

Check all the parts for wear and renew any that are defective. If there is difficulty in removing the cup for the lower bearing of the steering shaft, tap the housing onto a clean piece of wood to dislodge the cup. Renew both bearings if they are worn or show signs of fretting or breaking up.

1 If the sector shaft bearing and oil seal have been removed, press new parts back into place. Lubricate the oil seal lightly. Install a new oil seal in the bearing adjuster nut if the old seal is damaged or worn.

2 Install the lower bearing cup back into the housing. Lubricate all the bearings and install both bearings onto the steering shaft. Carefully slide the steering shaft and nut assembly back into the housing. Lightly smear the threads of the bearing adjuster with sealing compound. Do not allow sealing compound onto the shaft bearings and do not apply it to the female threads. Clamp the housing gently in a vice so that sector shaft axis is horizontal and the splined end of the steering shaft is vertically upwards. Install the bearing cup, adjuster with oil seal, and finally the locknut with the marking S outwards. Leave the locknut slack. Fit a torque spanner that measures in lb inch to the splined end of the steering shaft and tighten down the adjuster nut until the correct torque is required to turn the shaft. Hold the adjuster and tighten the locknut to keep it in this position.

3 Use feeler gauges to check the clearance between the adjuster and the sector shaft, with the shim in position. If the clearance exceeds .002 inch then a new shim must be selectively fitted to reduce the clearance to a figure below this limit. Lightly smear the adjuster threads with sealing compound and partially screw it back into the cover. Install the sector shaft and shim onto the adjuster. Rotate the steering shaft until the ball nut is correctly positioned for meshing with the sector teeth. Lower the sector shaft down into the

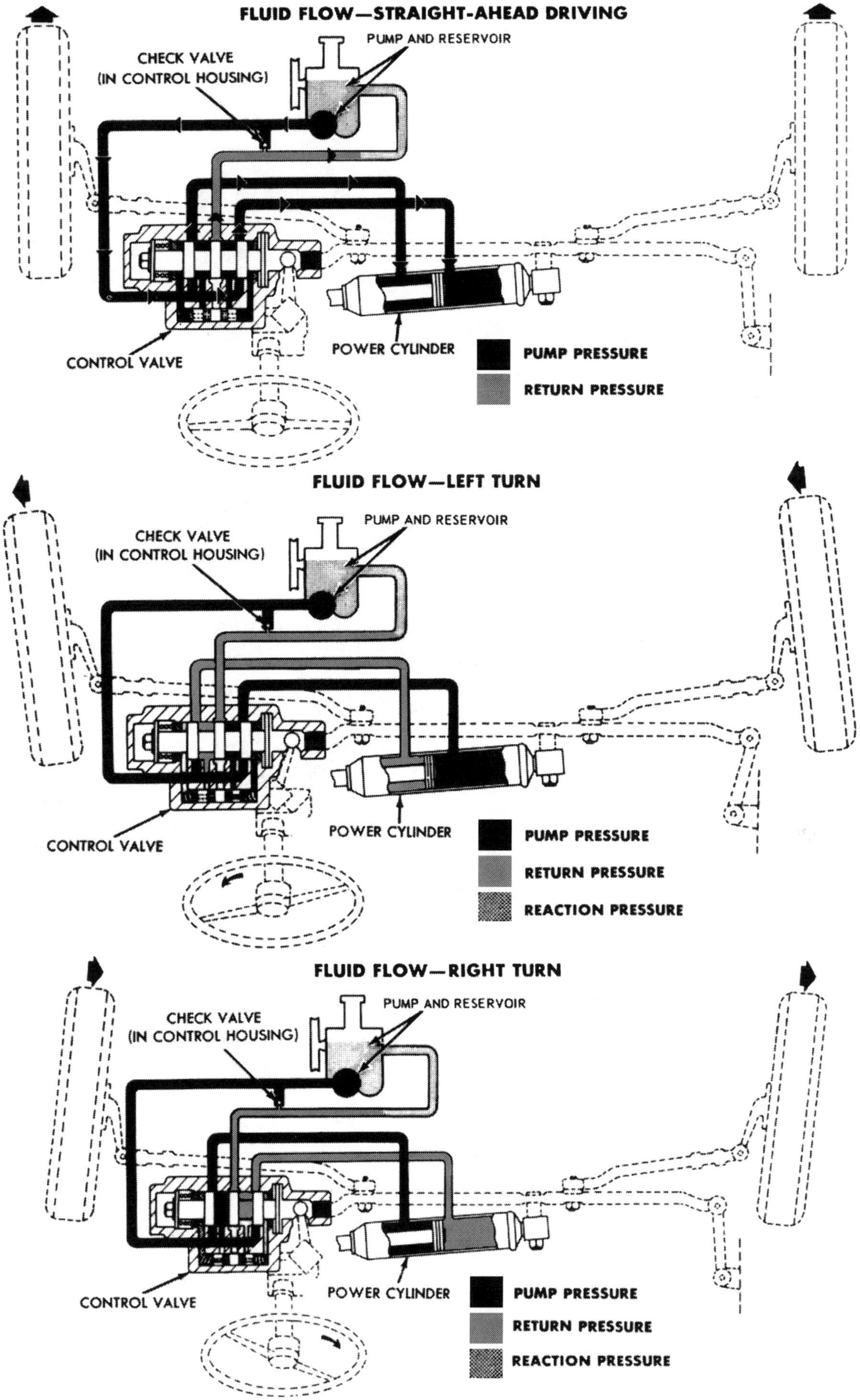

FIG 10:16 Power steering flow diagram

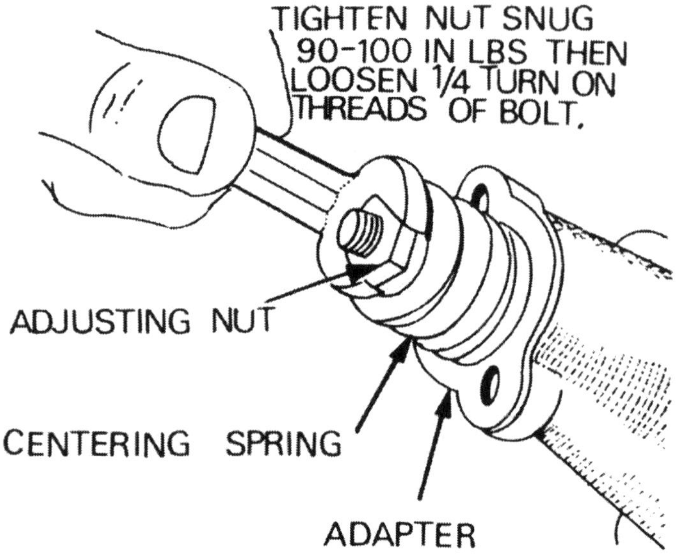

FIG 10:17 Adjusting the power steering valve centring spring

housing. Lightly smear the threads of the cover attachment bolts with sealing compound and use them to secure the cover, but do not tighten the bolts at this stage. Turn the sector adjuster counterclockwise until there is end clearance and then fully tighten the cover bolts. Set the mesh adjuster until the correct torque is required to turn the steering shaft and then lock the adjuster with the locknut.

10:7 The power steering

Operation:

The operation of the system is best shown in **FIG 10:16.** When the steering wheel is rotated and the force exerted on it exceeds 4 lb approximately, the valve spool overcomes the pressure of the centering spring and the spool moves to allow pressure to be fed to the appropriate side of the power cylinder. At the same time fluid returns from the other side so that the pressure differential moves the power cylinder piston, and therefore body, so as to assist the steering wheel effort.

A check valve is fitted between the pressure and return line so that when excessive pressure is reached, such as when the steering is at its stops, the check valve opens and limits the maximum pressure in the system.

Adjusting centering spring:

An incorrectly adjusted centering spring can cause stiff steering. The method of adjustment, after removing the cover from the cover from the valve assembly is shown in **FIG 10:17.** Tighten the nut until there is a snug fit and then slacken it off by $\frac{1}{4}$ turn. Do not overtighten the nut otherwise the stop pin may break at the travel regulator stop. Install the cover and apply a little silicone grease at the sealing areas.

Pump:

A defective pump cannot be repaired and a new one must be fitted in its place. On the earlier models only the pulley can be renewed if it is damaged but on later models the reservoir can be renewed as well.

A pressure gauge, T-piece, and shut-off valve are required to check the operation of the pump, as well as a tachometer. Disconnect the pressure line at the pump and reconnect it using the T-piece and shut off valve with the gauge on the third arm of the T-piece. **Make sure that the shut-off valve is open.** Start the engine and top up to replace any fluid loss. Allow the engine to run at idle for at least two minutes to warm up the fluid in the system. Cycle the steering from stop to stop several times and switch off the engine. Top up with more fluid to replace any air that was in the system. Restart the engine, leave the steering untouched and the engine at idle. The pressure in the system should be approximately 50 lb/sq inch.

Speed the engine up to 1000 rev/min and carefully close the shut-off valve. Note the maximum pressure reading and open the shut-off valve. **The shut-off valve should not be held closed for more than 5 seconds otherwise the fluid in the pump will overheat and cause undue wear in the parts.** The pressure should be a minimum of 750 lb/sq inch. A low pressure will be caused by a defective pump, but make sure that belt slip is not the cause of poor pump operation. If the pressure is above normal but the steering is stiff then it is most likely that the valve assembly is at fault.

The pump is removed by freeing the drive belt tension, disconnecting the hoses (and draining the fluid into a suitable container) and then removing the two pivot bolts and two adjuster bolts.

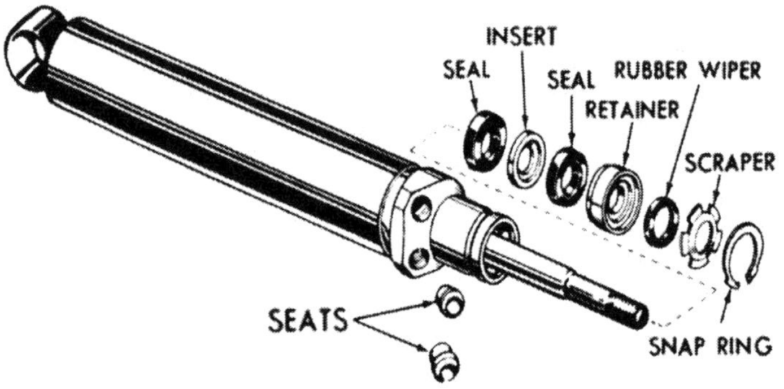

FIG 10:18 The power cylinder seals

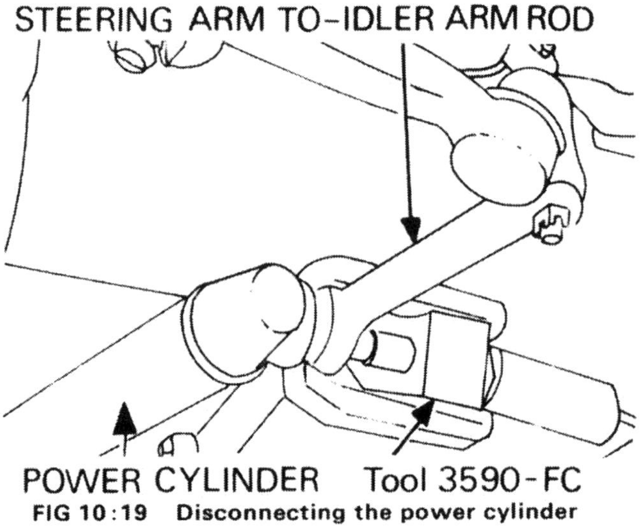

FIG 10:19 Disconnecting the power cylinder

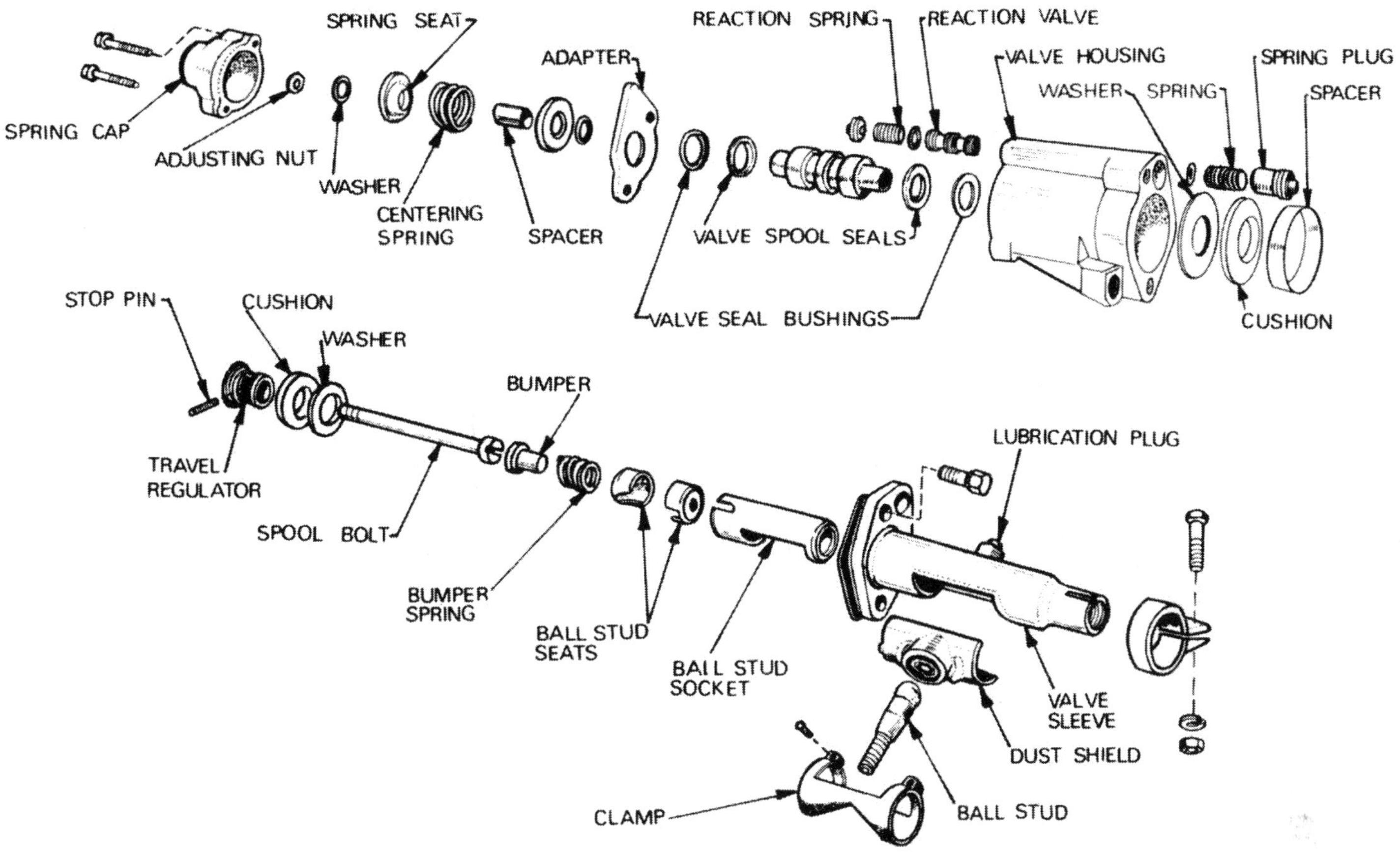

FIG 10:20 The control valve parts

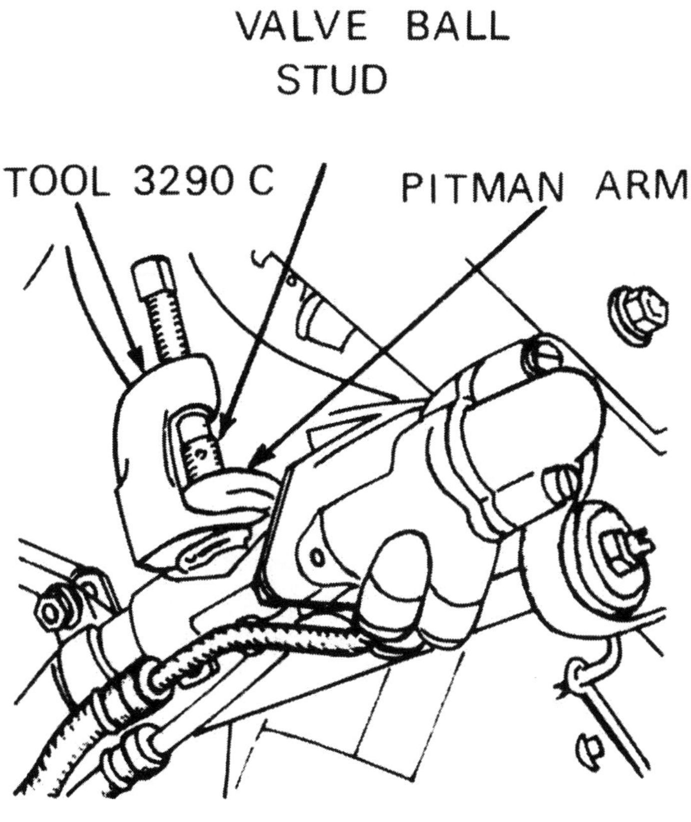

FIG 10:21 Removing the control valve ball stud

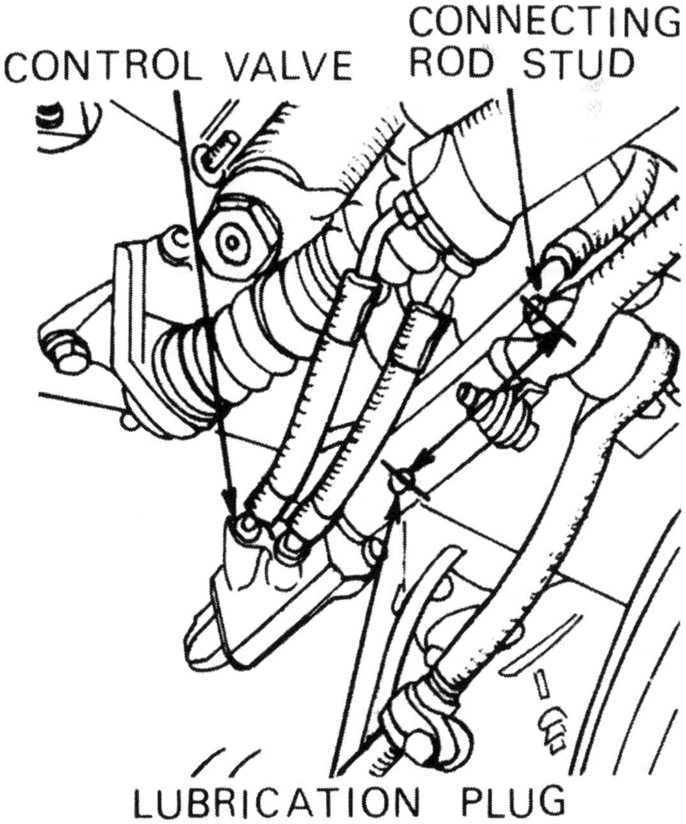

FIG 10:22 Installing the control valve

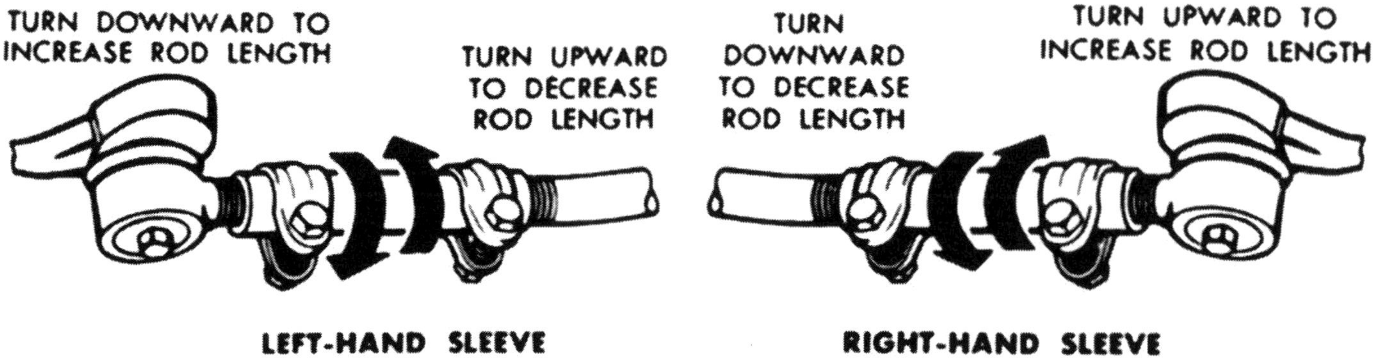

FIG 10:23 Adjusting the connecting spindles

When the pump has been installed and the correct belt tension set, fill the reservoir to the correct level with automatic transmission fluid and start the engine. Cycle the steering several times to expel any air in the system and top up the reservoir to the correct level.

Power cylinder:

The components of a power cylinder sealing are shown in **FIG 10:18.** New seals are available in kits of service spares. When installing new seals, use a deep socket or suitable piece of tubing to press the seals back into position.

To remove the power cylinder, first disconnect the hoses from it and allow the fluid to drain into a suitable container. It is advisable to mark the hoses so that they are correctly reconnected. **If the hoses are incorrectly connected the steering wheel is liable to kick back with considerable force when it is turned.** Remove the Pal nut which secures the cylinder rod to the frame. Take off the washer and insulator to free the rod. Remove the castellated nut, after extracting the cotterpin, that secures the cylinder to the center link and then disconnect the cylinder from the link using the tool shown in **FIG 10:19.**

Install the cylinder in the reverse order of removal. Fill the fluid reservoir and top it up while cycling the steering to expel air from the system. Top up to the correct level in the reservoir.

Control valve:

The components of the control valve are shown in **FIG 10:20.** The valve can be dismantled and new seals fitted as well as removing slight burrs from the valve spool but if there is a major defect then a new valve assembly should be fitted. If Crocus paper is used to remove burrs, **take great care not to round off the sharp edges on the spool.**

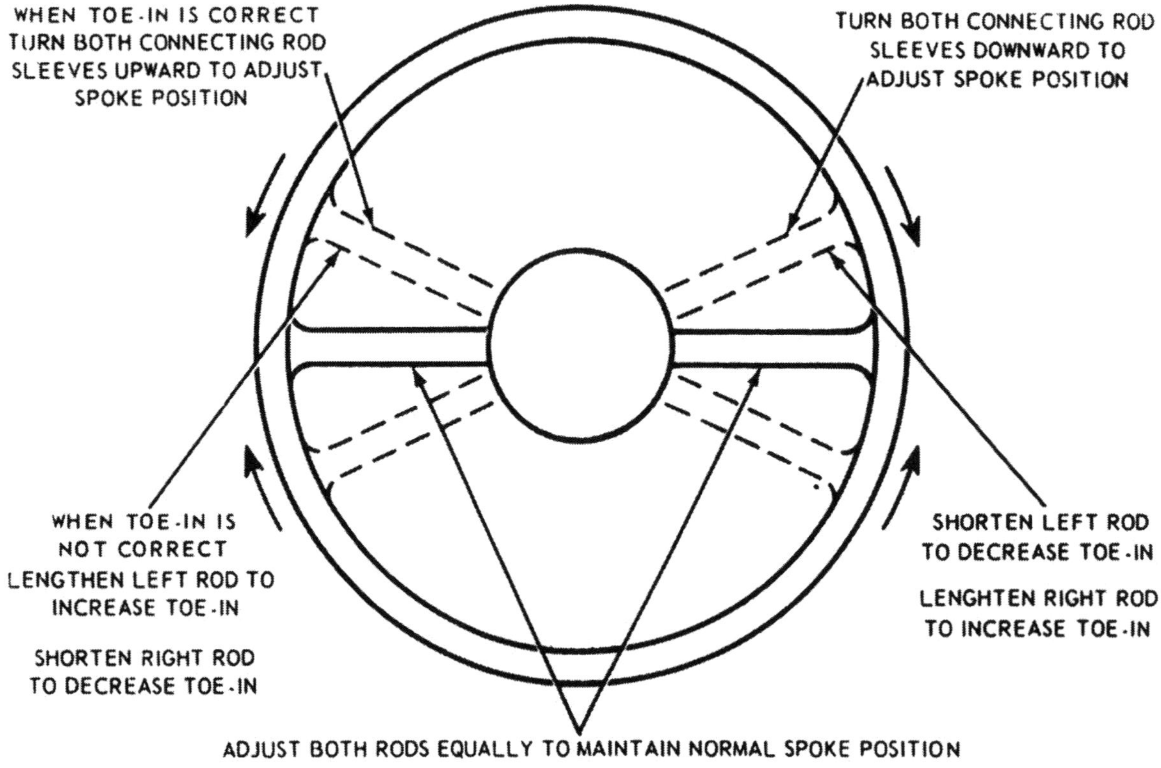

FIG 10:24 Adjusting the toe-in or spoke alignment

To remove the unit, first disconnect all four hoses from it and allow the fluid to drain into a container. Label the hoses as they are disconnected to ensure that they will be correctly installed. Turn the steering gently fully left and right to force out fluid. Loosen the clamping nut and bolt at the righthand end of the sleeve. Remove the roll-pin that releases the valve through the slot in the sleeve. Remove the nut and then free the ball stud as shown in **FIG 10:21** from the sector shaft arm. Turn the steering fully to the left and unscrew the valve assembly.

Screw the valve back into position until about four threads are left visible and position the ball stud in the sector shaft arm. Measure the distance between the center of the grease plug and center of the stud, as shown in **FIG 10:22**. On all models, except 1964 and 1965, the valve should be screwed in or out until the dimension shown is correct at $5\frac{5}{8}$ inch. On 1964 and 1965 models the dimension should be set between $4\frac{1}{4}$ and $4\frac{3}{8}$ inch. When the distance is correct, align the hole in the steering arm to the idler arm rod with the slot in the sleeve and install a new roll-pin to lock the valve in position. Tighten the clamp and install the remainder of the parts in the reverse order of disassembly. Fill the reservoir to the correct level and top up as required after cycling the system and checking for leaks.

10:8 Steering goemetry

The front wheel alignment should be checked by a service garage using accurate gauges. Incorrect alignment will cause excessive tire wear leaving the tread of the tire worn with a characteristic feathered edge on one side.

The toe-in of the wheels and the setting of the steering wheel spokes is carried out by altering the effective lengths of the spindle connecting rods, as shown in **FIG 10:23**. The various adjustments that can be carried out are shown in **FIG 10:24**.

10:9 Trouble shooting guide

(a) Wheel tramp or shimmy

1 Worn hub bearings
2 Defective or worn suspension
3 Unbalanced wheels and tires
4 Slack or worn steering connections
5 Incorrect suspension geometry

(b) Wander

1 Check 2, 4 and 5 in (a)
2 Front and rear suspensions out of alignment (frame damage)
3 Uneven tire pressures or wear

(c) Hard steering

1 Check 5 in (a)
2 Neglected lubrication
3 Steering gear incorrectly adjusted
4 Steering column bent or misaligned
5 Defective bearings in steering column
6 *Defective pump
7 *Defective valve assembly
8 *Slipping drive belt
9 *Control valve spool centering not accurately set
10 *Low fluid level

(d) Lost motion

1 Check 1 and 4 in (a) and 3 and 9 in (c)
2 Defective column coupling
3 Loose steering wheel or badly worn splines

(e) Noisy operation*

1 Check 6, 8 and 10 in (c)
2 Incorrect hoses used or hoses incorrectly routed

*Power steering only

Inches	Decimals	Millimetres	Inches to Millimetres: Inches	Inches to Millimetres: mm	Millimetres to Inches: mm	Millimetres to Inches: Inches
1/64	.015625	.3969	.001	.0254	.01	.00039
1/32	.03125	.7937	.002	.0508	.02	.00079
3/64	.046875	1.1906	.003	.0762	.03	.00118
1/16	.0625	1.5875	.004	.1016	.04	.00157
5/64	.078125	1.9844	.005	.1270	.05	.00197
3/32	.09375	2.3812	.006	.1524	.06	.00236
7/64	.109375	2.7781	.007	.1778	.07	.00276
1/8	.125	3.1750	.008	.2032	.08	.00315
9/64	.140625	3.5719	.009	.2286	.09	.00354
5/32	.15625	3.9687	.01	.254	.1	.00394
11/64	.171875	4.3656	.02	.508	.2	.00787
3/16	.1875	4.7625	.03	.762	.3	.01181
13/64	.203125	5·1594	.04	1.016	.4	.01575
7/32	.21875	5.5562	.05	1.270	.5	.01969
15/64	.234375	5.9531	.06	1.524	.6	.02362
1/4	.25	6.3500	.07	1.778	.7	.02756
17/64	.265625	6.7469	.08	2.032	.8	.03150
9/32	.28125	7.1437	.09	2.286	.9	.03543
19/64	.296875	7.5406	.1	2.54	1	.03937
5/16	.3125	7.9375	.2	5.08	2	.07874
21/64	.328125	8.3344	.3	7.62	3	.11811
11/32	.34375	8.7312	.4	10.16	4	.15748
23/64	.359375	9.1281	.5	12.70	5	.19685
3/8	.375	9.5250	.6	15.24	6	.23622
25/64	.390625	9.9219	.7	17.78	7	.27559
13/32	.40625	10.3187	.8	20.32	8	.31496
27/64	.421875	10.7156	.9	22.86	9	.35433
7/16	.4375	11.1125	1	25.4	10	.39370
29/64	.453125	11.5094	2	50.8	11	.43307
15/32	.46875	11.9062	3	76.2	12	.47244
31/64	.484375	12.3031	4	101.6	13	.51181
1/2	.5	12.7000	5	127.0	14	.55118
33/64	.515625	13.0969	6	152.4	15	.59055
17/32	.53125	13.4937	7	177.8	16	.62992
35/64	.546875	13.8906	8	203.2	17	.66929
9/16	.5625	14.2875	9	228.6	18	.70866
37/64	.578125	14.6844	10	254.0	19	.74803
19/32	.59375	15.0812	11	279.4	20	.78740
39/64	.609375	15.4781	12	304.8	21	.82677
5/8	.625	15.8750	13	330.2	22	.86614
41/64	.640625	16.2719	14	355.6	23	.90551
21/32	.65625	16.6687	15	381.0	24	.94488
43/64	.671875	17.0656	16	406.4	25	.98425
11/16	.6875	17.4625	17	431.8	26	1.02362
45/64	.703125	17.8594	18	457.2	27	1.06299
23/32	.71875	18.2562	19	482.6	28	1.10236
47/64	.734375	18.6531	20	508.0	29	1.14173
3/4	.75	19.0500	21	533.4	30	1.18110
49/64	.765625	19.4469	22	558.8	31	1.22047
25/32	.78125	19.8437	23	584.2	32	1.25984
51/64	.796875	20.2406	24	609.6	33	1.29921
13/16	.8125	20.6375	25	635.0	34	1.33858
53/64	.828125	21.0344	26	660.4	35	1.37795
27/32	.84375	21.4312	27	685.8	36	1.41732
55/64	.859375	21.8281	28	711.2	37	1.4567
7/8	.875	22.2250	29	736.6	38	1.4961
57/64	.890625	22.6219	30	762.0	39	1.5354
29/32	.90625	23.0187	31	787.4	40	1.5748
59/64	.921875	23.4156	32	812.8	41	1.6142
15/16	.9375	23.8125	33	838.2	42	1.6535
61/64	.953125	24.2094	34	863.6	43	1.6929
31/32	.96875	24.6062	35	889.0	44	1.7323
63/64	.984375	25.0031	36	914.4	45	1.7717

UNITS	Pints to Litres	Gallons to Litres	Litres to Pints	Litres to Gallons	Miles to Kilometres	Kilometres to Miles	Lbs. per sq. In. to Kg. per sq. Cm.	Kg. per sq. Cm. to Lbs. per sq. In.
1	.57	4.55	1.76	.22	1.61	.62	.07	14.22
2	1.14	9.09	3.52	.44	3.22	1.24	.14	28.50
3	1.70	13.64	5.28	.66	4.83	1.86	.21	42.67
4	2.27	18.18	7.04	.88	6.44	2.49	.28	56.89
5	2.84	22.73	8.80	1.10	8.05	3.11	.35	71.12
6	3.41	27.28	10.56	1.32	9.66	3.73	.42	85.34
7	3.98	31.82	12.32	1.54	11.27	4.35	.49	99.56
8	4.55	36.37	14.08	1.76	12.88	4.97	.56	113.79
9		40.91	15.84	1.98	14.48	5.59	.63	128.00
10		45.46	17.60	2.20	16.09	6.21	.70	142.23
20				4.40	32.19	12.43	1.41	284.47
30				6.60	48.28	18.64	2.11	426.70
40				8.80	64.37	24.85		
50					80.47	31.07		
60					96.56	37.28		
70					112.65	43.50		
80					128.75	49.71		
90					144.84	55.92		
100					160.93	62.14		

UNITS	Lb ft to kgm	Kgm to lb ft	UNITS	Lb ft to kgm	Kgm to lb ft
1	.138	7.233	7	.967	50.631
2	.276	14.466	8	1.106	57.864
3	.414	21.699	9	1.244	65.097
4	.553	28.932	10	1.382	72.330
5	.691	36.165	20	2.765	144.660
6	.829	43.398	30	4.147	216.990

CHAPTER 11

THE BRAKING SYSTEM

11 :1 Description

Hydraulically operated brakes are fitted to all four wheels. Drum brakes are fitted on all rear wheels but on later models, disc brakes can be fitted as an option to the front wheels instead of conventional drum brakes which are standard.

The rear brakes are mechanically operated by the park brake mechanism, and a system of cables transmits the pull to the rear brakes. A system of metal lines and flexible hose transmits the hydraulic pressure generated by the master cylinder to the brakes, and a typical system is shown in **FIG 11 :1**.

On the early models a single master cylinder is fitted and all four brakes are hydraulically interconnected. On later models a tandem master cylinder is fitted. The later master cylinder contains two pistons in series and the front and rear brakes are hydraulically independent from one another so that a failure in one half of the braking system will still allow the other half to operate at full efficiency, though pedal travel will be longer and the overall stopping distance of the car will be increased.

A power unit can be fitted as an optional extra between the brake pedal and master cylinder. The unit is vacuum operated, from the intake manifold, and proportionally increases the load applied to the master cylinder. For a given retardation, less pressure needs to be applied to the brake pedal. The unit is designed so that the brakes will still operate if the unit fails but there will be no vacuum assistance. This point must be born in mind when towing the vehicle as the reservoir suction will soon be used up and a heavier brake pedal pressure will be required.

11 :2 Maintenance

Fluid level:

At regular intervals check the fluid level in the master cylinder reservoir. Wipe the filler cap clean and remove it from the master cylinder. On the dual master cylinder, the diaphragm should come away with the filler cap. Top up, if required, to the correct level using Rotunda Super Duty Brake fluid (Specification SAE.70R3).

On models fitted with disc brakes the level will steadily drop as the disc brake pads wear and this is normal. Any undue drop requires immediate investigation of the braking system as it can be caused by a fluid leak in the lines. Leaks must be repaired as soon as possible as they can lead to brake failure.

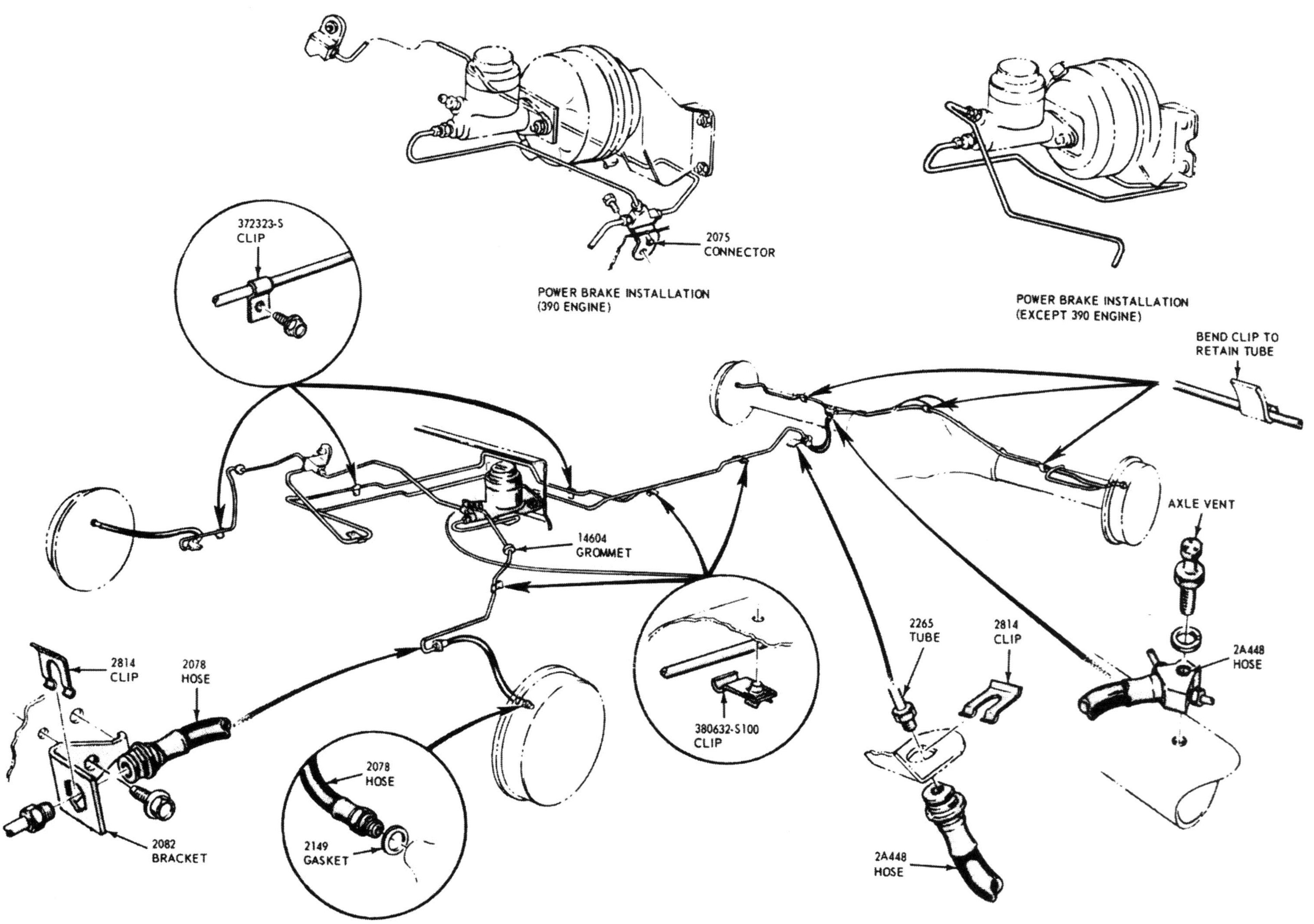

FIG 11:1 Typical service brake hydraulic system

Adjustments:

All brakes are automatically adjusting with use. Disc brakes have a seal around the piston which automatically sets the clearance between the pad and rotor. Drum brakes have an internal mechanism that operates an adjusting lever when the clearance between the lining of the shoe and drum becomes excessive. At regular intervals the brake drums should be removed, loose dust and dirt blown out, and the linings checked for wear. Drum brake linings must be renewed when there is less than $\frac{1}{32}$ inch of lining material left above the rivet heads.

The adjustment of the park brake is dealt with in **Section 11:11.**

Preventative maintenance:

Regularly examine the friction linings on the brakes as well as all lines and flexible hoses. Lines which are corroding or dented by stones should be renewed as soon as possible as the damage will cause a weak spot which may fail under the stress of heavy braking.

At intervals of 18 months, drain the system and fill and bleed it using fresh hydraulic fluid. Alternately bleed the system using sufficient fresh fluid to ensure that all the fluid is bled out and replaced by fresh fluid. The fluid gradually accumulates moisture from the atmosphere and, coupled with the heat from the brakes, this makes fluid lose efficiency. It is for these reasons that the fluid should be regularly changed.

At intervals of three years, drain out the fluid and dis-assemble all the parts of the braking system. Discard all the old seals, fitting all new ones on reassembly, and check all components for wear or damage, renewing those that are defective.

11:3 Flexible hoses

When removing or installing a flexible hose, the flexible portion must not be twisted. When the hose is installed the line on the hose must be straight.

Hoses which show signs of defect, such as cracking, perishing or chafing, must be renewed. If a hose is blocked, try clearing it with gentle air pressure and if this does not remove the obstruction fit a new hose. **Do not attempt to clear a blocked hose by poking wire through it.**

Removal:

Hold the hexagons on the adjacent end of the hose with a wrench to prevent the hose from turning. Undo the union nut that secures the brake line to the hose, using another wrench. Remove the clip or locknut that secures the hose to the bracket, still holding the hose with a wrench, and then free the hose from the bracket. The other end of the hose can then be unscrewed, allowing the flexible portion to rotate freely.

Install the hose in the reverse order of removal. If seals are fitted at the end of the hose, they should be discarded and new seals fitted on installation. Check that the hose is clear of any obstructions, or the muffler, when the suspension and steering moves through its range of movement.

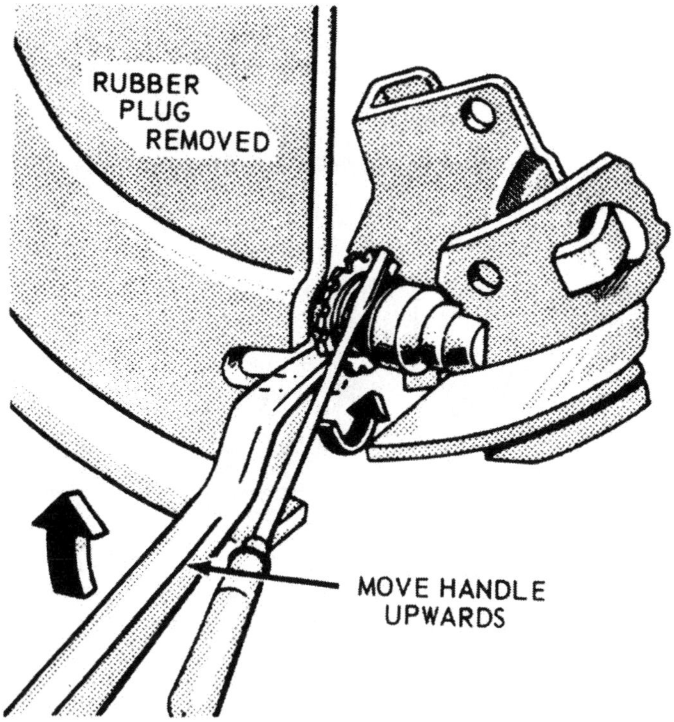

FIG 11:2 Backing off the brake adjustment

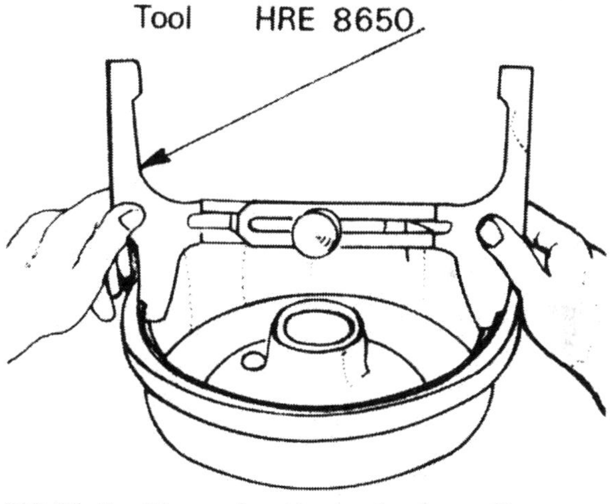

FIG 11:3 Measuring the brake drum diameter

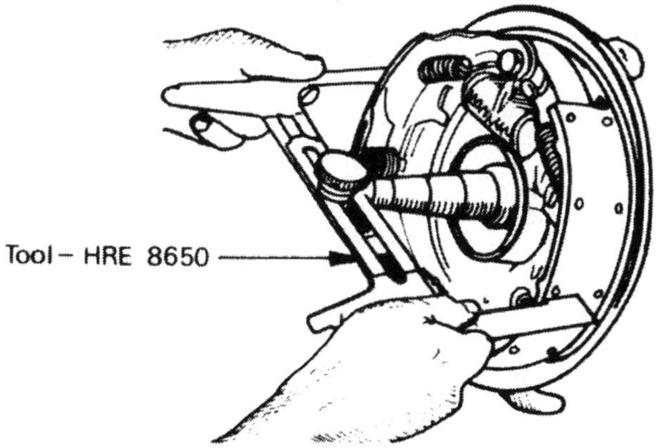

FIG 11:4 Adjusting the shoes to the brake drum

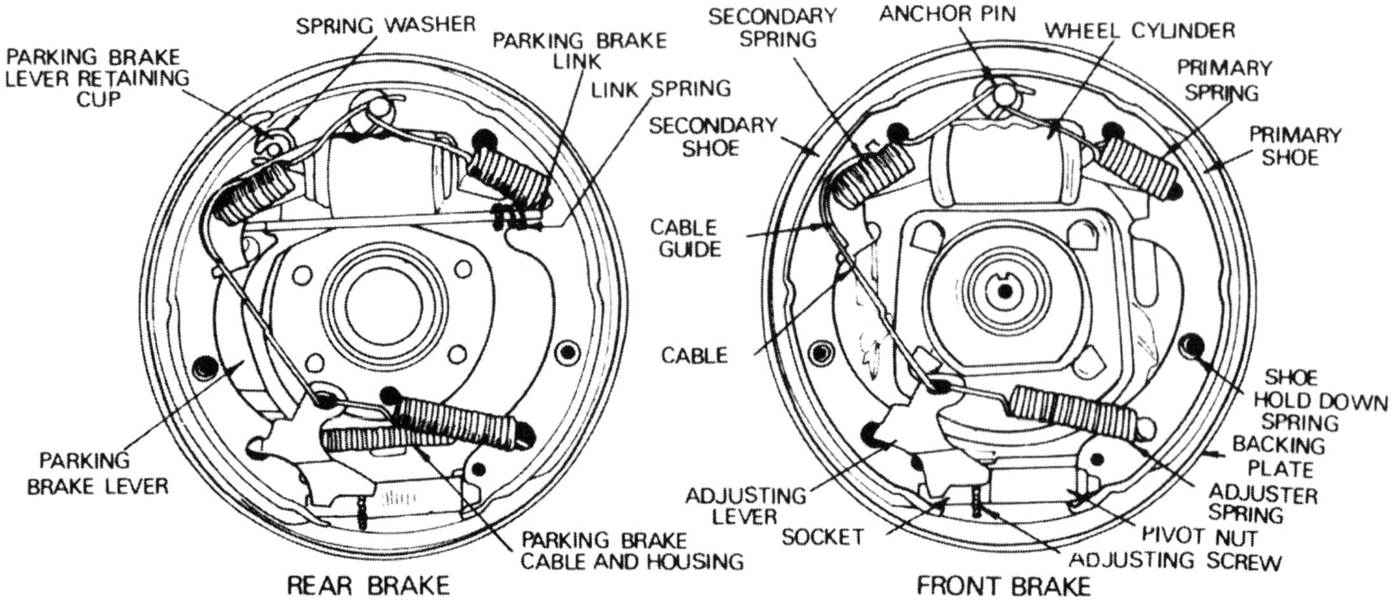

FIG 11 :5 The drum brake assemblies

11 :4 Brake drums

Removal:

Slacken the road wheel attachment nuts slightly and jack up the appropriate side of the car. Remove the road wheel. On the front wheels it is best to remove the complete front hub assembly (see **Chapter 9, Section 9 :3**) and then separate the drum from the hub. On the rear wheels, remove the three Timmerman nuts and draw the drum off.

If the drum is held in place by the brake shoes, slacken off the shoes to give clearance. Knock out the access slug at the back of the carrier plate and insert a thin screwdriver to disengage the adjusting lever from the adjusting screw, as shown in **FIG 11 :2**. Use the brake adjusting tool to back off the adjusting screw as shown. **Take great care not to damage or burr the teeth on the adjusting screw otherwise the automatic adjuster may not work correctly.**

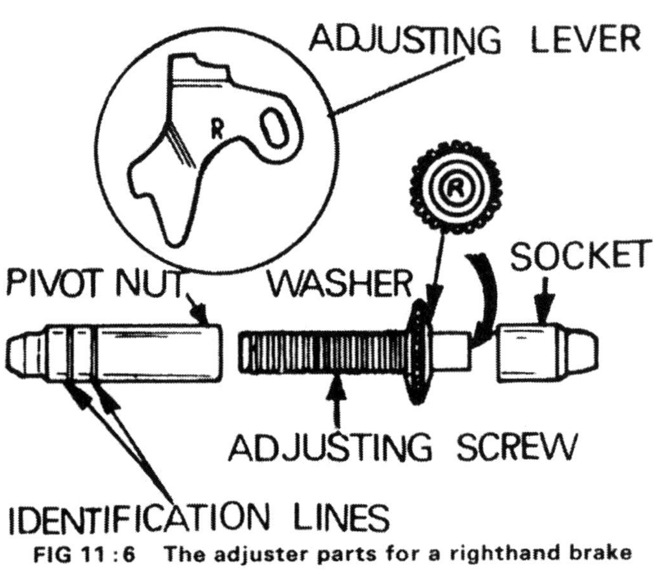

FIG 11 :6 The adjuster parts for a righthand brake

Examination:

Wash down the carrier plate assembly using denatured alcohol and then brush and blow away any loose dirt. Wash the brake drum in any suitable degreasing solvent. On models where the drum is integral with the hub, avoid washing the grease retainer with solvent and wipe out the operating surface of the brake drum with a separate clean cloth and solvent to ensure that grease from the hub is not spread onto the operating surface. **Do not allow any solvent onto the brake linings themselves.** Take the opportunity to examine the wheel cylinder and hub assembly for leaks as well as checking the lining thickness.

Examine the operating surface of the drum for scores. Light damage may be machined off provided that not more than .060 inch of stock is removed. When machining a drum, be sure to mount it on an arbor and not in a chuck, as a chuck will slightly distort the drum. If the scoring is too deep to remove by machining, a new drum should be fitted. If the linings are only partially worn, leave the scored drum in place, provided that the scores are circumferential, and renew both drum and shoes when the linings are worn. Fitting a new drum to shoes that have mated in to the scores in the old drum will reduce braking efficiency and probably cause scoring in the new drum as well. Oversize shoes will have to be fitted if the diameter is greater than +.030 inch.

Check the drum for cracks. Hang the drum on a wooden handle and tap it with a light metal tool. A flat note shows that the drum is cracked while if it rings true then the drum is satisfactory. Do not use a cracked drum but discard it and fit a new one in its place. If the cracked drum is also found to be scored, new brake shoes must be fitted—irrespective of the wear left in the old linings.

Installation:

Measure the internal diameter of the brake drum with the Rotunda calipers, as shown in **FIG 11 :3**. Clamp the calipers so that they do not alter and use them to set the brake shoes, as shown in **FIG 11 :4**. The brake shoes

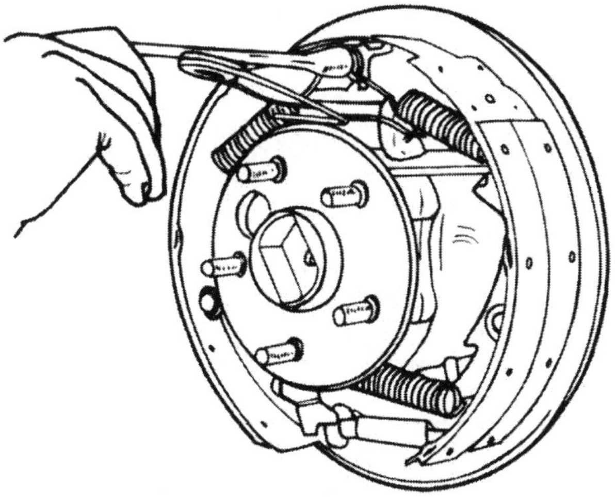

FIG 11 : 7 Removing the shoe return springs

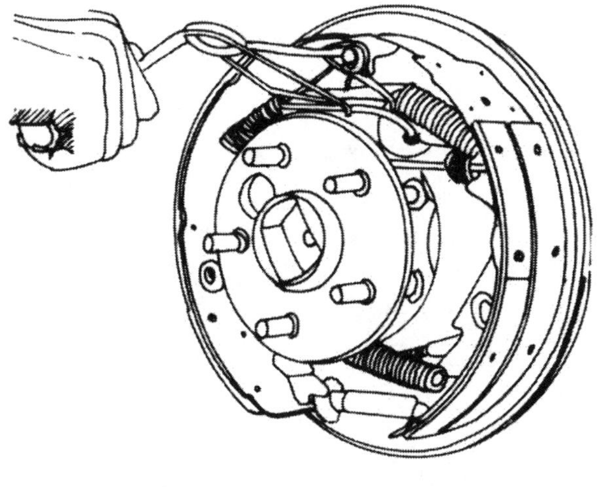

FIG 11 : 8 Installing the shoe return springs

are adjusted while holding the adjusting lever off the adjusting screw, as shown in **FIG 11 : 2.** Check that the adjusting screw rotates freely and if necessary lubricate it with a thin film of C1Az-19590-B grease on the threads. Apply very small quantities of high temperature grease to points where the brake shoes contact the carrier plate and anchor pin. **Avoid getting any lubricant onto the linings themselves and avoid touching the shoes with dirty or greasy hands.**

When the shoes have been set, install the drum in the reverse order of removal. On the front wheels, set the correct hub clearance (see **Chapter 9, Section 9 : 3**). Install the road wheel.

When all the parts are installed and all drums checked, drive the vehicle forwards to ensure that the brakes are not binding. Accurately set the adjustments by backing the car and stopping it with a pedal pressure greater than 50 lb. After every stop drive the vehicle forwards and apply the brakes before stopping the vehicle while backing.

Do not apply the brake pedal while a brake drum is off, otherwise the pressure will force out the pistons from the wheel cylinder.

11 : 5 Drum brake shoes and adjusting screws

The front and rear drum brakes are shown in **FIG 11 : 5.** The wheel hub assemblies are removed for illustrative purposes only. As the car backs and the brakes are applied, the movement of the shoes causes the anchor pin to pull the cable which acts on adjusting lever. If there is sufficient play, the lever will move the adjusting

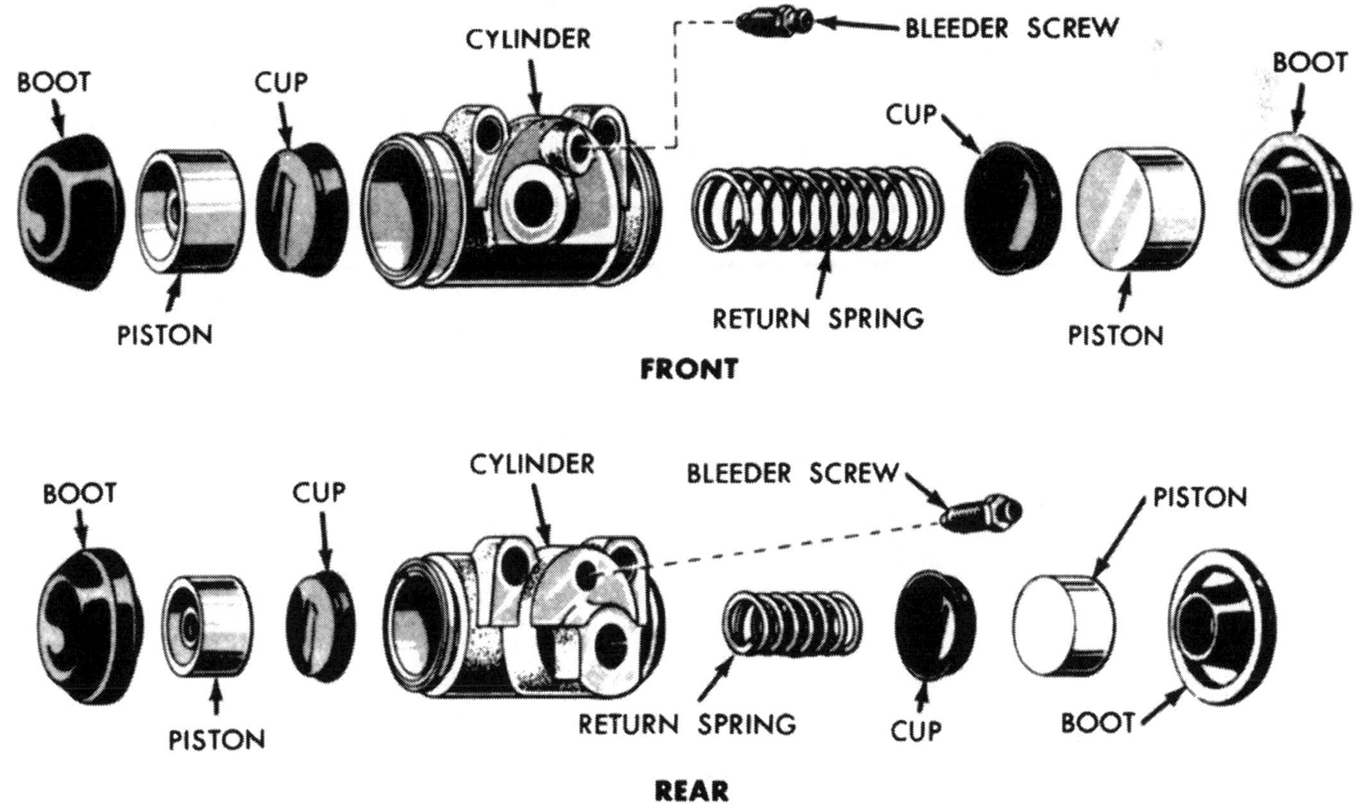

FIG 11 : 9 Wheel cylinder components

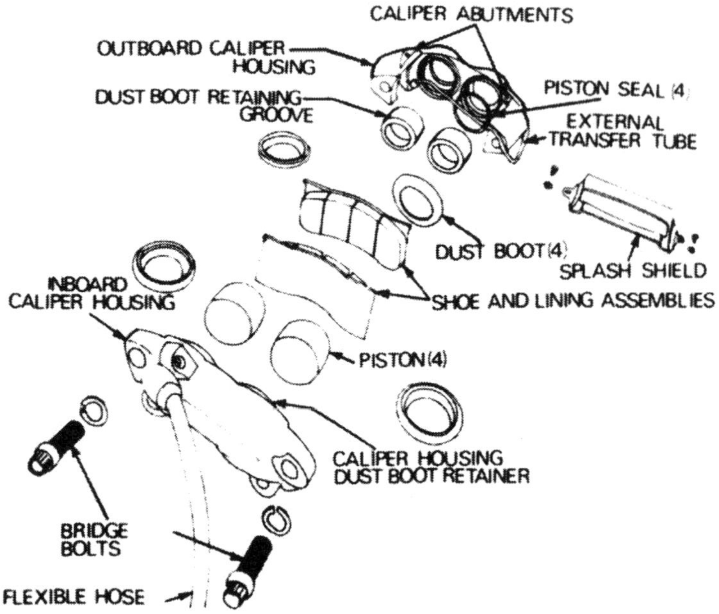

FIG 11:10 Typical early disc brake components

screw by one tooth to take up the play. When the brakes are in adjustment there is insufficient movement for the adjusting lever to grip the tooth and turn it. To ensure that the parts actually adjust the brake, as fitting the parts to the incorrect side would cause them to slacken the shoes, the parts of the adjuster are marked; R for righthand and L for lefthand. The righthand side adjuster parts are shown in **FIG 11:6** and the position nuts are also identified; two grooves for a righthand side nut and a single groove for a lefthand side one.

Removing parts:

1 Remove the brake drum as described in the previous section. Remove the shoe hold-down spring parts. Hold the pin in place with a finger behind the brake carrier plate and grip the washer with a pair of pliers. Press the washer in against the pressure of the spring and turn it through 90 deg. so that the washer can be slipped off the head of the retaining pin. Collect the spring and withdraw the retaining pin through the carrier plate.

2 As a safety precaution to keep the pistons in the wheel cylinder, clamp them in place. Free the return springs from the anchor pin, using the tool shown in **FIG 11:7.** If the special tools for removing and installing the springs are not available use a strong length of cord through the end of the spring to extend it and to lift it off or onto the anchor pin.

3 Remove the shoes and adjusting screw as an assembly and then dismantle the parts on the bench. On the rear brakes, remove the parking brake link and spring and also disconnect the parking brake cable from the parking brake lever. After the shoes have been removed from the brake, remove the retaining clip and spring washers to free the parking brake lever from the secondary shoe.

Inspection:

Disassemble the components of the adjusting screw. Check the shoe linings for wear. New linings must be fitted if the old ones are contaminated with oil or grease or if the linings are worn down to $\frac{1}{32}$ inch above the rivet heads.

New linings must not be fitted to one wheel only as this will lead to unbalanced braking. Always renew linings in complete axle sets. **When contaminated linings are found they must be renewed and no attempt made to restore the linings by baking or washing with solvents.**

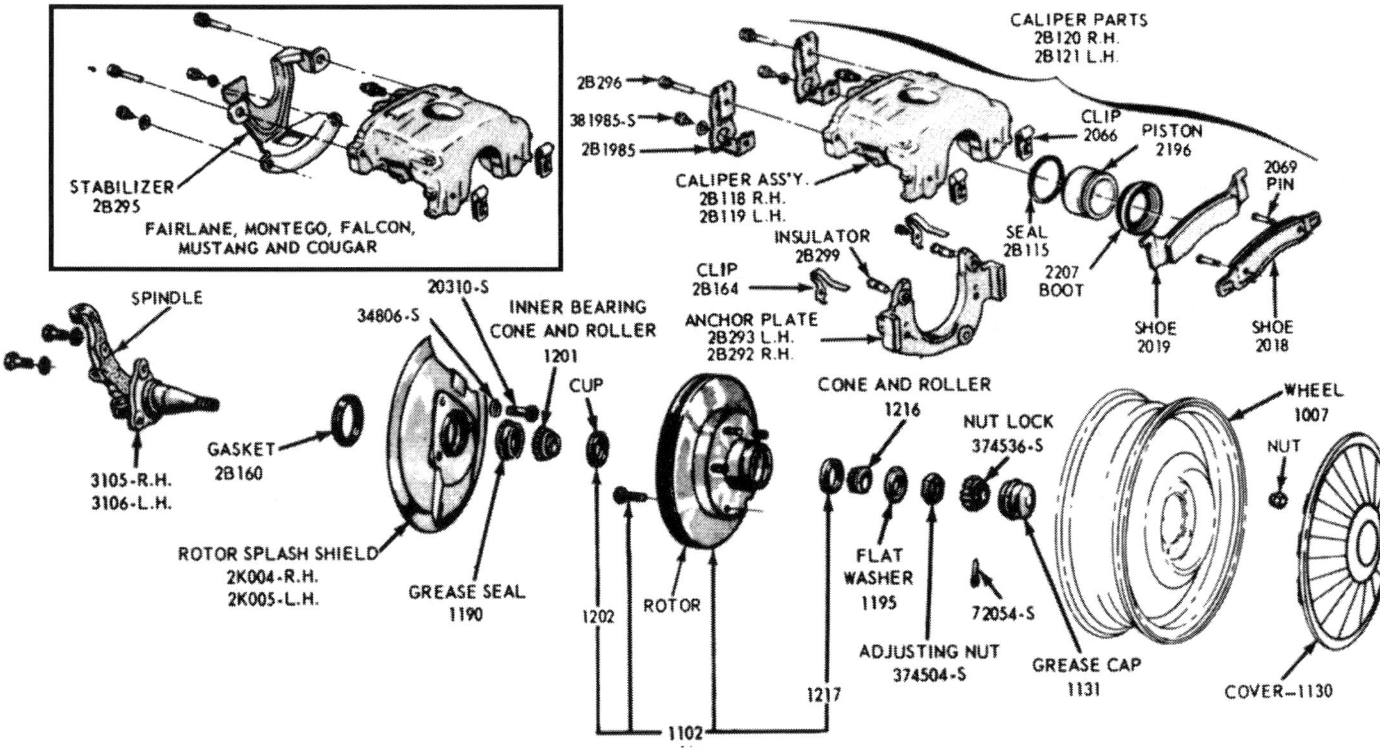

FIG 11:11 Fixed caliper disc brake

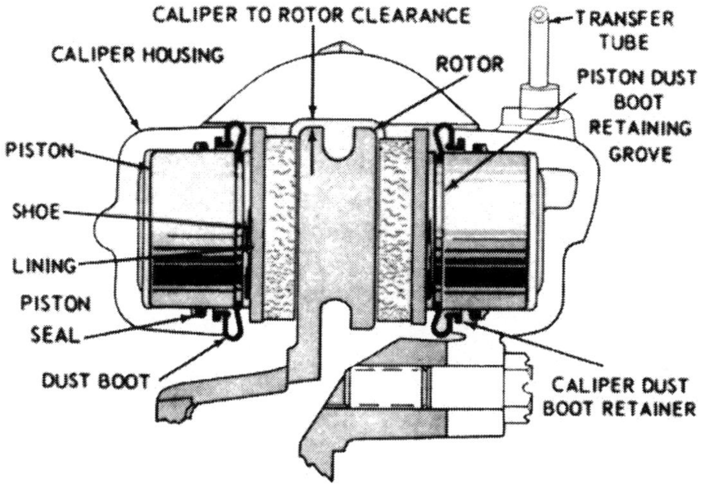

FIG 11:12 Sectioned view through a swinging caliper type brake

New linings are available already ground to shape but if the owner has not had experience of rivetting it is best to obtain exchange shoes and fit these. If new linings are fitted to the old shoes, use only the rivets supplied to secure them and make sure that the rivets are well clinched, leaving no gaps between shoe and lining. File the leading and trailing edges of the linings to a chamfer as this will help to prevent brake squeal.

Check the adjusting cable for damage or weakening. The cable must completely fill or extend beyond the crimped fittings. If the cable appears weak or damaged, discard it and fit a new cable assembly, as a poor cable can be one cause of the automatic adjuster failing to operate correctly. Check the cable guide for damage. The groove for the cable should be parallel to the shoe web and the body of the guide should lie flat against the web.

Check the pivot hook on the adjusting lever. The hook surfaces should be square with the body of the lever.

Renew any damaged or defective parts.

Installation:

The parts are installed in the reverse order of disassembly. Lubricate bearing points lightly with high temperature grease. The cable must be fitted with the crimped side of its eye towards the carrier plate and the cable must be routed in the guide and not between the guide and carrier plate. Be sure that the parts all lie flat on the anchor pin and install the return springs as shown in **FIG 11:8**, noting that a length of cord can be used to extend the springs into position.

When the parts are in place, the adjusting lever should contact the adjusting screw teeth $\frac{3}{16} \pm \frac{1}{32}$ inch above the center line of the adjusting screw.

Check the operation of the adjuster. Pull on the adjuster cable, between the cable guide and anchor pin, toward the secondary shoe web far enough to lift the adjusting lever past a tooth on the wheel. The lever should spring into position behind the tooth and when the cable is released the return action of the lever should move the screw by one tooth.

Adjust the brakes and install the brake drums and road wheels. Set the final adjustment by backing the car and applying the service brake firmly.

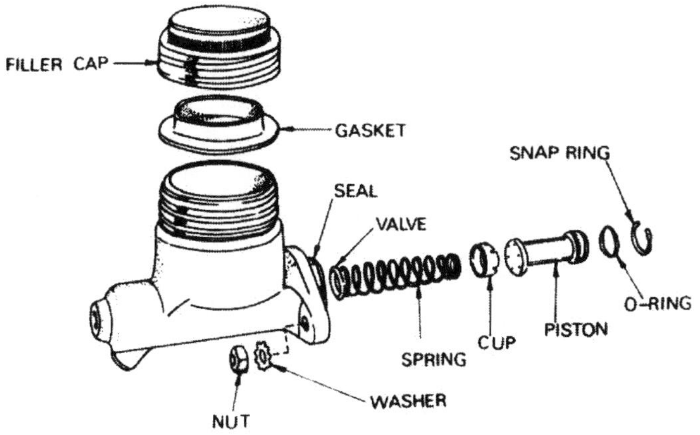

FIG 11:13 Single master cylinder fitted with a standard system

11:6 Drum brake wheel cylinders

These are accessible after the brake drum has been removed (see **Section 11:4**) and the shoe assembly has been removed (see previous section). The components of the wheel cylinders are shown in **FIG 11:9.**

Disassembly and removal:

The parts can be disassembled without having to remove the wheel cylinder from the carrier plate. Only if the wheel cylinder bore is found to be scored or damaged need it be removed from the carrier plate.

Remove the links and boots from the wheel cylinder and then take out the cups, pistons and return spring from the cylinder.

If the cylinder requires removal, take off the bleeder screw. On front wheels, disconnect the flexible hose at the frame and then unscrew it at the wheel cylinder. On rear wheels, undo the union nut that secures the metal pipe to the wheel cylinder but do **not** pull the pipe away from the cylinder. Pulling the pipe will distort it and the pipe will free as the cylinder is removed from the back-plate. Remove the wheel cylinder attaching bolts and take out the wheel cylinder.

Inspection and cleaning:

Discard the old cup seals as well as rubber boots and fit new ones on assembly. Wash the parts with denatured alcohol. If the bore of the cylinder is scored, worn or pitted, a complete new assembly must be fitted.

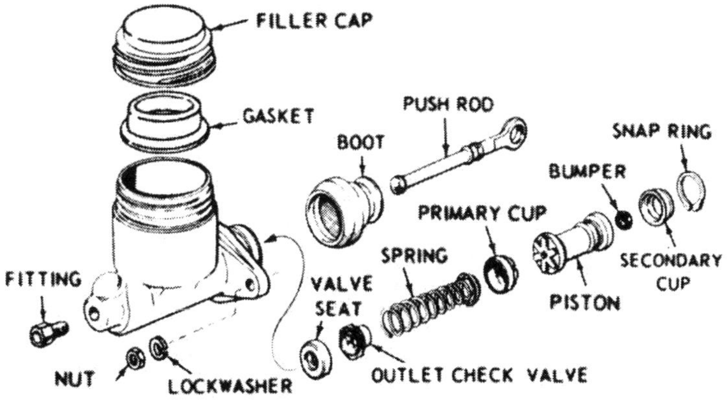

FIG 11:14 Single master cylinder fitted in conjunction with a booster

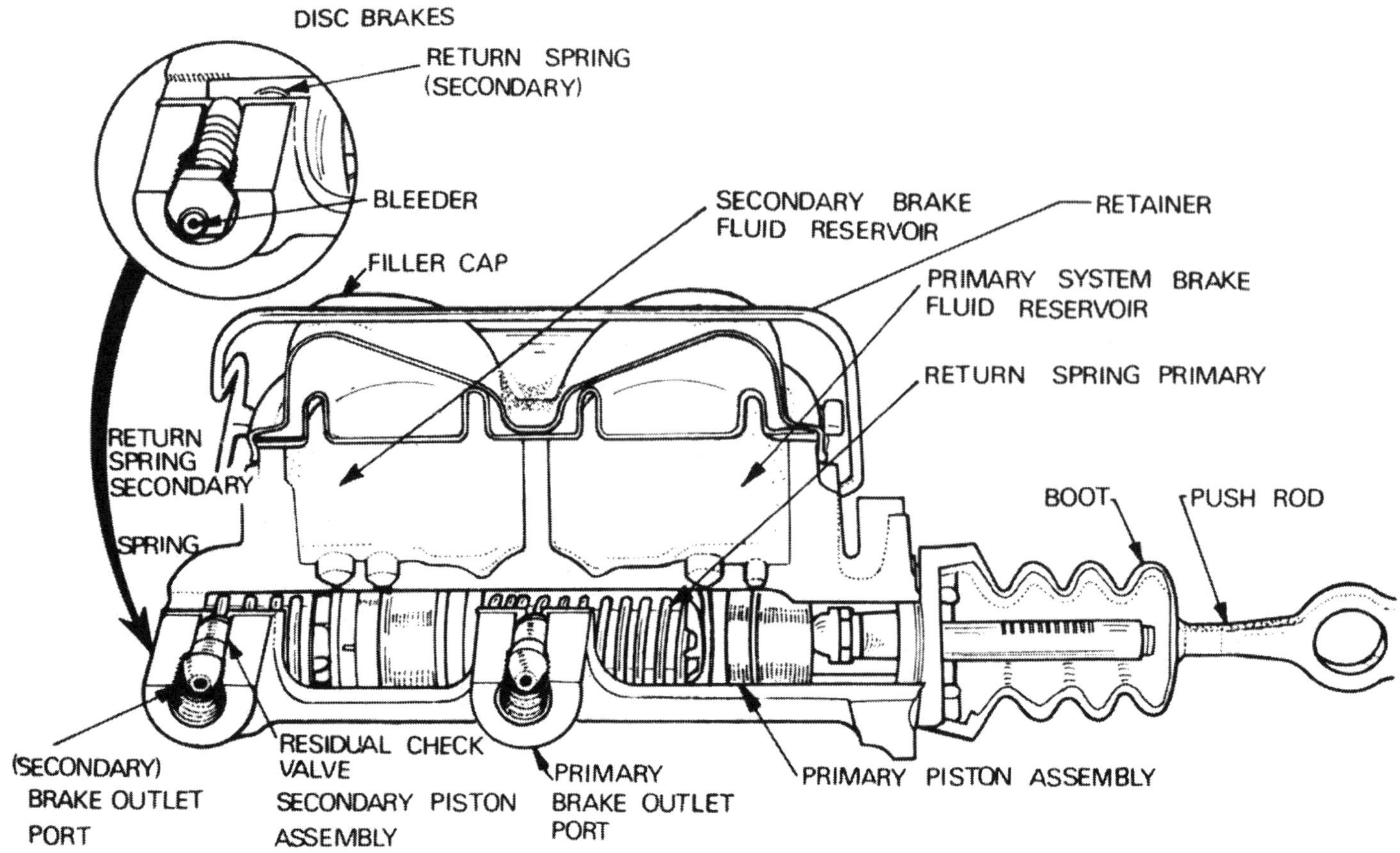

FIG 11 :15 Dual master cylinder fitted with drum brakes

Assembly and installation:

Wet the internal parts with hydraulic fluid. Fit one cup seal back into the bore, **taking great care not to bend back or damage the lips of the seal as it enters the bore.** From the open end, install the return spring and other cup seal, taking care not to bend or damage the lips. Be sure that the return spring is seated between the seals and insert the pistons, flat faces leading. Secure the parts in place with the rubber boots.

The components are installed in the reverse order of disassembly. Once the brake drums have been installed, fill the reservoir and bleed the system as described in **Section 11 :10.**

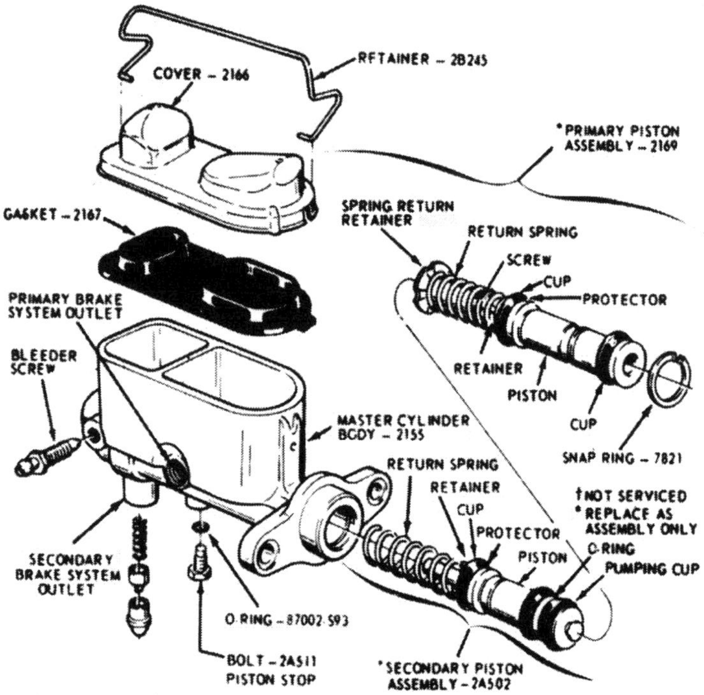

FIG 11 :16 Dual master cylinder fitted with disc brakes

Brake carrier plates:

Removal of the carrier plates entails removing the hub assembly. The complete brake assembly, apart from drum, can be removed as a unit after disconnecting the line to the wheel cylinder. The hydraulic system must be bled after the parts have been installed.

11 :7 Disc brakes

A typical disc brake is shown disassembled in **FIG 11 :10.** On later models a fixed caliper brake may be fitted and the parts are shown in **FIG 11 :11. The parts of the fixed caliper brake are shown fully disassembled but under no circumstances may the bolts that secure the two housing pieces together be removed or slackened and all work must be carried out with the housing correctly bolted together.**

A sectional view through a caliper is shown in **FIG 11 :12.** The piston seal ensures that there is always a slight running clearance between the friction linings and rotor. When pressure is applied the piston moves in the bore to apply the friction linings to the rotor. As the piston moves it drags the outer edge of the seal with it so that the seal is deformed. If there is excessive clearance the piston slides through the seal, leaving the seal still distorted. As soon as the pressure is released, the seal returns to its

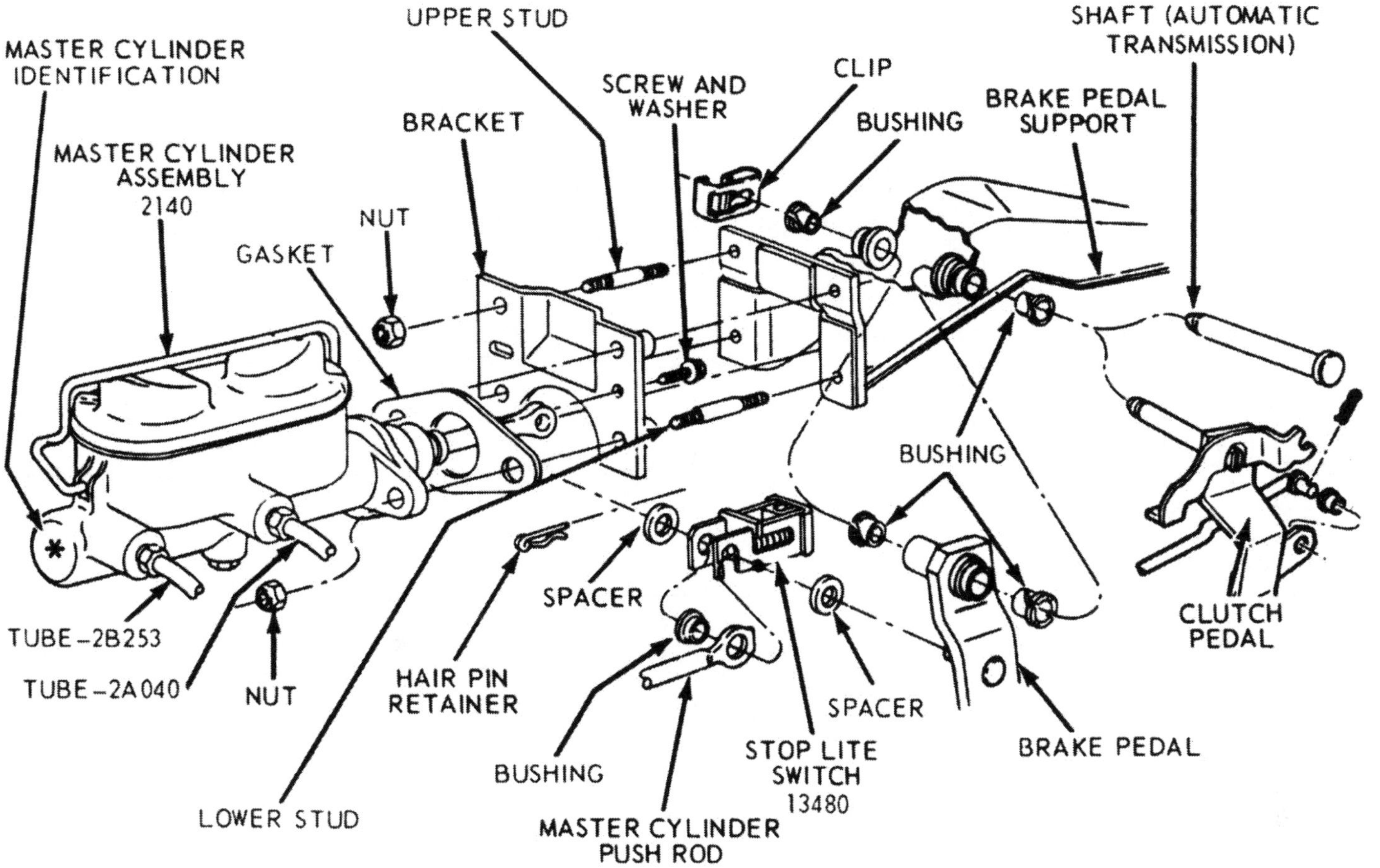

FIG 11:17 Installation of dual master cylinder

normal shape and draws the piston slightly back with it.

A pressure control valve is fitted into the system to ensure that the hydraulic pressure to the rear brakes is limited when high pressures are required on the disc brakes. This valve ensures that the rear wheels do not lock under heavy braking. The proportioning valve cannot be dismantled or serviced and a new valve must be fitted if the old one is defective. The hydraulic system must be bled after the valve has been removed and refitted.

Caliper removal:

If the caliper is to be removed completely, disconnect the hydraulic line. If the caliper is being removed for the servicing of parts of the front suspension, leave the caliper attached to the line, and when it has been freed from the suspension hang it safely out of the way so that the flexible hose is not strained or twisted.

Raise the front of the car on a jack and remove the road wheel. Take out the two bolts that secure the caliper to the suspension and slide the caliper off from the rotor.

Install the caliper in the reverse order of removal. The upper bolt should be tightened first. If the bolts are so equipped, lock them using a new length of safety wire after they have been torque-loaded.

If the flexible hose has been disconnected, the hydraulic system must be filled and bled as described in **Section 11:10.**

Rotor:

Removal of the rotor entails removing the hub assembly and then separating the rotor from the hub (see **Chapter 9, Section 9:3**). Before separating the rotor from hub, make alignment marks so that the rotor and hub will be reassembled into the original alignment.

If the rotor is scored or has excessive runout (measured with a dial indicator after the hub assembly has been correctly installed) the rotor may be machined down on a

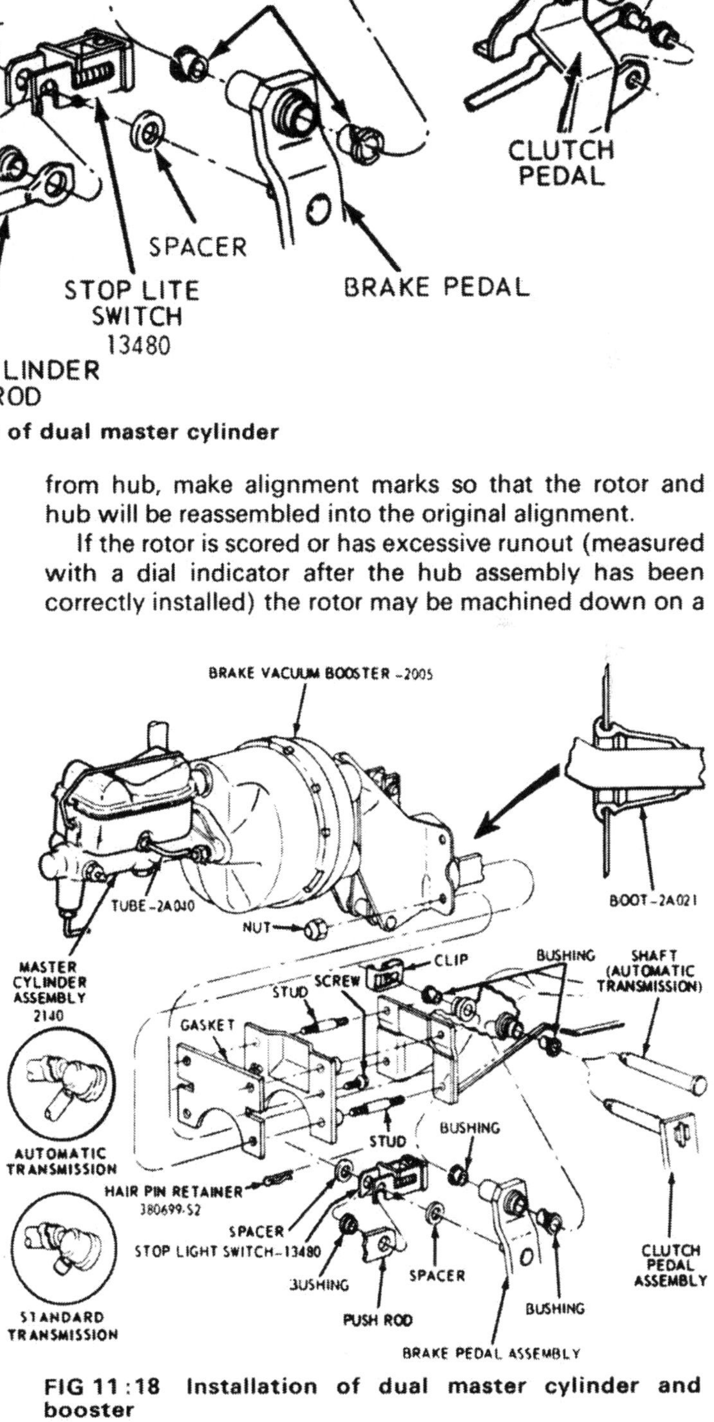

FIG 11:18 Installation of dual master cylinder and booster

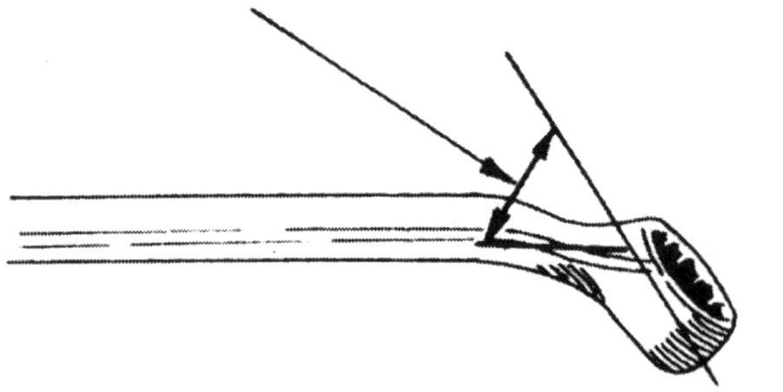

FIG 11 : 19 Modified wrench for bleeding brakes

Rotunda rotor grinding machine. The rotor must be machined so that its operating faces are parallel and the thickness between them does not vary by more than .0007 inch. The rotor must not be machined below its minimum thickness, but this can be checked using a special ball and gauge. If the runout or scores cannot be cured by machining, a new rotor must be fitted.

Dirt and deposits, which affect braking efficiency, can be removed from the rotor faces by washing with trichlorethylene and gentle scrubbing with emerycloth.

Before installing the rotor to the hub, be sure that the mating faces are scrupulously clean, as specks of dirt will cause excessive runout on the rotor.

When the assembly has been installed on the spindle and the wheel bearings correctly set, use a dial indicator to measure the runout of the outer operating face near to the outer edge. If the runout is excessive try rotating the rotor with respect to the hub to minimize the runout, but if this fails the rotor must either be machined or renewed.

Renewing brake shoes and friction linings:

Exchange shoes are supplied with the linings bonded to them. **The owner should not attempt to bond new linings to the old shoes.**

New shoes must be fitted if the lining material has worn down to .030 inch at the front edge or if the lining material is contaminated with oil or grease.

1 Jack up the front of the vehicle and remove the road wheel.
2 Syphon out sufficient fluid from the reservoir of the master cylinder connected to the disc brakes so that the level is at least half-way down the reservoir. If this precaution is not taken, fluid may overflow from the reservoir as the pistons are pressed back into the caliper. Discard the old fluid siphoned out, and use fresh fluid for topping up.
3 Remove the bolts and clips that secure the shoes. Press the shoes carefully but firmly back into the caliper so that the pistons return into their bores. Pressure will need to be maintained for at least a minute and take great care not to damage or score either the caliper or the rotor.
4 Grip the lugs on each shoe with two pairs of pliers and pull each shoe in turn out of the caliper. Brush and blow out all loose dirt. If a solvent has to be used to wash out dirt use only denatured alcohol or hydraulic cleaning fluid, as any other solvent will attack the material of the seals.
5 Install the new shoes in the reverse order of removal, taking care not to allow grease, oil or solvent onto the friction linings and not handling the parts with greasy or dirty hands. Secure the shoes in place.
6 Install the road wheel. Pump the brake hard several times to take up the adjustment before using the vehicle on the road. Top up the reservoir with fresh fluid until the level is $\frac{1}{4}$ inch below the top of the reservoir and fit the cap and diaphragm assembly.

Servicing brake caliper:

Remove the shoes from the caliper and the caliper from the suspension. Lay a piece of rag into the caliper and use air pressure through the inlet to gently blow out the pistons. Remove the dust covers from the pistons and carefully pry out the piston seals from the caliper bores. Discard the old dust covers and seals, fitting new ones on reassembly.

Wash all the parts in denatured alcohol or hydraulic cleaning fluid. Check the bores of the caliper for wear, scoring or pitting. If the bore is defective a complete new caliper assembly must be fitted.

Wet the piston seals with hydraulic fluid and insert them back into their recesses. The seals should be refitted using only the fingers and they must seat fully and squarely into the recesses.

Install the dust covers onto the pistons. Wet the pistons with hydraulic fluid. **Carefully press the pistons back into their bores, taking great care to enter each piston squarely and make sure that it does not tilt.** Secure the dust covers back in place on the caliper. The caliper is now ready for installation. After the hydraulic system has been bled, pump the brakes hard several times to take up the adjustment of the disc brakes.

11 :8 The master cylinder

Early models are fitted with a single master cylinder where the one chamber operates all four brakes. On later models a dual master cylinder is fitted, in which one chamber operates the front brakes and the other chamber operates the rear brakes so that the hydraulic system is divided into two independent parts.

Care must be taken when ordering spare parts as the size and type of master cylinder varies with year and application.

The single master cylinder for models without power brakes is shown in **FIG 11 :13** and that for a system with power brakes in **FIG 11 :14**. A dual master cylinder for models with drum brakes is shown in **FIG 11 :15**, while a dual master cylinder without disc brakes is shown in **FIG 11 :16**.

Removal:

The installation of a dual master cylinder with power brakes is shown in **FIG 11 :17** and that without power brakes in **FIG 11 ;18**. The single master cylinders are installed in a very similar manner.

Siphon out all the fluid from the reservoir(s). Disconnect the hydraulic lines from the master cylinder, using rags to catch any spillage. Hydraulic fluid will rapidly damage paintwork if it is spilt. On models not fitted with power brakes, the pushrod must be disconnected from the brake pedal. Disconnect the

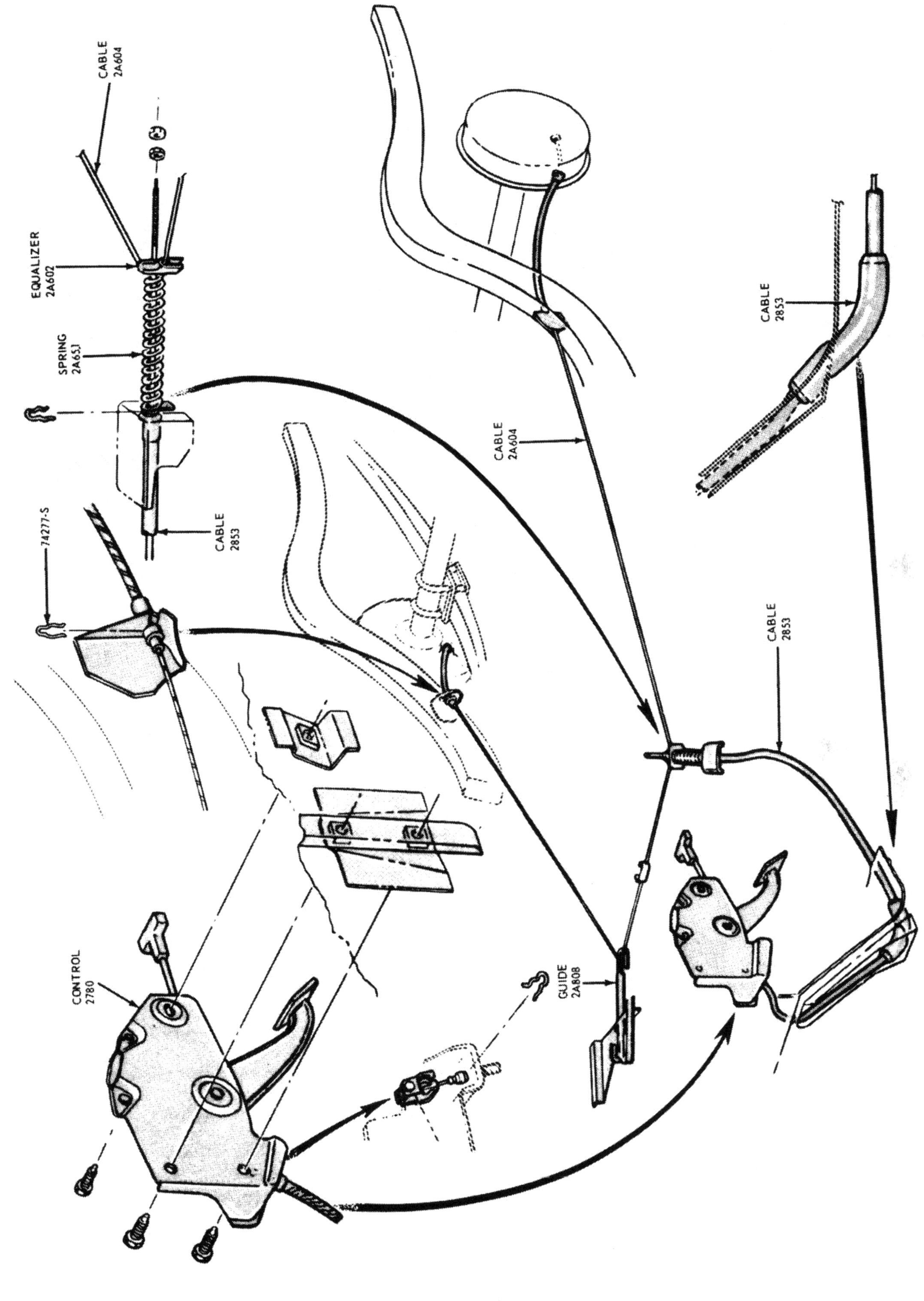

FIG 11:20 Early parking brake linkage

connectors at the stoplight switch, remove the hairpin retainer and slide the switch off the pivot and up and out, taking care not to damage the switch. Remove the two attachment nuts and withdraw the master cylinder from the vehicle.

The master cylinder is installed in the reverse order of removal. Once the unit is in place the hydraulic system must be filled and bled as described in **Section 11 :10.**

Servicing:

Wash the outside of the unit with denatured alcohol and pour out any remaining fluid. Pull back the rubber boot. Gently press the piston down the bore and remove the snap ring with snap ring pliers. Tap the open end of the master cylinder bore against the palm of the hand to free the internal parts. Service kits are available and these contain all the parts necessary. On dual master cylinders, the primary piston parts must not be disassembled.

Wash the parts in denatured alcohol and blow through all passages to clean them. Lay the clean parts onto clean paper to prevent them picking up dirt. Inspect the bore. If there is only light etching, wear, or pitting, the bore may be honed out by a specialist but if the damage is deep a new master cylinder must be fitted.

Wet the internal parts with hydraulic fluid and install them in the reverse order of disassembly. **Take great care when inserting the piston(s) not to damage or bend back the lips of the seals.** Secure the parts in place with the snap ring and install the rubber boot.

Pressure differential valve:

This is only fitted on models with dual master cylinders. If one half of the system fails, there will be a pressure differential and this pressure differential will move the plunger of the valve causing it to operate the switch and illuminate the warning light. The unit cannot be serviced and if it is defective a new valve must be fitted in its place.

The method of centralizing the plunger if it has moved is dealt with in **Section 11 :10.**

Removal of the valve is straightforward after disconnecting the electrical leads and hydraulic lines. When the valve has been installed, both halves of the brake system must be bled.

11 :9 The power brake booster unit

Testing:

Apply the parking brake and pump the service brake several times to use up any vacuum. Apply the service brake, noting the pressure required. Start the engine and if the booster is functioning satisfactorily the pedal will gradually sink down and the pressure required to hold the brake reduce.

If the unit does not operate, check the vacuum supply and the non-return valve on the unit. Disconnect the hose (two hoses on models fitted with automatic transmission) and remove the check valve from the unit. Renew the valve if it is defective. Check the hoses for blockages or kinks.

If the unit still fails to operate, either fit a new unit or disassemble the old one and service it using a service kit.

Removal:

The installation with a dual master cylinder is shown in **FIG 11 :17.** Drain the master cylinder reservoir and disconnect the hydraulic lines from the master cylinder. The master cylinder can then be removed from the booster unit. Disconnect the pushrod of the booster unit from the pedal, as described for removal of a master cylinder without booster. Take off the nuts that secure the unit and remove it from the vehicle, after disconnecting the vacuum lines.

Install the unit in the reverse order of removal. Bleed the hydraulic system after installation of the master cylinder.

Servicing:

A service kit of seals and parts is available for servicing the unit.

Routine maintenance is confined to changing the air filter at the inlet (accessible after removing the unit from the car) at regular but long intervals.

Before installing the master cylinder or booster unit, check the protrusion of the booster unit pushrod at the front end of the unit. A special gauge is made for checking the protrusion, and the gauge bridges over the end of the unit. The pushrod is then adjusted so that it just touches the gauge.

11 :10 Bleeding the brakes

This is required when air has entered the system, either by allowing the fluid level to fall so low that air is drawn into the master cylinder or by dismantling the system.

On the models fitted with dual master cylinders, only the part of the system that has been dismantled requires bleeding as the other half of the system is hydraulically independent. On these systems, both systems must be bled if a component which is common to both systems has been removed and installed.

Fluid that has passed through the system should be discarded. Only if it is perfectly clean may it be used again but it must not be used until it has been stored for at least 24 hours in a sealed clean container. The storage ensures that all air has risen out of the fluid. **Never return fluid directly from bleeding to the master cylinder reservoir. Never use fluid that has been stored in an open or dirty container.**

Before commencing the operation, remove the filler cap from the master cylinder and top up the reservoir as far as it will possibly go without spillage. Keep an eye on the level as the operation progresses and top up with more fluid when the level reaches the halfway mark. If the level is allowed to drop too far, air will be drawn into the master cylinder.

Bleed the brakes in the order of decreasing pipe runs, starting at the brake with the longest run. Disc brakes will require more bleeding than conventional drum brakes to remove all trapped air. The master cylinder itself should be bled first, if it is fitted with a bleeder screw. On the models fitted with disc brakes, it will be necessary to jack up the front of the vehicle and remove the road wheels in order to gain access to the disc brake bleeder screws.

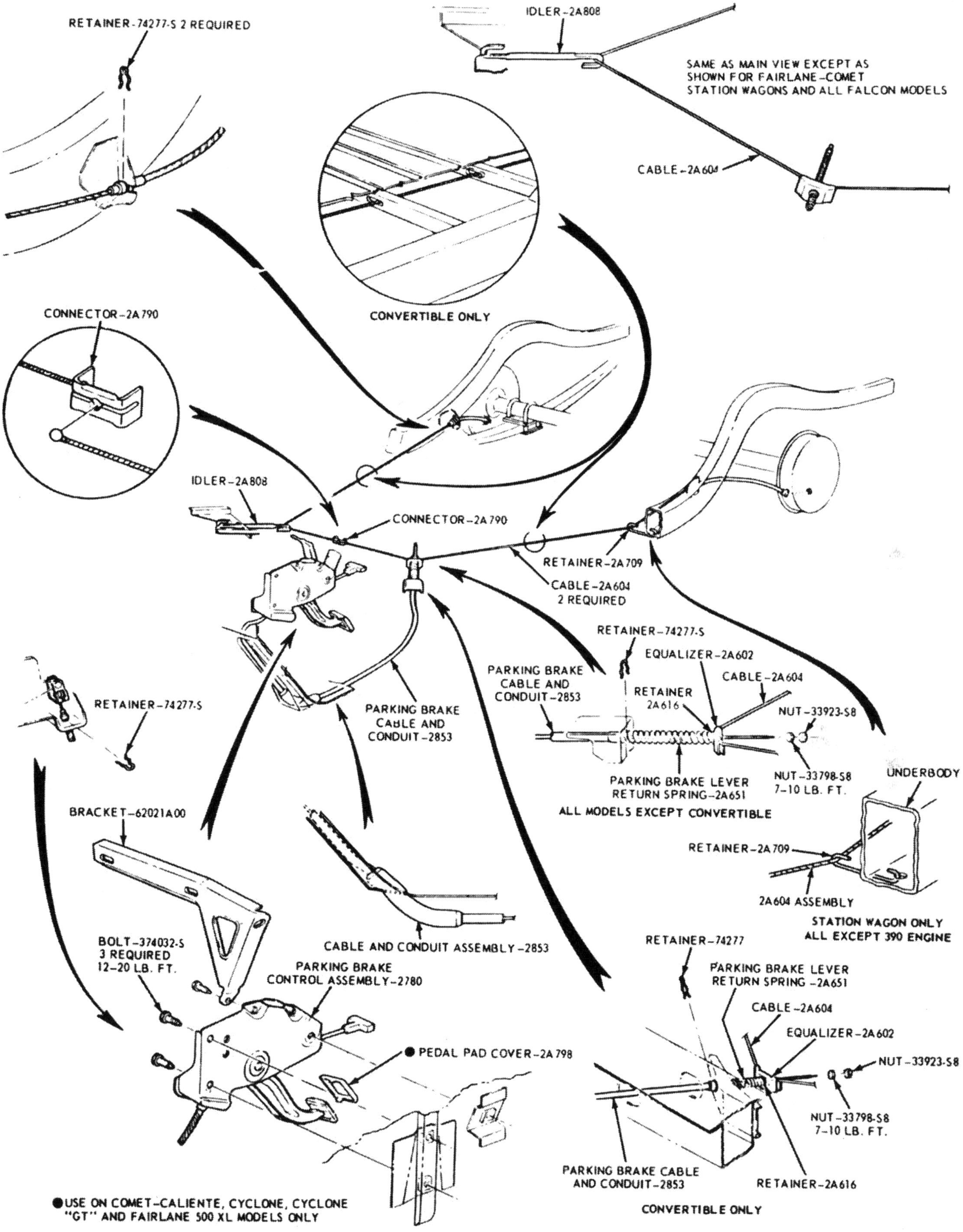

FIG 11 : 21 Later parking brake linkage

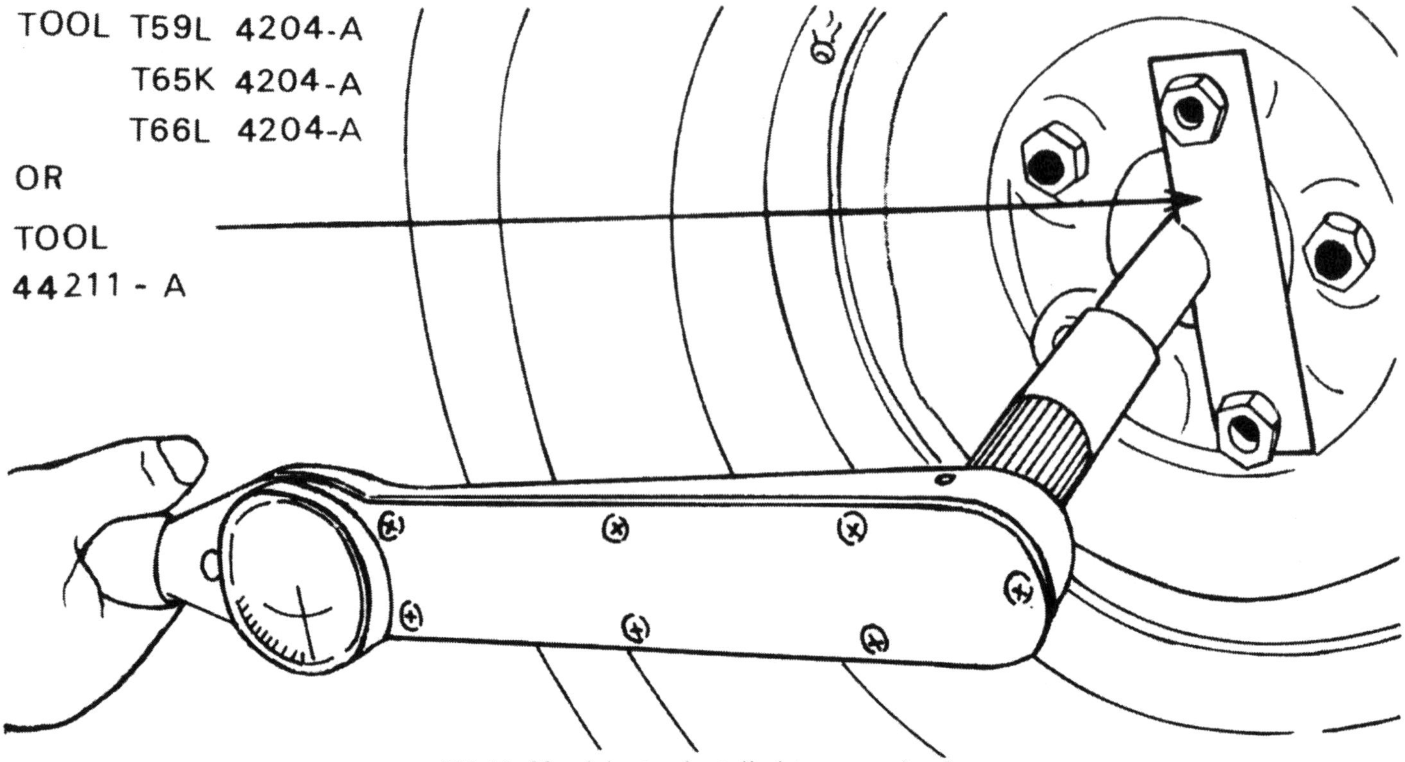

FIG 11 : 22 Adaptor installed to rear wheel

Attach a length of plastic or rubber tube to the bleeder screw of the brake to be bled. Be sure that the tube is a snug fit on the bleeder screw. Dip the free end of the tube into a little clean hydraulic fluid in a clean glass container.

Open the bleeder screw by $\frac{3}{4}$ turn. The task will be easier if a $\frac{3}{8}$ inch box wrench is bent to the angle shown in **FIG 11 : 19** and slipped onto the bleeder screw before installing the bleeder tube. Have an assistant slowly press the brake pedal right down through its full range of travel. If a power brake is fitted, exhaust the unit by pumping the service brake several times and carry out the bleeding without the engine running. When the pedal has reached the end of its travel, close the bleeder screw and have the pedal released. Repeat this cycle of opening, pressing down pedal, closing and releasing pedal until the fluid comes out of the bleeder tube in an unbroken stream with no air bubbles in it. Once the fluid is air-free, transfer the bleeder tube to the next bleeder screw to be bled and repeat the process.

If the system has been fully dismantled, it will be difficult to remove all the air. Bleed the system until a firm pedal pressure is obtained and the brakes are safe to use. Use the vehicle normally for a few days and then rebleed the complete system. All the minute air bubbles will then have had a chance to coalesce into larger bubbles which are easier to bleed out.

In extreme cases, take the vehicle to a service garage for bleeding using pressure equipment.

Centralizing pressure differential valve:

If the valve is accidentally moved over during the bleeding operation, or the valve operated because of a leak in one system, it can be centralized to turn off the warning light. Make sure that the fault that caused the valve to move has been repaired.

A reverse pressure will be required to move the valve spool back into the central position. Switch on the ignition to ACC or On position but do not start the engine. Open a bleed screw on the other half of the system to the half that leaked and caused the spool to move. Apply steadily increasing pressure to the brake pedal and stop further movement of the pedal the instant the warning light goes out. Close the bleed screw and release the brake pedal. If excessive pressure is applied to the pedal, the warning lamp will go out and then come on again as the spool moves past the center position. The original system will then have to have a bleeder screw opened and the pressure applied to move the spool back again.

Before using the vehicle, be sure that a firm pedal pressure is obtained.

11 : 11 The parking brake

On the 1964 to 1966 models a hand operated parking brake is fitted. The brake is applied by pulling on the handle in the vehicle, and released by turning the handle clockwise and allowing it to return. The linkage is shown in **FIG 11 : 20.** To adjust the cables, pull the handle out by three notches and raise the rear of the vehicle off the ground so that the rear wheels are free. Slacken the nuts on the equalizer rod and screw them forwards along the rod, so that the rear nut pushes the equalizer forwards, until moderate drag is felt on the rear wheels when they are turned in a forward direction. Tighten back the front nut to the equalizer. Release the handle and check that both rear wheels rotate freely, allowing for the drag of the rear axle.

On later models, a foot-operated parking brake is fitted. The brake is applied by pressing down on the parking brake pedal and it is released by pulling on the release handle. A typical linkage is shown in **FIG 11 : 21.**

To adjust the later linkage, jack up the rear of the vehicle and fit an adaptor to each rear wheel, as shown in **FIG 11:22**.

Check that the wheels rotate freely, allowing for the drag of the rear axle, when the parking control is fully released. Apply the parking brake pedal to the first notch. Slacken the nuts on the equalizer, shown in **FIG 11:21**, and wind the rear nut forwards to tension the rear cables. Use a torque spanner to check the force required to rotate the rear wheels. The adjustment at the equalizer is correct when a torque of 100 lb ft at breakaway. Tighten the front nut at the equalizer to hold the adjustment.

Release the parking brake and check that the rear wheels rotate freely. Depress the parking brake pedal to the third notch and check that this holds the vehicle. Release the brake and check that the brake shoes return fully. If the shoes do not fully release, check the cables for bends or kinks.

11:12 Trouble shooting guide

(a) Spongy pedal

1 Air in the hydraulic system
2 Fluid leak in the hydraulic system
3 Gap between lining and shoes on drum brakes
4 Excessive hub end play on disc brakes

(b) Excessive pedal movement

1 Check 1, 2 and 4 in (a)
2 Excessively worn friction linings
3 Defective automatic adjusters
4 Very low fluid level in reservoir

(c) Brakes grab or pull to one side

1 Wet or oily friction linings
2 Distorted or damaged brake rotor
3 Excessive rotor runout
4 Seized piston in wheel cylinder or caliper
5 Seized parking brake cable
6 Mixed linings of different grades
7 Uneven tire pressures or wear
8 Defective steering or suspension

Inches	Decimals	Millimetres	Inches to Millimetres: Inches	Inches to Millimetres: mm	Millimetres to Inches: mm	Millimetres to Inches: Inches
1/64	.015625	.3969	.001	.0254	.01	.00039
1/32	.03125	.7937	.002	.0508	.02	.00079
3/64	.046875	1.1906	.003	.0762	.03	.00118
1/16	.0625	1.5875	.004	.1016	.04	.00157
5/64	.078125	1.9844	.005	.1270	.05	.00197
3/32	.09375	2.3812	.006	.1524	.06	.00236
7/64	.109375	2.7781	.007	.1778	.07	.00276
1/8	.125	3.1750	.008	.2032	.08	.00315
9/64	.140625	3.5719	.009	.2286	.09	.00354
5/32	.15625	3.9687	.01	.254	.1	.00394
11/64	.171875	4.3656	.02	.508	.2	.00787
3/16	.1875	4.7625	.03	.762	.3	.01181
13/64	.203125	5·1594	.04	1.016	.4	.01575
7/32	.21875	5.5562	.05	1.270	.5	.01969
15/64	.234375	5.9531	.06	1.524	.6	.02362
1/4	.25	6.3500	.07	1.778	.7	.02756
17/64	.265625	6.7469	.08	2.032	.8	.03150
9/32	.28125	7.1437	.09	2.286	.9	.03543
19/64	.296875	7.5406	.1	2.54	1	.03937
5/16	.3125	7.9375	.2	5.08	2	.07874
21/64	.328125	8.3344	.3	7.62	3	.11811
11/32	.34375	8.7312	.4	10.16	4	.15748
23/64	.359375	9.1281	.5	12.70	5	.19685
3/8	.375	9.5250	.6	15.24	6	.23622
25/64	.390625	9.9219	.7	17.78	7	.27559
13/32	.40625	10.3187	.8	20.32	8	.31496
27/64	.421875	10.7156	.9	22.86	9	.35433
7/16	.4375	11.1125	1	25.4	10	.39370
29/64	.453125	11.5094	2	50.8	11	.43307
15/32	.46875	11.9062	3	76.2	12	.47244
31/64	.484375	12.3031	4	101.6	13	.51181
1/2	.5	12.7000	5	127.0	14	.55118
33/64	.515625	13.0969	6	152.4	15	.59055
17/32	.53125	13.4937	7	177.8	16	.62992
35/64	.546875	13.8906	8	203.2	17	.66929
9/16	.5625	14.2875	9	228.6	18	.70866
37/64	.578125	14.6844	10	254.0	19	.74803
19/32	.59375	15.0812	11	279.4	20	.78740
39/64	.609375	15.4781	12	304.8	21	.82677
5/8	.625	15.8750	13	330.2	22	.86614
41/64	.640625	16.2719	14	355.6	23	.90551
21/32	.65625	16.6687	15	381.0	24	.94488
43/64	.671875	17.0656	16	406.4	25	.98425
11/16	.6875	17.4625	17	431.8	26	1.02362
45/64	.703125	17.8594	18	457.2	27	1.06299
23/32	.71875	18.2562	19	482.6	28	1.10236
47/64	.734375	18.6531	20	508.0	29	1.14173
3/4	.75	19.0500	21	533.4	30	1.18110
49/64	.765625	19.4469	22	558.8	31	1.22047
25/32	.78125	19.8437	23	584.2	32	1.25984
51/64	.796875	20.2406	24	609.6	33	1.29921
13/16	.8125	20.6375	25	635.0	34	1.33858
53/64	.828125	21.0344	26	660.4	35	1.37795
27/32	.84375	21.4312	27	685.8	36	1.41732
55/64	.859375	21.8281	28	711.2	37	1.4567
7/8	.875	22.2250	29	736.6	38	1.4961
57/64	.890625	22.6219	30	762.0	39	1.5354
29/32	.90625	23.0187	31	787.4	40	1.5748
59/64	.921875	23.4156	32	812.8	41	1.6142
15/16	.9375	23.8125	33	838.2	42	1.6535
61/64	.953125	24.2094	34	863.6	43	1.6929
31/32	.96875	24.6062	35	889.0	44	1.7323
63/64	.984375	25.0031	36	914.4	45	1.7717

UNITS	Pints to Litres	Gallons to Litres	Litres to Pints	Litres to Gallons	Miles to Kilometres	Kilometres to Miles	Lbs. per sq. In. to Kg. per sq. Cm.	Kg. per sq. Cm. to Lbs. per sq. In.
1	.57	4.55	1.76	.22	1.61	.62	.07	14.22
2	1.14	9.09	3.52	.44	3.22	1.24	.14	28.50
3	1.70	13.64	5.28	.66	4.83	1.86	.21	42.67
4	2.27	18.18	7.04	.88	6.44	2.49	.28	56.89
5	2.84	22.73	8.80	1.10	8.05	3.11	.35	71.12
6	3.41	27.28	10.56	1.32	9.66	3.73	.42	85.34
7	3.98	31.82	12.32	1.54	11.27	4.35	.49	99.56
8	4.55	36.37	14.08	1.76	12.88	4.97	.56	113.79
9		40.91	15.84	1.98	14.48	5.59	.63	128.00
10		45.46	17.60	2.20	16.09	6.21	.70	142.23
20				4.40	32.19	12.43	1.41	284.47
30				6.60	48.28	18.64	2.11	426.70
40				8.80	64.37	24.85		
50					80.47	31.07		
60					96.56	37.28		
70					112.65	43.50		
80					128.75	49.71		
90					144.84	55.92		
100					160.93	62.14		

UNITS	Lb ft to kgm	Kgm to lb ft	UNITS	Lb ft to kgm	Kgm to lb ft
1	.138	7.233	7	.967	50.631
2	.276	14.466	8	1.106	57.864
3	.414	21.699	9	1.244	65.097
4	.553	28.932	10	1.382	72.330
5	.691	36.165	20	2.765	144.660
6	.829	43.398	30	4.147	216.990

CHAPTER 12

THE ELECTRICAL SYSTEM

12:1 Description

A 12-volt electrical system is fitted on all models covered by this manual. The correct polarity must be observed otherwise some of the components, such as alternator and radio, may be irreparably damaged.

Power for the system is produced by an alternator driven by a belt from the engine and the battery acts as a storage. A standard generator is fitted to 1964 models. The battery supplies the very heavy current required by the starter motor for starting the engine and it also supplies current to the accessories when the engine is not running. A lead/acid battery is used. It is essential that the charging circuit is disconnected when boost-charging the battery or carrying out arc welding repairs to the frame or bodywork.

A test lamp and leads, or a 0-20 voltmeter, may be used to test the continuity of a circuit, but cheap instruments must not be used for checking the performance of, or adjusting components in, the electrical system. Cheap instruments will not measure to the accuracy required and for such tests high-grade, preferably moving-coil, instruments should be used.

Typical schematic wiring diagrams are given in **Technical Data** at the end of the manual to enable those with electrical experience to trace and rectify wiring faults.

Instructions for servicing electrical components are given in this chapter but it must be pointed out that it is a waste of time and money to try to repair items which are seriously defective, either mechanically or electrically. Such items should either be exchanged for new or reconditioned units, or be discarded and new units fitted in their place.

12:2 The battery

Regular battery maintenance is essential for long battery life and to ensure satisfactory engine starting. If the battery is neglected, it will rapidly fail and, as well as producing starting difficulties, the performance of the whole electrical system will suffer.

In cold weather, the battery loses efficiency and at the same time more demands are made on it as the engine is stiffer to turn and the accessories are used more often. A battery that has limped through the summer will not be able to cope with the winter. If the battery is satisfactory, efficiency can be restored in extreme cold weather by removing the battery and storing it in a warm room overnight.

The alternator produces a higher charging current at all speeds compared with a conventional generator, but if the vehicle is used for short-mileage journeys only in

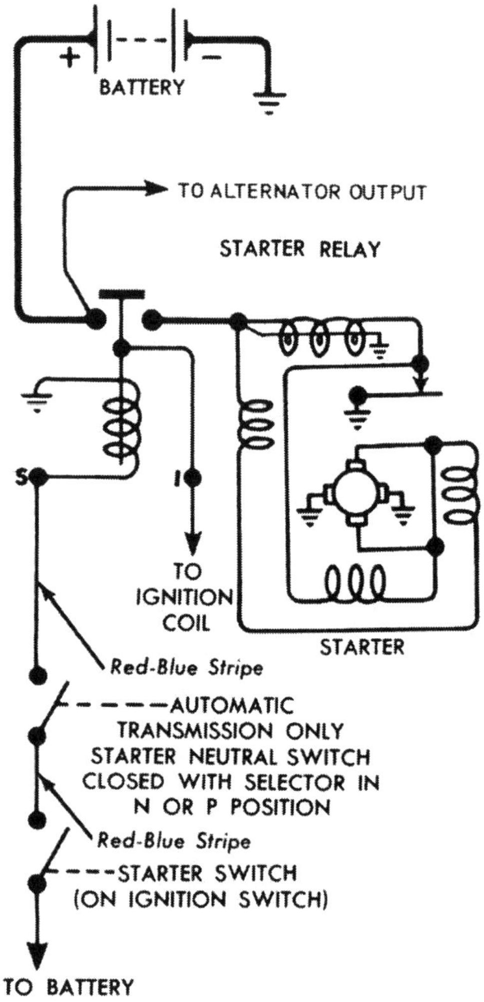

FIG 12:1 The starter circuit

winter, a boost charge from a trickle charger at monthly intervals will keep the battery at full charge and make starting easier.

Maintenance:

Always keep the top of the battery clean and dry. Wipe away any spillage when refilling the cells with distilled water and at regular intervals wipe away any dirt or moisture from the top.

If spillage has caused corrosion to form on the battery surrounds, remove the battery and clean the area. Wash with a dilute solution of ammonia or baking powder and then flush liberally with clean water. When the parts are dry, paint them with anti-sulphuric paint.

Keep the battery terminal connectors clean and tight. High resistance at this point is often a cause of difficult starting. Disconnect the connectors. Use special wire brushes to clean the connectors and battery terminal posts. Corrosion may be washed off using dilute ammonia followed by flushing with clean water. Smear the connectors and posts with a non-metallic grease (such as petroleum jelly) to prevent further corrosion and make sure that the connectors are tight on the posts.

At regular intervals, approximately monthly, check the electrolyte level in the cells. Remove the cell vent caps and top up with pure distilled water until the level is just over the ring in the filler well. Do not inspect the level with a naked light and do not add anything but pure distilled water. Ready-mixed electrolyte should only be added to replace spillage or leakage.

Testing:

A hydrometer provides the easiest method available to the owner of testing the battery. If it is suspected that the battery or starter circuit is defective, take the vehicle

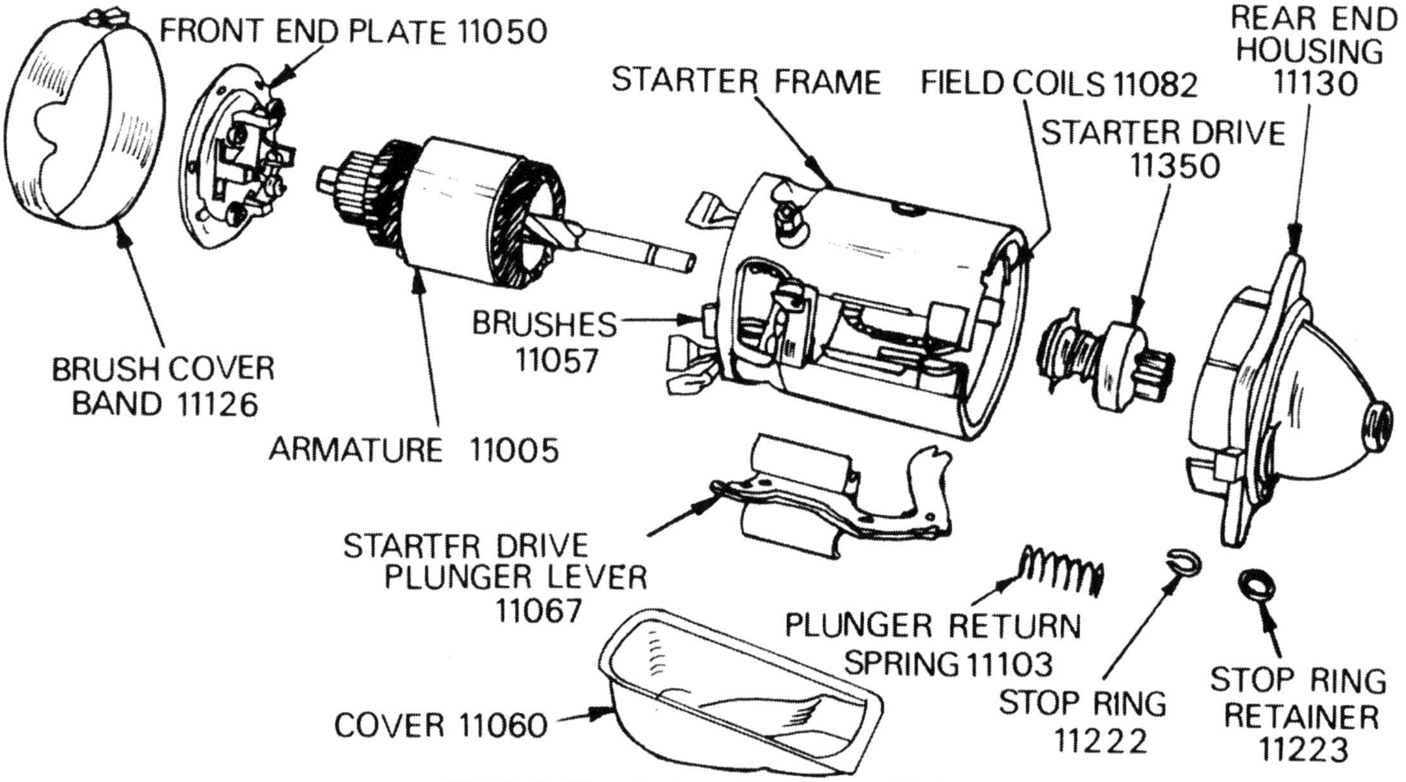

FIG 12:2 The starter motor disassembled

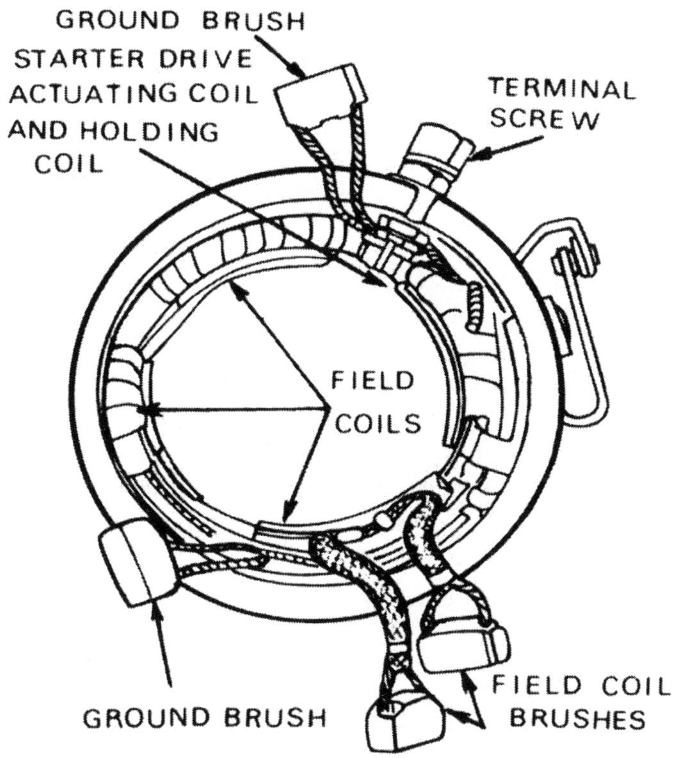

FIG 12:3 The starter motor field coil assembly

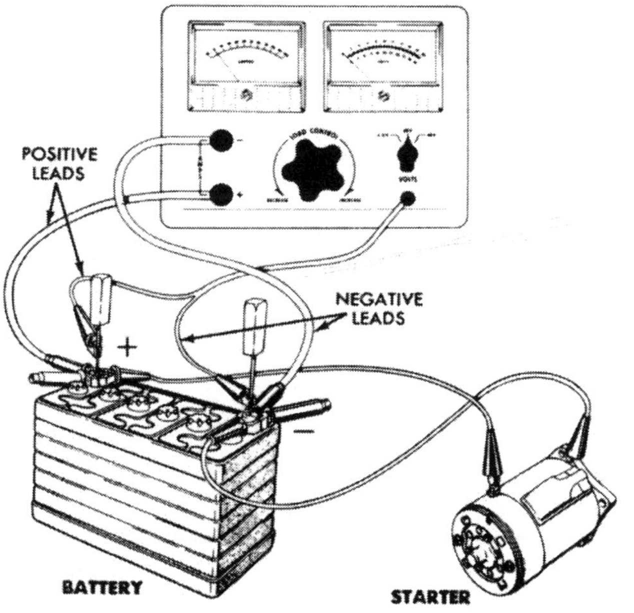

FIG 12:4 The starter motor no-load bench test

to a service garage where they will check the battery under heavy discharge and also measure the voltage drop in the heavy duty starter motor leads.

Draw sufficient electrolyte up from each cell in turn to ensure that the float of the instrument is clear of the sides and bottom. Take the reading at eye level, check the color of the electrolyte, and return it to its original cell before testing the next cell. If the electrolyte appears cloudy or filled with specks, the cell is probably defective and this will be confirmed by the reading of that cell varying by more than .050 from the readings of the other cells.

If the level in the cell is so low that electrolyte cannot be drawn up into the instrument, top up with distilled water and charge the battery for ½ hour at 35 amps to thoroughly mix the electrolyte.

The readings of the float give the following indications:

For climates below 32°C (90°F):

1.270 to 1.290	Cell fully charged
1.190 to 1.210	Cell half-charged
1.110 to 1.130	Cell discharged

Replace spillage with electrolyte of 1.270 specific gravity.

For climates above 32°C (90°F):

1.120 to 1.230	Cell fully charged
1.130 to 1.150	Cell half-charged
1.050 to 1.070	Cell discharged

Replace spillage with electrolyte of 1.120 specific gravity.

The figures are given assuming a standard electrolyte temperature of 16°C (60°F). To convert the actual reading to standard, add .002 for every 3°C (5°F) increment above standard temperature and subtract for every increment below.

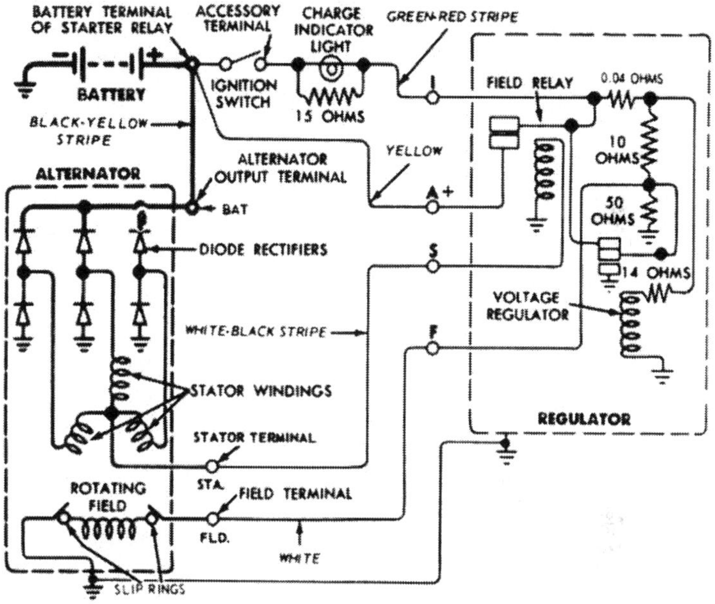

FIG 12:5 Alternator system with indicator light

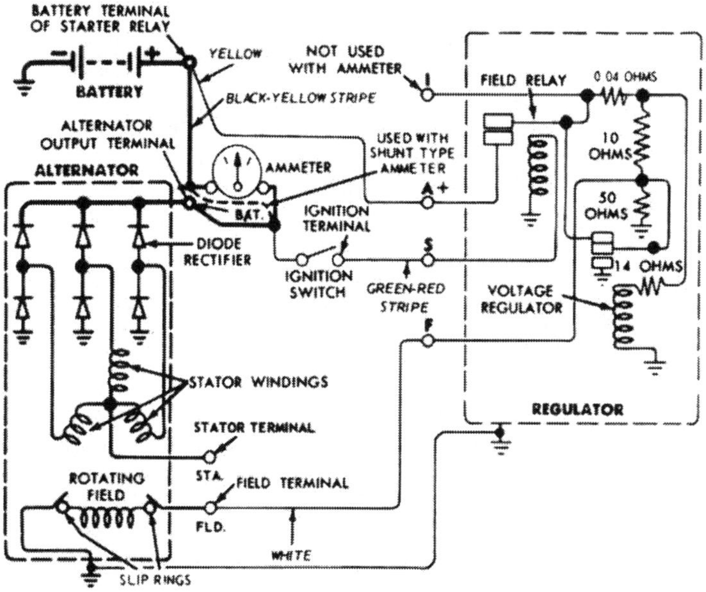

FIG 12:6 Alternator system with ammeter

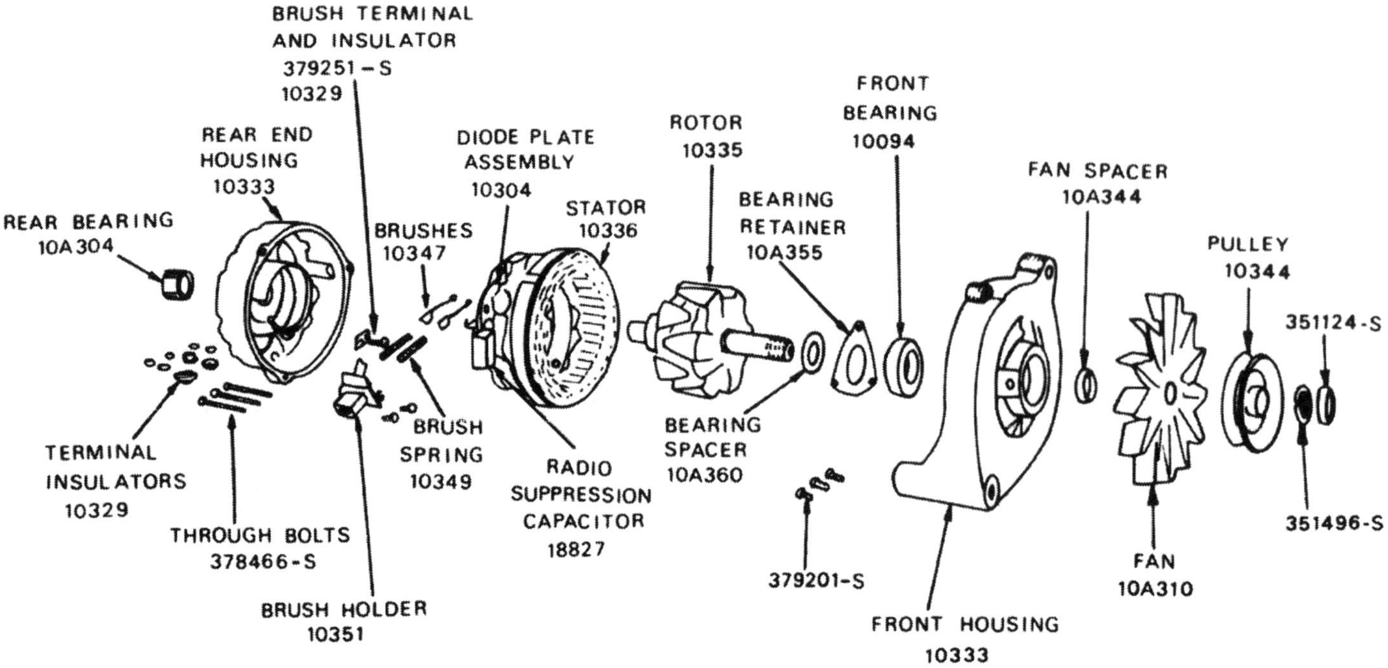

FIG 12:7 The alternator disassembled

Charging:

The maximum charging current and period for the various battery capacities and state of charge are given in the following table:

Specific gravity reading	*Charge rate amperes*	*Battery capacity—ampere hours* 45	55	70	80	85
1.125 to 1.150 (1)	35	65 min.	80 min.	100 min.	115 min.	125 min.
1.150 to 1.175	35	50 min.	65 min.	80 min.	95 min.	105 min.
1.175 to 1.200	35	40 min.	50 min.	60 min.	70 min.	75 min.
1.200 to 1.225	35	30 min.	35 min.	45 min.	50 min.	55 min.
Above 1 225	5	(2)	(2)	(2)	(2)	(2)

(1) If the specific gravity is below 1.125, use the indicated high rate of charge for the 1.125 specific gravity, then charge at 5 amperes until the specific gravity reaches 1.250 at 80°F.
Charge at 5 ampere rate only until the specfic gravity reaches 1.250 at 80°F.

(2) At no time during the charging operation should the electrolyte temperature exceed 130°F.

(2) (1)

The rates given are the maximum and no harm will be done if the battery is charged at a low rate for a correspondingly longer time.

A battery is fully charged when it is gassing freely and the specific gravity does not rise in three consecutive readings taken at hourly intervals. **The gases given off by the battery when charging are explosive so do not bring naked lights near the battery and avoid sparks when installing a newly charged battery.**

A poor battery may sometimes be reclaimed partially by giving it three or four cycles of discharge and charge. Discharge the battery down to a cell voltage of 1.8 volts by connecting it to a lamp bank taking approximately 5 amps (approximately 60 watts). Recharge the battery at a 5 amp rate and then repeat the cycle. At the end of a few cycles charge the battery up ready for use.

Storage:

Short term storage of a well maintained battery presents no problems.

If the battery is to be stored for long periods, be sure that it is fully charged, the top dry, and all exposed metal well covered with non-metallic grease. Store the battery in a cool dry room well away from any temperature extremes.

At monthly intervals give the battery a freshening-up charge, and at intervals of three months discharge it on a lamp bank and fully recharge it. If this precaution is not taken, the battery will slowly self-discharge and the plates sulphate up.

11:4 Servicing electrical motors

All electrical motors for automobiles follow the same general design principles and therefore there are many similar constructional features. To save repetition, general instructions for servicing motors are gathered into this section. It should be noted that some motors may be sealed on manufacture and therefore cannot be serviced but must be renewed if defective.

Brushgear:

Dismantle the motor sufficiently to expose the brushes. If the armature has to be removed, or the brush holder assembly pulled off the commutator, take great care not to allow the brushes to be damaged on the edge of the commutator.

Check the brushes for wear and renew them if they are excessively worn. Check that the brushes move freely in their holders and that the springs press them firmly down onto the commutator. Renew the springs if they are weak or damaged. If the brushes stick, remove them from their holders and lightly polish the sides of the brushes with a fine file. Clean the brush holders with a piece of cloth dipped into fuel. Install the brushes into their original positions so that the bedding-in is not altered.

Brushes are usually supplied with their ends ground to shape but if further bedding-in is required, make up a wooden mandrel the same diameter as the commutator. The commutator itself may be used. Wrap the mandrel with glasspaper (never emerycloth as this leaves particles embedded in the copper) and rotate the mandrel with the brushes lightly pressed against the glasspaper.

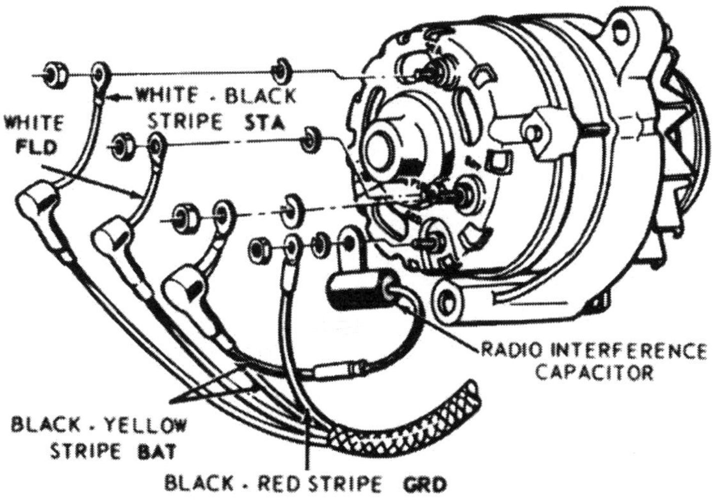

FIG 12:8 Typical early alternator wiring connections

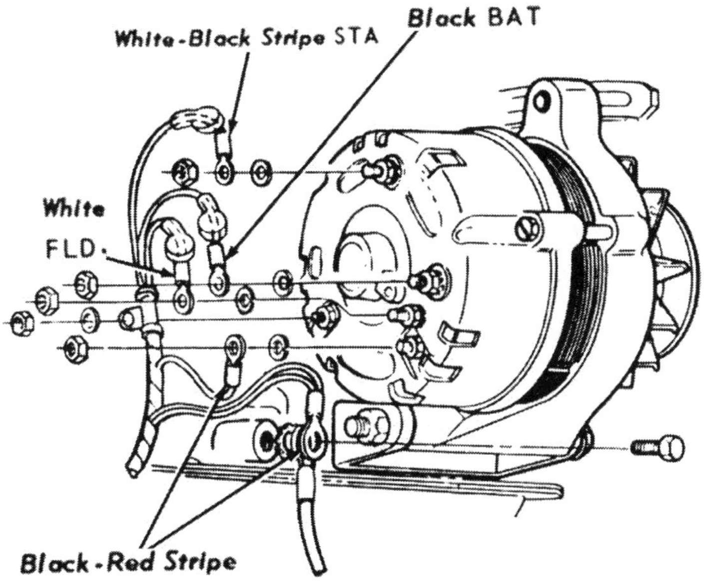

FIG 12:9 Typical late alternator wiring connections

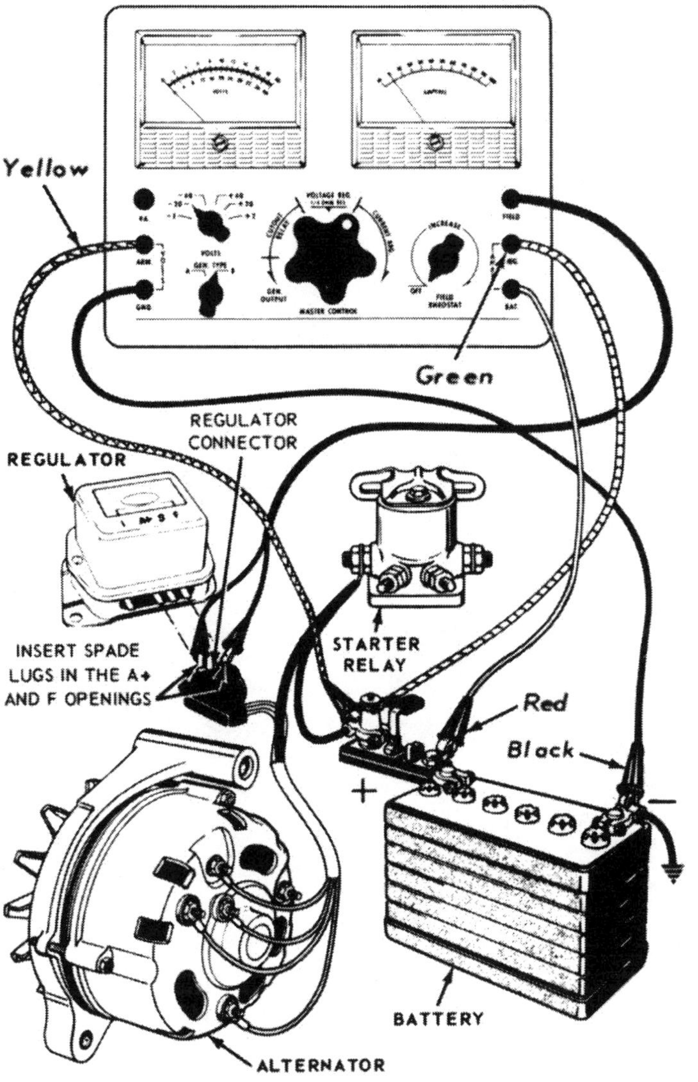

FIG 12:10 Alternator output test

When new brushes are to be attached by soldering, grip the connector with a pair of needle-nosed pliers to act as a heat sink and prevent the solder from creeping up the connector and making it stiff.

Commutator:

Normally the commutator should have a darkened surface that is polished and free from scores or burn marks. Cleaning can be carried out by wiping the commutator with a piece of lint-free cloth dipped in denatured alcohol or gasoline.

Light burn marks or scores can be polished out using a strip of fine-grade glasspaper. Never use emerycloth as this leaves hard particles embedded in the copper.

If the damage is deeper, it may be machined off in a lathe. Use the highest speed possible and a very sharp tool and remove only sufficient stock to take out the damage. On some motors the insulation between the segments is undercut. If undercutting is required, either use a special tool or grind a hacksaw blade to the exact width of the insulation and use it to undercut the insulation squarely to a depth of $\frac{1}{32}$ inch. Take a light final cut using a diamond or carbide-tipped tool and clean off metal dust before installing the armature. If a diamond-tipped tool is not available, the commutator may be polished instead, using a very fine grade of glasspaper.

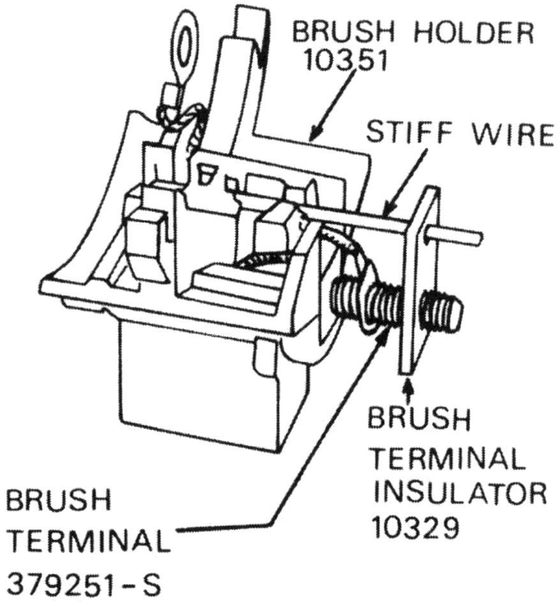

FIG 12:11 Alternator brush holder assembly

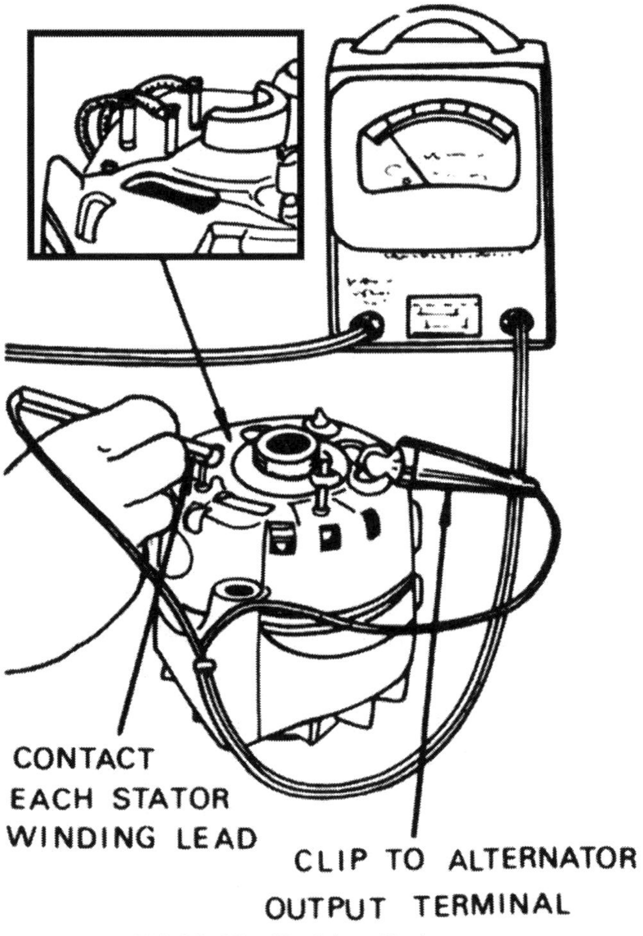

FIG 12:12 Positive diode test

Armature:

Apart from skimming the commutator, very little can be done to the armature apart from the visual examination. A service garage can check the windings for short-circuits and these can be suspected if indivudal commutator segments are badly burnt.

Check the laminations for scoring or similar damage. Possible causes of scoring are loose polepieces, excessively worn bearings, or a bent armature shaft. **Do not attempt to machine or straighten the armature.**

Field coils:

On some smaller motors a permanent magnet may be fitted in place of field coils. Care must be taken in dismantling such motors as the armature will be drawn out with the cover, with possible damage to the brushes.

The field coils may be tested for continuity by connecting a test lamp and battery across them. The resistance can be measured using a suitable resistance meter. It should be noted that the resistance can also be measured using a 12-volt battery and ammeter. Measure the current flow and use Ohm's law to calculate the resistance (Resistance=Voltage divided by Current).

The field coils on the starter motor are held in place by the polepieces and the polepieces in turn are secured by special screws. These screws must only be removed and installed using a special machine screwdriver and therefore if the field coils are defective renewal should be left to a service garage.

Insulation:

Wipe away all carbon and metal dust. If necessary use a cloth moistened with denatured alcohol or gasoline. The insulation may be checked using a 12-volt battery and test lamp. The lamp should not light when connected across the insulation. A better method is to use a 110 AC supply and neon test bulb.

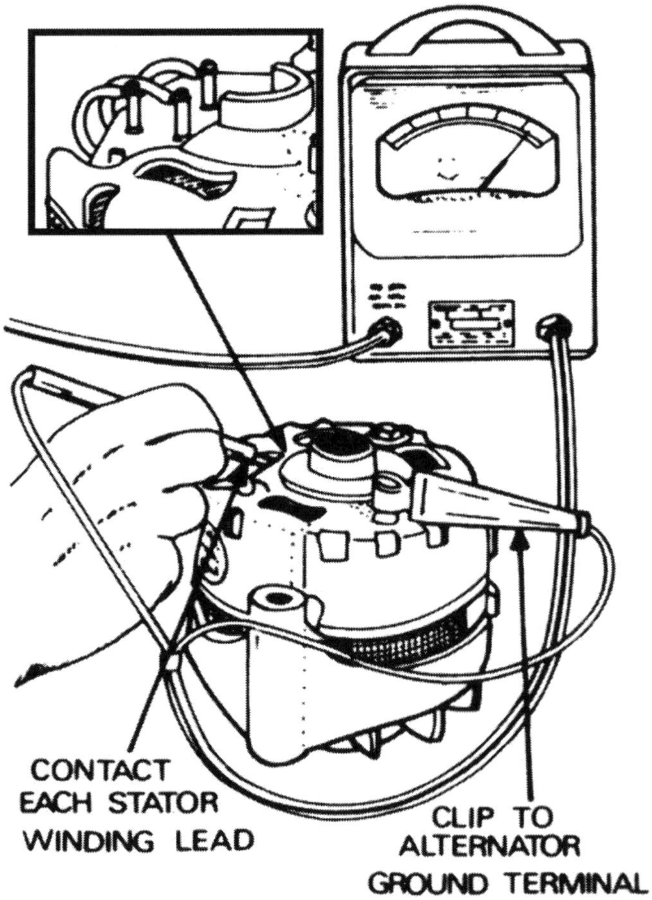

FIG 12:13 Negative diode test

12:4 The starter motor

The starting circuit is shown in **FIG 12:1** and the components of the motor in **FIG 12:2**. The starter relay is only energized when the ignition switch is turned to the START position and on models fitted with automatic transmission the selector is in the N or P position. When the relay is energized the coils draw in the plunger and close the heavy duty contacts, allowing current to flow to the starter motor. One of the field coils is connected directly to earth and therefore a heavy current will flow through this coil. The coil pulls across the moveable pole shoe and this presses the drive pinion into engagement with the teeth on the flywheel ring gear. When the moveable polepiece is fully seated, the linkage opens the contacts which ground the field coil. A holding coil keeps the moveable polepiece in position, against the pressure of its return spring, while current is flowing through the motor.

Starter fails to operate or operates sluggishly:

1 Check the condition and charge of the battery, paying special attention to the connectors and battery posts. Check the grounding points for corrosion or dirt that will cause a high resistance. If need be, take the vehicle to a service garage for a voltage-drop check on the starter circuit.

2 The starter may fail to operate because the pinion is jammed in mesh. On a vehicle fitted with manual transmission, select a gear and rock the vehicle backwards and forwards. If this fails, remove the cover over the moveable pole and try to tap the pinion out of engagement. If all fails, remove the starter motor for further checks and repair. Worn teeth on pinion and ring gear can cause the pinion to jam in mesh.

3 If the motor fails to operate at all, switch on some lights that can be seen from the drivers seat. Select Start on the ignition switch and if the starter motor is taking current the lights will go dim. If the lights do not go dim then there is a fault in the wiring, starter relay, switch, or starter motor itself. Trace through using a test lamp or voltmeter to find the point at which current is no longer available. If current is available up to the starter relay short across the heavy duty terminals of the relay with a thick piece of metal, and if the starter motor operates correctly then the contacts are burnt and a new relay should be fitted. If current is reaching the starter motor but it fails to operate, the motor must be removed for further examination.

Removal:

Disconnect the battery and raise the front of the vehicle. Disconnect the starter cable at the starter motor terminal. Remove the starter motor attachment bolts, withdraw the motor and lower it down and out of the vehicle.

Install the motor in the reverse order of removal. Hold the motor firmly in place while snugging up the bolts, to ensure that the motor is mounted squarely against its mounting surface.

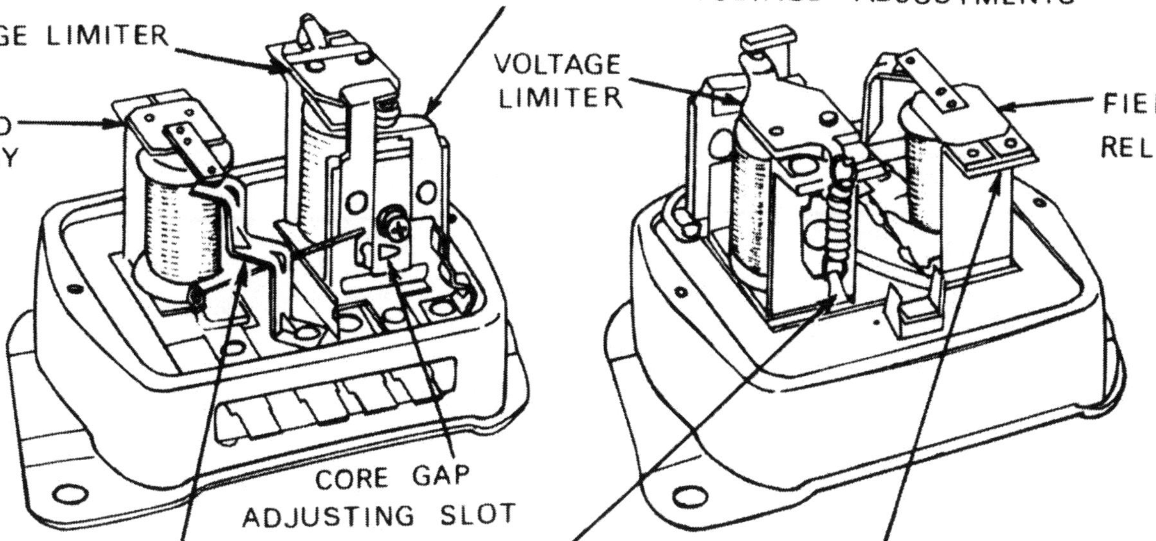

FIG 12:14 Typical regulator adjustments

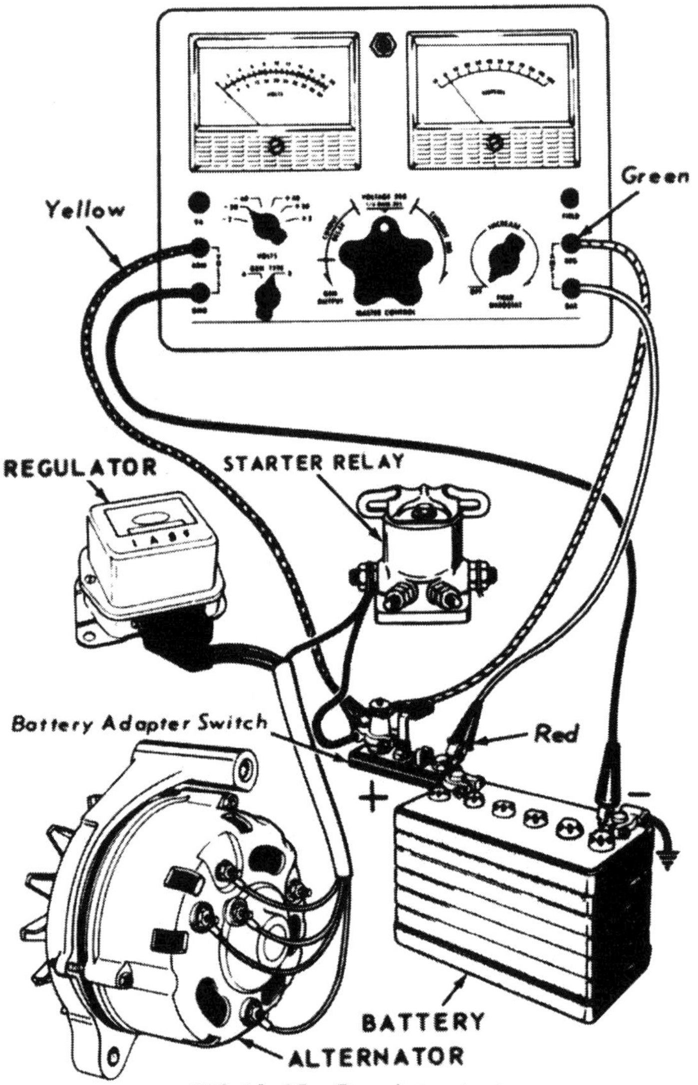

FIG 12:15 Regulator test

Disassembly:

The parts of the starter motor are shown in **FIG 12:2.** Slacken off the brush coverband retaining screw and remove the brush cover and starter drive cover. Note the positions of the brushes and remove them from the brush holders. Remove the throughbolts (not shown in the figure) and take off the drive end housing and plunger return spring. If needle roller bearings are fitted, insert a dummy shaft into the bearing to keep the needle rollers in place. Drive out the pivot pin for the starter drive plunger lever and remove the lever as well as the armature. Remove the brush end plate from the starter frame.

If the starter drive requires removal, remove the stop ring retainer so that the stop ring can be taken off and discarded.

The motor is assembled in the reverse order of disassembly. Lubricate the armature shaft splines with a thin coating of Lubriplate 777 and apply a little grease to needle roller bearings. Note that a fiber thrust washer is fitted to the armature shaft between the commutator and the starter frame. The brushes fitted are shown in **FIG 12:3.** Note they must all be removed from their brush holders before removing the brush end plate.

Starter drive:

Check that the starter drive moves freely along the splines of the armature shaft. Dirt will cause the drive to jam. The teeth must mesh to a depth of at least $\frac{1}{2}$ inch to ensure trouble-free operation. Light damage to the teeth may be smoothed off with an oil stone or smooth file but if the teeth are milled, worn, or broken a new drive must be fitted. Examine the teeth on the ring gear through the starter motor mounting hole while slowly turning the engine.

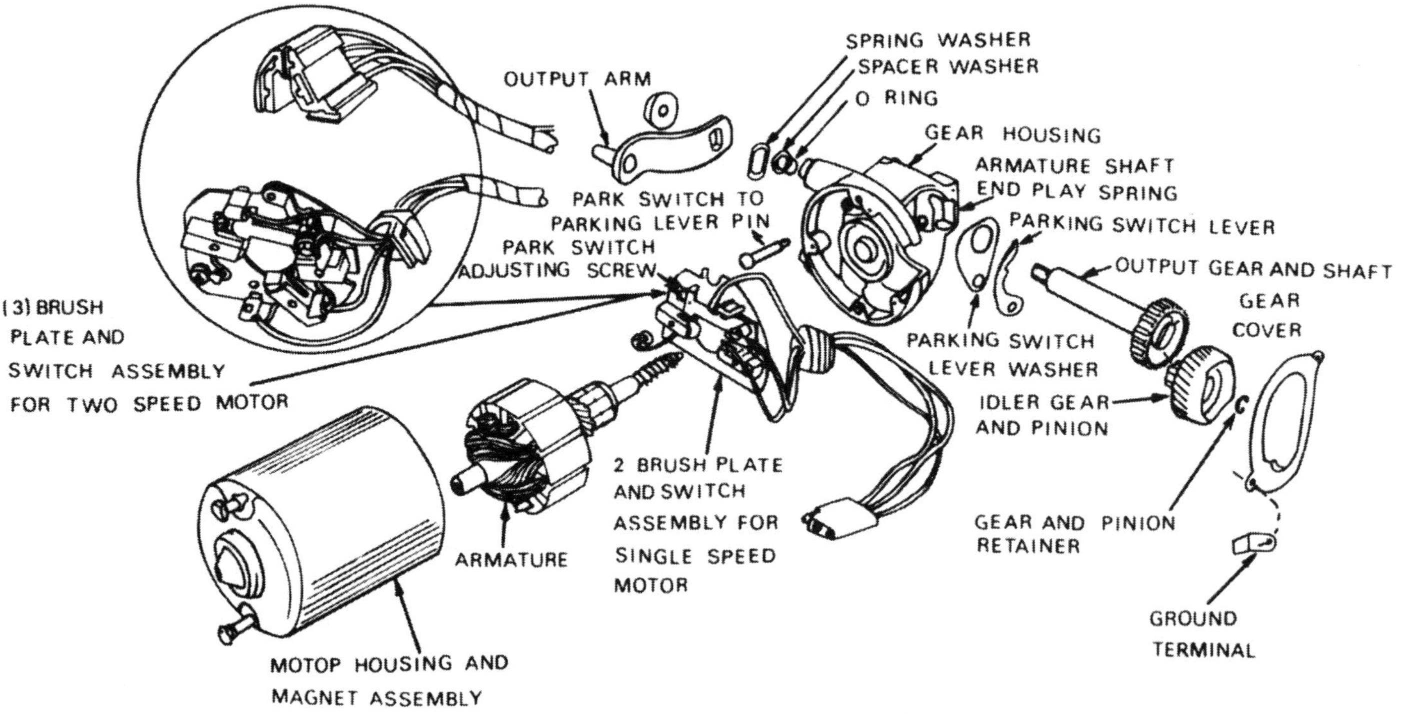

FIG 12:16 Permanent magnet type windshield wiper motor

To check the drive with the motor fitted, flood the engine by pumping the accelerator cable 8 to 10 times. Set the ignition switch to Start and hold it there. The engine should fire but not start while the starter motor should continue to crank the engine. If possible, remove the cover and observe the operation of the drive plunger. If the engine stops turning and the engine spins at high speed then there is a fault in the drive.

Starter motor tests:

The circuit for testing the motor on the bench is shown in **FIG 12:4.** The starter motor should be firmly held down as it will kick vigorously on start-up. With the battery connected as shown the starter motor should spin freely and at high speed. Heavy duty leads must be used because even on the no-load condition the starter motor will take 70 amps at least.

12:5 The alternator

Typical alternator charging circuits are shown in **FIGS 12:5** and **12:6.** Typical alternator components are shown in **FIG 12:7.**

The alternator produces an AC current. Heavy stator windings carry the main current, and as these are stationary no allowance need be made for centrifugal forces, so heavy wiring can be safely used. The rotor carries the light excitation current. The current does not require to change directions and the varying magnetic fields are set up by the polepieces, which give alternate North and South poles. The rotor windings are therefore light and the rotor can be constructed for high rotational speeds. Slip rings and light brushes carry the excitation current to and from the rotor windings and these are far more reliable than the commutator and brushes on a conventional generator, which not only carry a heavy current but also have to change the direction of flow.

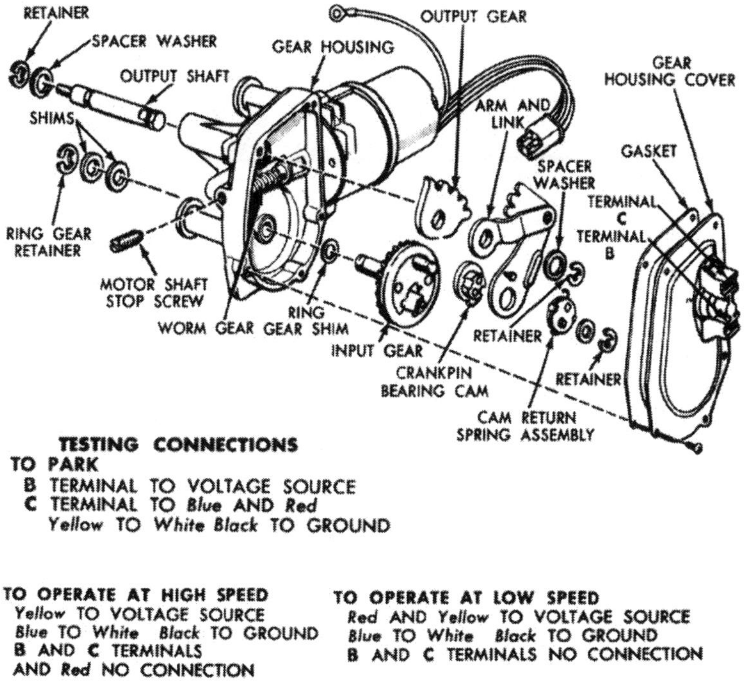

FIG 12:17 Two-speed cranked wiper motor

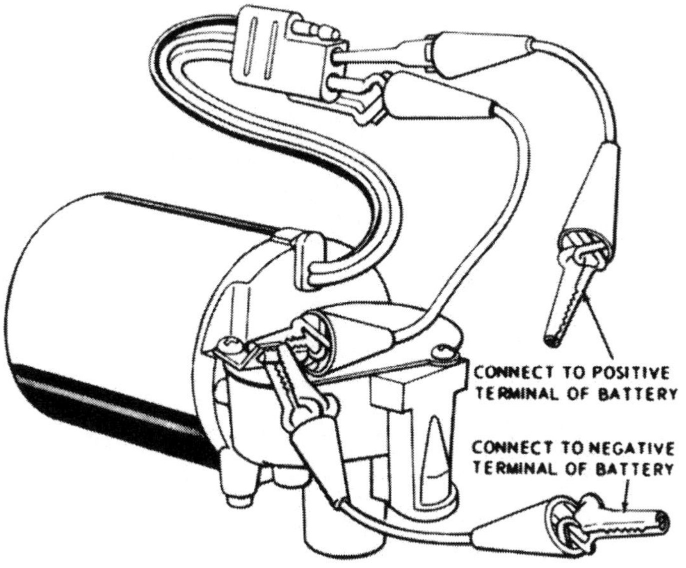

FIG 12:18 Wiper motor park switch test

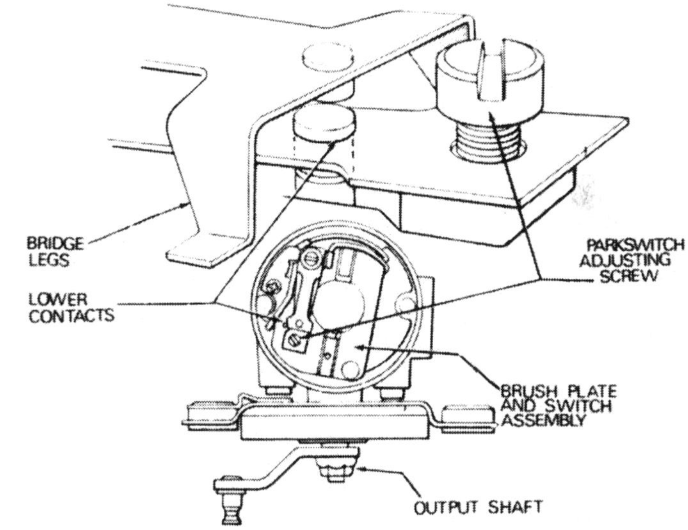

FIG 12:19 Wiper motor park switch adjustment

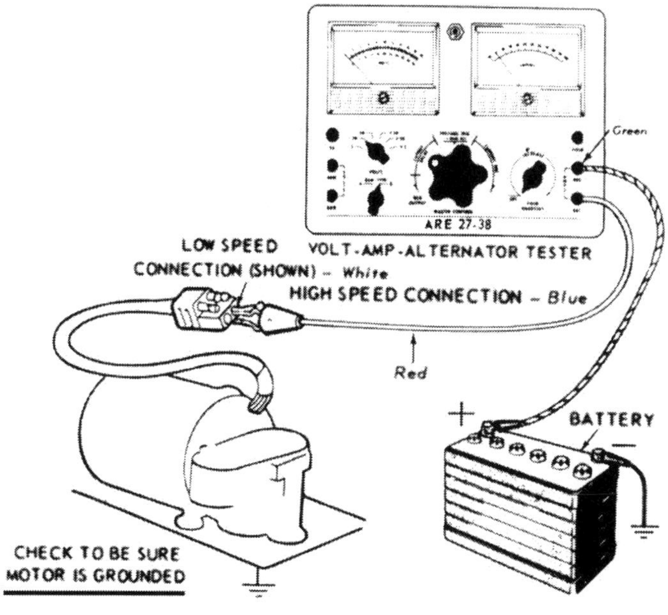

FIG 12:20 Wiper motor test running circuit

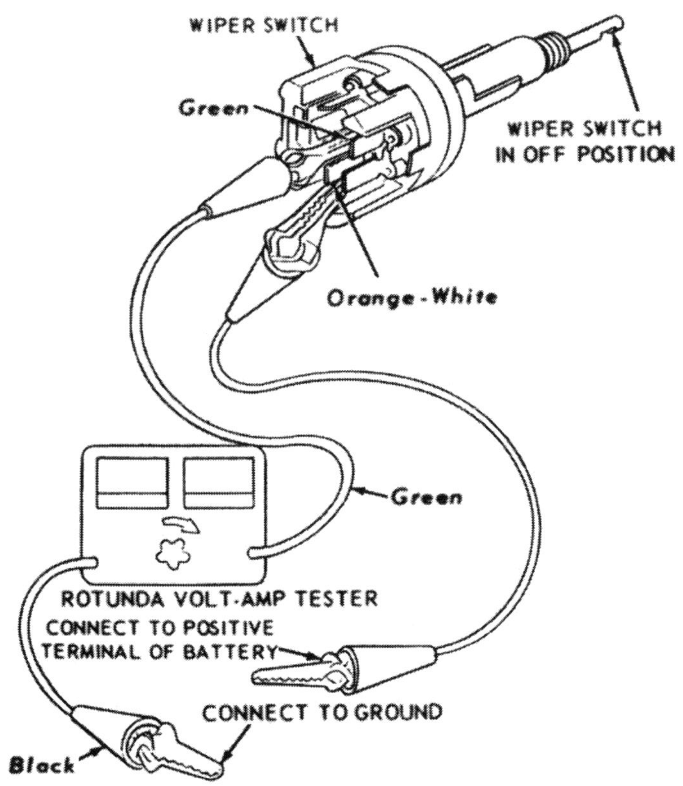

FIG 12:21 Wiper switch circuit breaker test

Diode rectifiers change the AC current produced by the alternator to a DC current acceptable to the battery. The diodes also prevent a reverse current from flowing, so no cut-out is required. A separate regulator limits the maximum voltage to the demands of the system. It should be noted that a defective diode will lower the maximum current output by approximately 10 amps and cause the alternator to operate with a whining noise.

Alternator fails to charge:

1 Check the drive belt tension (see **Chapter 4, Section 4:4**) and be sure that the belt is not excessively glazed or slipping.
2 Be sure that all the connections are clean and tight. **The charging circuit connections must not be made or broken while the engine is running.** Typical earlier alternator connections are shown in **FIG 12:8** and those on later models in **FIG 12:9**. It should be noted that if the alternator is removed from the engine, the battery must be disconnected.
3 A typical output test circuit is shown in **FIG 12:10**. If an ammeter and voltmeter are used instead of the proper test equipment, apply load to the alternator by switching on the headlamps. The engine should be run for at least 20 minutes to allow the alternator to reach its normal operating temperature. Adjust the load until the voltmeter reads 15 volts with the engine running at 2500 rev/min. The alternator should then be delivering its rated output.

Disassembly and checking:

Remove the alternator from the engine, after slackening the drive belt and disconnecting the battery.

Mark both end housings and stator with light aligning scribe marks. Take out the three through-bolts and remove the front end housing and rotor from the stator and rear end housing. Take off the nuts and washers so that the rear end housing can be removed. Take out the brush holder mounting screws and remove the brush holder, shown in **FIG 12:11**.

Brush gear and slip rings:

Check the brushes for wear and renew them if they are short. Check the springs and renew the assembly if the springs are weak.

The slip rings should only require cleaning with a fine grade of glasspaper or cloth dipped in denatured alcohol. They may be machined down if badly worn or burnt but do not go beyond the minimum diameter of 1.22 inch.

Rotor and stator:

Check the continuity of the stator, using a test lamp and 12-volt battery between any pair of stator wires, noting that the wires must be unsoldered from the diode assembly. Again check for continuity between one of the wires tested and the third wire.

Measure the resistance of the field coils between the slip rings. Wipe off dirt from both stator and rotor using a rag moistened with denatured alcohol. Check the insulation using a 110 AC supply and neon test lamp.

Diode tests:

The test circuit for the diodes is shown in **FIGS 12:12** and **12:13**. A test lamp may be used in place of the meter. Press the probe firmly through the varnish to make good contact. With the polarity in one direction the meter should read or the test lamp illuminate, but with the polarity reversed there should be no reading. No reading in either direction or a reading in both directions indicates a faulty diode.

Voltage regulator:

A typical regulator is shown in **FIG 12:14**. Dirt on the contacts can be a cause of trouble. If the contacts require cleaning, take a sheet of 400 grade silicone carbide paper and partially wear it by folding and rubbing together. Fold a strip so that the abrasive is outwards and clean the contacts by pulling the strip between them. Blow away all loose dust and dirt. If the contact gaps are being set, use hospital clean feeler gauges. The circuit for testing the regulator is shown in **FIG 12:15**.

Too high a voltage setting will cause bulbs to blow, fuses to fail, and the battery to use excessive amounts of distilled water. Too low a setting will prevent the alternator from charging the battery correctly.

12:6 The windshield wiper

A typical single-speed wiper motor is shown disassembled in **FIG 12:16** and a two-speed wiper motor in **FIG 12:17**.

Parking:

The motor is self-parking and the wiper arms should be lying flat along the bottom edge of the windshield. If the arms are not flush along the lower edge when the motor stops, remove the arms from the wheelbox spindles and install them so that they lie in the correct position. With the arms removed, operate the wiper switch briefly so that the motor runs and returns to its stop position.

Typical connections for testing the motor are shown in **FIG 12:18.** The motor should run to the park position and stop. If the motor operates satisfactorily there is a fault in the switch or wiring. If the motor is defective, remove the through-bolts and take off the motor cup and armature. Rotate the output shaft until the contacts, shown in **FIG 12:19,** are firmly closed. Rotate the adjusting screw clockwise until the contacts just open and then rotate the adjusting screw counterclockwise by one complete turn. Check the bridge to ensure that the legs are contacting the brush plate when the contacts are closed.

Install the armature and motor cup in reverse order of removal and check that it now operates satisfactorily.

Poor wiper operation:

Note that the wiper blades will drag on a dry windshield, so wet the windshield before carrying out tests.

Disconnect the linkage from the motor and run the motor so that it can be checked separately. Move the linkage by hand and check for binding or defects.

Check that full battery voltage is reaching the terminals of the motor. The test circuits for the motor are shown in **FIG 12:20** and the test circuit of the switch circuit breaker is shown in **FIG 12:21.** If the motor operates satisfactorily on the test circuit then the wiring or switch is at fault. **If the motor is operated with the gearbox cover removed, take great care as the gears have sufficient torque to cause injuries to the fingers.**

Likely faults of poor motor operation are defective brushgear, insufficient armature end float (.003 inch approximately), insufficient lubricant, and worn or damaged gears.

12:7 The temperature and fuel gauges

Both these gauges operate by the bending of a bimetallic strip when it is heated by current passing through it. To ensure that voltage fluctuations in the electrical system do not cause the gauges to read incorrectly they take their supply from a voltage stabilizer unit. This unit makes and breaks the circuit by the action of another bimetallic strip to give, over a period of time, the same current flow as would a constant voltage over the same period of time. For accurate testing a special test unit is required and the sender units must

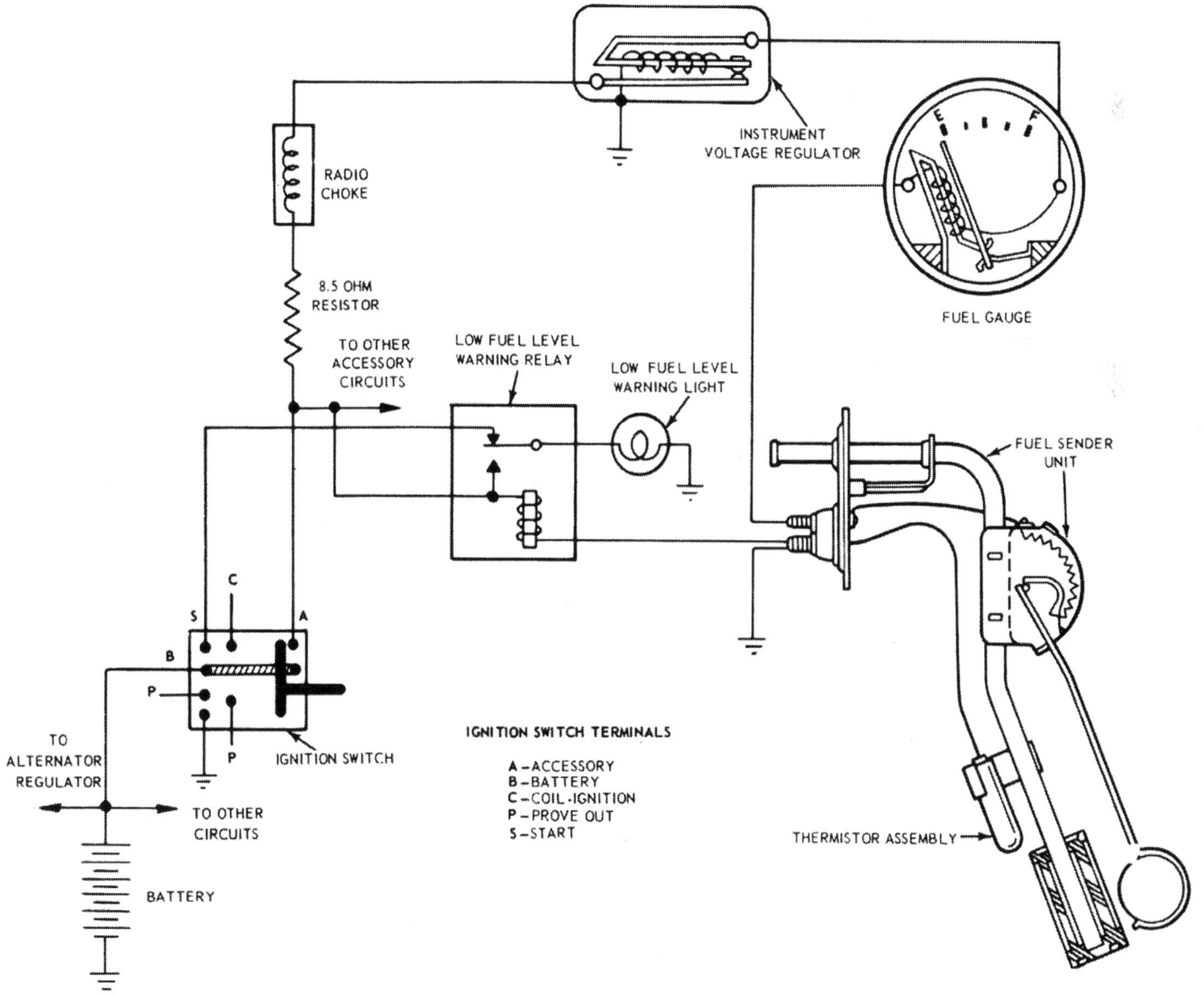

FIG 12:22 Low fuel level indicator sender assembly

not be connected directly to full battery voltage. If a voltmeter is connected to the output of the voltage stabilizer it should oscillate between 10 and zero volts. If the supply is steady the unit is defective.

Intermittent or incorrect readings can be caused by wiring faults so trace through the wires and check for breaks, faulty insulation, or poor connections. No readings indicate a defective voltage stabilizer unit (mounted on the back of the instrument cluster).

If the fuel gauge operates incorrectly, borrow a new tank sender unit. Disconnect the lead from the sender unit fitted to the tank and connect it to the new sender unit. Connect a jumper lead between a good ground and the new sender unit. Switch on the ignition and leave the float arm at its lowest position. The needle on the gauge should settle on the Empty mark. Raise the float arm and wait a minute for the needle to settle, when it should read Full. If the reading is incorrect and the wiring is satisfactory then the fault lies with the gauge itself. If the reading is satisfactory the sender unit in the tank is defective and should be exchanged for a new unit.

If only warning lights are fitted, they operate from full battery voltage.

If temperature warning lights are fitted, they operate from full battery voltage. The sender unit contains a bi-metallic strip which closes one contact when the temperature is low and the other contact when the temperature is high.

It should be noted that on some models a low fuel level warning light is fitted. The system operates when there is approximately $4\frac{1}{2}$ gallons of fuel left in the tank. The tank unit is shown in **FIG 12:22.** The thermistor is normally covered by fuel and kept cool. When the level is low the thermistor is uncovered, warms up and operates a relay to illuminate the warning light.

If a temperature gauge is fitted, the system can be checked using two resistors. Disconnect the lead from the sender unit and connect it to ground in series with a 10 ohm resistor. The gauge should then give a high reading. Substitute a 75 ohm resistor and the gauge should give a low reading. If the gauge still reads incorrectly, try substituting another voltage stabilizer unit. The sender unit must be sealed in place with a waterproof sealant that is also electrically conducting.

12:8 The turn indicators

The turn indicators use the dual-filament parking lights to indicate the direction of turn. A flasher unit located in the back of the instrument panel makes and breaks the circuit to provide the flashing at the indicator lamps. A separate emergency flasher unit comes into action when the emergency button is pressed and this makes all four turn indicator lamps flash.

If the indicators do not operate correctly, check the bulbs for blown filaments as if a filament is blown the flash rate will alter.

Reach under the instrument panel, push in and then twist the flasher unit to free it from its mounting. Disconnect the wires. Switch on the ignition and use a test lamp or voltmeter to check that battery voltage is reaching the flasher unit. Switch off the ignition and connect the wires together. Switch on the ignition and operate the turn indicator switch. The appropriate lamps should come on and stay on. If the lamps do not light,

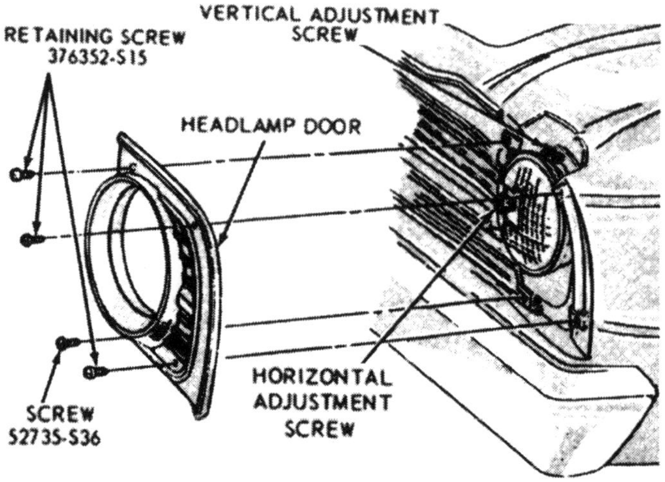

FIG 12:23 Typical headlight attachments

check through the wiring until the fault is found. If current is passing through the wires and switch, check the earth points on the lamps. If the lamps operate correctly, the flasher unit is defective and must be renewed.

Handle the flasher unit with care as it is delicate and will be damaged by mishandling.

12:9 The headlights

Two sealed-beam headlamps are fitted. A typical attachment is shown in **FIG 12:23** and access is made after removing the headlamp door.

To remove the headlamp, take out the screws which secure the door. Hold the headlamp assembly by hand and unhook the spring from the retaining ring. Remove the retaining ring by unhooking it from the clip on the opposite side to the spring. Pull the headlight bulb out and disconnect it from the wiring assembly plug.

Install the headlight in the reverse order of removal.

Beam setting:

Two adjusting screws are fitted to each headlight and they are used to set the beams correctly. The vehicle should be taken to a service garage for accurate setting using special equipment, as the owner cannot set the beams with accuracy.

12:10 Fuses

A fuse panel containing the fuses is fitted. The panel is marked with the circuits protected by the individual fuses as well as the correct rating of the fuse to be used.

A fuse blowing is not necessarily an indication of a fault in the system, as fuses do weaken with age. If a fuse does blow, briefly check the circuits for damage and fit a new fuse of the correct rating. If the new fuse blows immediately a circuit is switched on, check the circuit thoroughly. Shake the wires in case insulation has frayed and is causing a ground. Do not install a fuse of higher rating than called for, as this could prevent the fuse from blowing and the wiring may burn out instead.

Circuit breakers:

These are fitted to some circuits, such as headlights and windshield wipers. In the event of an excessively high current being taken the circuit breaker will operate.

The circuit breakers can be tested, using a 12-volt battery variable resistance and ammeter. The current is adjusted to the normal value and the circuit breaker should not operate. When the current is adjusted to a high value, the circuit breaker should operate within 30 seconds. Renew defective circuit breakers as they cannot be repaired.

12:11 Trouble shooting guide

(a) Battery discharged

1 Loose or dirty terminal connectors
2 Insufficient charging current
3 Shortcircuit in part of circuit not protected by fuse
4 Defective alternator diode
5 Battery internally defective
6 Accessories left on

(b) Battery will not hold charge

1 Low electrolyte level
2 Battery plates sulphated or separators ineffective
3 Electrolyte leakage or spillage

(c) Alternator output low or nil

1 Broken or slipping drive belt
2 Defective or incorrectly set regulator
3 Loose connectors
4 Defective diodes
5 No battery supply
6 Defective rotor or stator windings

(d) Starter motor lacks power or will not operate

1 Battery discharged or loose and dirty terminals
2 Starter pinion jammed in mesh
3 Defective ignition switch
4 Defective starter relay
5 Defective brushgear or commutator
6 Defective armature or field coil windings
7 Starter motor mechanically defective
8 Engine abnormally stiff

(e) Starter runs but does not turn engine

1 Badly worn or broken teeth on drive

Inches	Decimals	Millimetres	Inches to Millimetres: Inches	Inches to Millimetres: mm	Millimetres to Inches: mm	Millimetres to Inches: Inches
1/64	.015625	.3969	.001	.0254	.01	.00039
1/32	.03125	.7937	.002	.0508	.02	.00079
3/64	.046875	1.1906	.003	.0762	.03	.00118
1/16	.0625	1.5875	.004	.1016	.04	.00157
5/64	.078125	1.9844	.005	.1270	.05	.00197
3/32	.09375	2.3812	.006	.1524	.06	.00236
7/64	.109375	2.7781	.007	.1778	.07	.00276
1/8	.125	3.1750	.008	.2032	.08	.00315
9/64	.140625	3.5719	.009	.2286	.09	.00354
5/32	.15625	3.9687	.01	.254	.1	.00394
11/64	.171875	4.3656	.02	.508	.2	.00787
3/16	.1875	4.7625	.03	.762	.3	.01181
13/64	.203125	5·1594	.04	1.016	.4	.01575
7/32	.21875	5.5562	.05	1.270	.5	.01969
15/64	.234375	5.9531	.06	1.524	.6	.02362
1/4	.25	6.3500	.07	1.778	.7	.02756
17/64	.265625	6.7469	.08	2.032	.8	.03150
9/32	.28125	7.1437	.09	2.286	.9	.03543
19/64	.296875	7.5406	.1	2.54	1	.03937
5/16	.3125	7.9375	.2	5.08	2	.07874
21/64	.328125	8.3344	.3	7.62	3	.11811
11/32	.34375	8.7312	.4	10.16	4	.15748
23/64	.359375	9.1281	.5	12.70	5	.19685
3/8	.375	9.5250	.6	15.24	6	.23622
25/64	.390625	9.9219	.7	17.78	7	.27559
13/32	.40625	10.3187	.8	20.32	8	.31496
27/64	.421875	10.7156	.9	22.86	9	.35433
7/16	.4375	11.1125	1	25.4	10	.39370
29/64	.453125	11.5094	2	50.8	11	.43307
15/32	.46875	11.9062	3	76.2	12	.47244
31/64	.484375	12.3031	4	101.6	13	.51181
1/2	.5	12.7000	5	127.0	14	.55118
33/64	.515625	13.0969	6	152.4	15	.59055
17/32	.53125	13.4937	7	177.8	16	.62992
35/64	.546875	13.8906	8	203.2	17	.66929
9/16	.5625	14.2875	9	228.6	18	.70866
37/64	.578125	14.6844	10	254.0	19	.74803
19/32	.59375	15.0812	11	279.4	20	.78740
39/64	.609375	15.4781	12	304.8	21	.82677
5/8	.625	15.8750	13	330.2	22	.86614
41/64	.640625	16.2719	14	355.6	23	.90551
21/32	.65625	16.6687	15	381.0	24	.94488
43/64	.671875	17.0656	16	406.4	25	.98425
11/16	.6875	17.4625	17	431.8	26	1.02362
45/64	.703125	17.8594	18	457.2	27	1.06299
23/32	.71875	18.2562	19	482.6	28	1.10236
47/64	.734375	18.6531	20	508.0	29	1.14173
3/4	.75	19.0500	21	533.4	30	1.18110
49/64	.765625	19.4469	22	558.8	31	1.22047
25/32	.78125	19.8437	23	584.2	32	1.25984
51/64	.796875	20.2406	24	609.6	33	1.29921
13/16	.8125	20.6375	25	635.0	34	1.33858
53/64	.828125	21.0344	26	660.4	35	1.37795
27/32	.84375	21.4312	27	685.8	36	1.41732
55/64	.859375	21.8281	28	711.2	37	1.4567
7/8	.875	22.2250	29	736.6	38	1.4961
57/64	.890625	22.6219	30	762.0	39	1.5354
29/32	.90625	23.0187	31	787.4	40	1.5748
59/64	.921875	23.4156	32	812.8	41	1.6142
15/16	.9375	23.8125	33	838.2	42	1.6535
61/64	.953125	24.2094	34	863.6	43	1.6929
31/32	.96875	24.6062	35	889.0	44	1.7323
63/64	.984375	25.0031	36	914.4	45	1.7717

UNITS	Pints to Litres	Gallons to Litres	Litres to Pints	Litres to Gallons	Miles to Kilometres	Kilometres to Miles	Lbs. per sq. In. to Kg. per sq. Cm.	Kg. per sq. Cm. to Lbs. per sq. In.
1	.57	4.55	1.76	.22	1.61	.62	.07	14.22
2	1.14	9.09	3.52	.44	3.22	1.24	.14	28.50
3	1.70	13.64	5.28	.66	4.83	1.86	.21	42.67
4	2.27	18.18	7.04	.88	6.44	2.49	.28	56.89
5	2.84	22.73	8.80	1.10	8.05	3.11	.35	71.12
6	3.41	27.28	10.56	1.32	9.66	3.73	.42	85.34
7	3.98	31.82	12.32	1.54	11.27	4.35	.49	99.56
8	4.55	36.37	14.08	1.76	12.88	4.97	.56	113.79
9		40.91	15.84	1.98	14.48	5.59	.63	128.00
10		45.46	17.60	2.20	16.09	6.21	.70	142.23
20				4.40	32.19	12.43	1.41	284.47
30				6.60	48.28	18.64	2.11	426.70
40				8.80	64.37	24.85		
50					80.47	31.07		
60					96.56	37.28		
70					112.65	43.50		
80					128.75	49.71		
90					144.84	55.92		
100					160.93	62.14		

UNITS	Lb ft to kgm	Kgm to lb ft	UNITS	Lb ft to kgm	Kgm to lb ft
1	.138	7.233	7	.967	50.631
2	.276	14.466	8	1.106	57.864
3	.414	21.699	9	1.244	65.097
4	.553	28.932	10	1.382	72.330
5	.691	36.165	20	2.765	144.660
6	.829	43.398	30	4.147	216.990

CHAPTER 13

THE BODYWORK

13:1 Bodywork repairs

Large-scale repairs to bodywork should be left to experts. Even small dents can be tricky as injudicious hammering will cause the metal to stretch and make the dent worse instead of better.

Acrylic finishes are used as standard on the exterior of the vehicle and Nitrocellulose lacquers must not be used.

Spotting or minor blemishes on the paintwork can be removed by polishing. This method can also be used to restore the finish if it has suffered by weathering and lost its gloss. Machine polishing will make the task easier. Remove the defects using 600 grit paper with water or white spirits as a lubricant. Remove only sufficient paint to take out the defect.

Paint on a white or light coloured medium grit machine polishing compound (such as Sno-Flake No. 16) with a brush. Polish the entire surface at 1850 rev/min and a carpet pad ($\frac{5}{8}$ inch nap) or lambswool pad. Remove the compound and buff the surface with a clean lambswool pad.

Self-spraying cans of matching paint and touch-up kits are available. The touch-up kit should be used for small stone chips and minor scratches. Polishing after hardening can be carried out using a mild metal polish. Regular use of the touch-up kit will prevent corrosion forming on the metal under chips and scratches.

For deeper dents, filling and then spraying is the best method of repair available to the owner. It is better to spray a complete panel rather than just a patch as any mismatch in the paint will then not be so obvious. Remember that the finish will alter color slightly and fade with age. Remove all trim and handles from the area, as a better finish will be obtained than with masking them over. Wash off all wax polish using white spirits. If silicone-based polishes have been used it is essential to use a special cleansing fluid to remove them. Traces of polish will leave the finish with pits, craters and pop-ups. Lightly scuff the area to give a good key for new finish. If there are any traces of corrosion, rub down to bright metal and apply a coat of primer.

Apply paste stopper or filler as required to bring the damaged area just slightly proud of the original finish. Leave the filler to harden and then rub it down using 400 grit paper with water as a lubricant. Add further stopper or filler as required to achieve the best possible surface finish. Wash off all slurry and leave the surface to dry. A second wash will probably be required to remove

FIG 13:1 Window mechanism lubrication points

slurry missed in the first wash. When the surface is dry, wipe it over with a proprietary 'tacky' cloth to remove any remaining dust.

Mask off surrounding areas with paper and masking tape. Spray the surface evenly and spray over onto the masking to ensure that the edges of the panel are also evenly sprayed. Apply the lacquer in thin layers, gently rubbing down between each coat, rather than in one thick coat which will probably run. Remove the masking and leave the finish to dry for at least 24 hours. Polish the surface as described earlier to remove any runs or sags and then leave the lacquer to harden for several weeks before applying wax polish.

13:2 Door components

The door components vary slightly from model to model and year to year, so only typical assemblies will be shown. Whenever the door mechanism parts are removed, all bearing surfaces and pivots should be lubricated with Polyethylene grease. This grease is available in aerosol containers with a long nozzle for reaching awkward spots. The shaded areas in **FIG 13:1** show the points of application.

Electrically operated windows:

Before carrying out checks, make sure that the battery is fully charged and turn the ignition key to the accessory position. Connect a test lamp or voltmeter between ground and the 20 amp circuit breaker on the starter relay. Full battery voltage should be available. On models fitted with power operated seats and hood, these accessories use the same circuit breaker and if they do not operate as well then the fault most likely lies with the circuit breaker. Renew a defective circuit breaker.

Similarly test the supply from the relay and renew the relay if it is defective. Check the wires for breaks or poor connections if the components are satisfactory.

If only one window fails to operate, activate the switch and listen for the motor running free. If the motor is running but the window does not operate, the most likely fault is a sheared rubber drive. If the motor does not run at all, disconnect the wires and use a test lamp or voltmeter to check the feed, tracing back through the wiring until a fault is found. If feed is available to the motor and it does not run then it is likely that the motor itself is defective. It should be noted that if the window will operate in one direction only the motor is satisfactory and the fault lies in wiring or switch.

If a window operates sluggishly, first check the wiring for poor connections or partially grounded wires. Disconnect the supply to the motor and reconnect it through an ammeter. Under normal running the motor should take approximately 12 amps and 20 amps at stall. A low stall current indicates poor connections in the circuit. The motor should take approximately 5 amps when disconnected from the winder mechanism and running free.

Regulator mechanism:

The front door regulator mechanism on a 1970 Model 62 is shown in **FIG 13:2.** The rear door adjustment points are shown in **FIG 13:3** while the quarter window mechanism is shown in **FIG 13:4.** Before access can be gained to the parts, the interior trim panel and handles must first be removed and the plastic sealing sheet taken off.

Lock mechanism:

Typical lock mechanism components and their attachments are shown in **FIG 13:5.**

13:3 Windshield and back window

Two differing methods of securing and sealing the glass in place are used. Earlier models use a weatherstrip type seal where the glass fits into a slit in the rubber weatherstrip. On some later models Butyl tape is used.

If the glass has broken but is still in place, it will be found easiest to remove it after sticking paper onto both sides and laying cloth or paper over the hood or luggage locker lid to protect the paint. If the windshield has broken, make sure that all particles are removed from the windshield demist system, otherwise they may be blown into the faces of the front seat occupants by the heater.

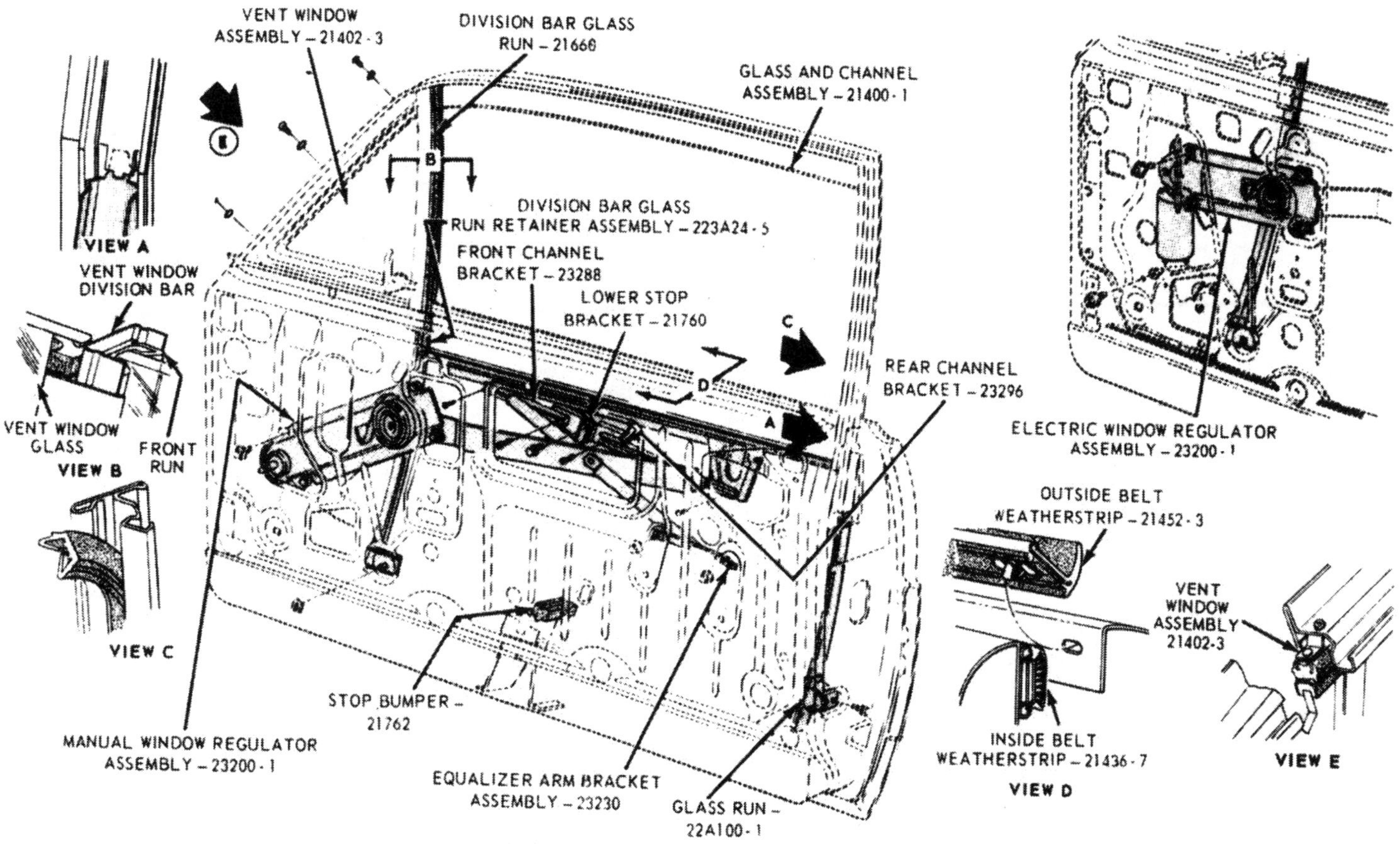

FIG 13:2 The front door window mechanism, typical

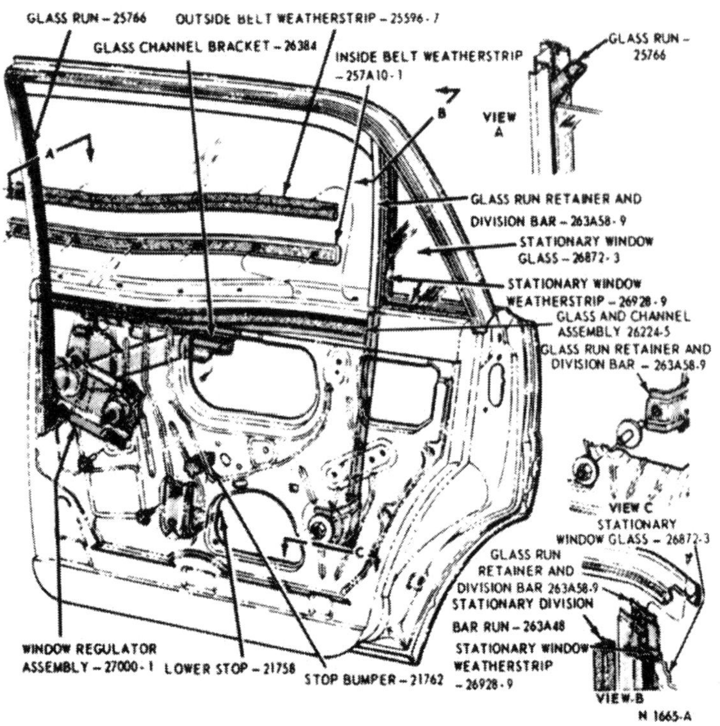

FIG 13:3 The rear door window mechanism, typical

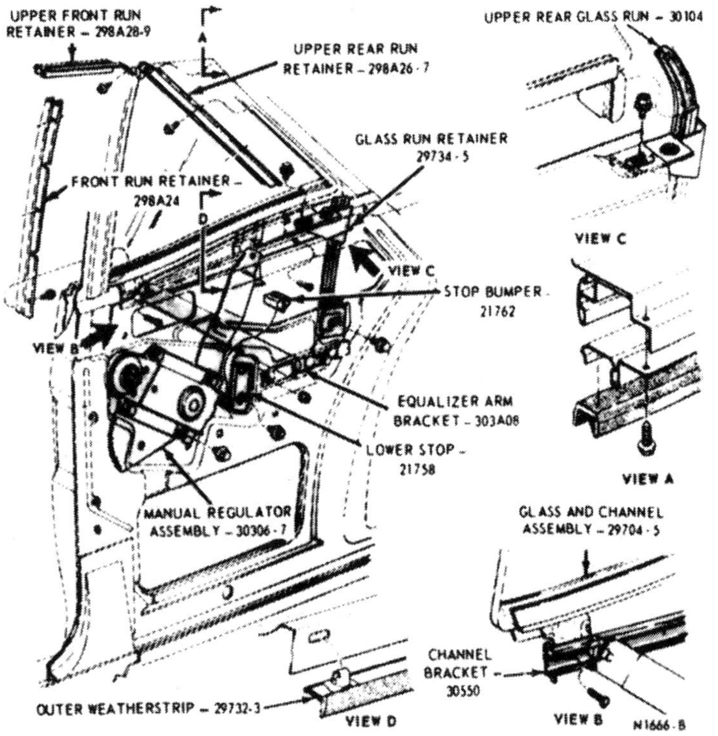

FIG 13:4 Quarter window operating mechanism, typical

Once the glass, mouldings and sealer have been removed, check the aperture flange for dents or protrusions. Dents should be dressed out using a block of metal on one side and a hammer to tap them flat, while protrusions should be smoothed down with a file. If the flange is left in a damaged state it may set up stress points in the glass which may cause it to fail in service.

Removal when Butyl strip is used:

Remove the moulding pieces. If the windshield is to be removed, first take off the wipers and their arms. A typical late model windshield installation is shown in **FIG 13:6**, and a back window installation in **FIG 13:7.** Insert a 2 foot length of thin piano wire between the flange and glass. Wrap each end of the wire around a piece of wooden dowel as a handle, and with the aid of an assistant cut through the seal all the way around the glass using a sawing motion. Remove the glass. Pry the remains of the Butyl strip away from one corner of the aperture, grasp the strip near to the flange and steadily pull it away from the flange. Collect the two spacers from the lower edge of the aperture.

Removal when weatherstrip is fitted:

Remove the moulding pieces. If the windshield is to be removed, take off the wipers and their arms. A typical early model windshield installation is shown in **FIG 13:8** and a back window installation in **FIG 13:9.** Remove any interior garnish mouldings. If a heated back window is fitted, disconnect the wires. If necessary use a thin wooden wedge to free the inner lip of the weatherstrip from the aperture. Apply pressure on a corner of the glass, using a well-padded foot or flat of hand, and press the glass and weatherstrip outwards from the aperture. Have an assistant outside the vehicle to take the glass as it comes free. Lay the glass onto a padded bench and strip off the weatherstrip. It is unlikely that the weatherstrip will be removed without causing damage to it so it is always best to fit a new weatherstrip on installation and discard the old one.

Installation using Butyl strip:

Install the spacers in the lower corners of the aperture. Check the moulding retainers and repair or renew as required if they are defective. Temporarily position the glass in the aperture and move it into the best position for even overlap. Handling of the glass will be easier if rubber suckers are attached to the outside and used as hand holds. When the glass is satisfactorily positioned, mark the location with wax crayon.

Remove the glass and apply primer (furnished with the kit) around the entire perimeter of the aperture flange and the edge of the glass where it will contact the Butyl strip. Allow the primer to dry for 10 minutes.

Position the Butyl tape around the aperture, so that its edge is within $\frac{1}{16}$ inch of the edge of the flange, but **do not allow the tape to overhang the edge of the flange. Do not stretch the tape at the corners but guide it gently around and into place.** Carefully splice the two ends of the tape where they meet.

Place the glass against the aperture, using the previously made crayon marks to align it accurately. When the glass is in position, press it firmly against the tape using hand pressure. If the tape appears dull through the glass the glass and tape are not in contact, so apply hand pressure at that point until the tape appears shiny. Remove surplus primer from the glass with a razor blade and wipe clean with a cloth moistened with naptha. Pour water over the outside, or use a hosepipe, to check for leaks. Leaks can be repaired by applying liquid Butyl around the edge of the glass.

Install the mouldings and any other parts removed.

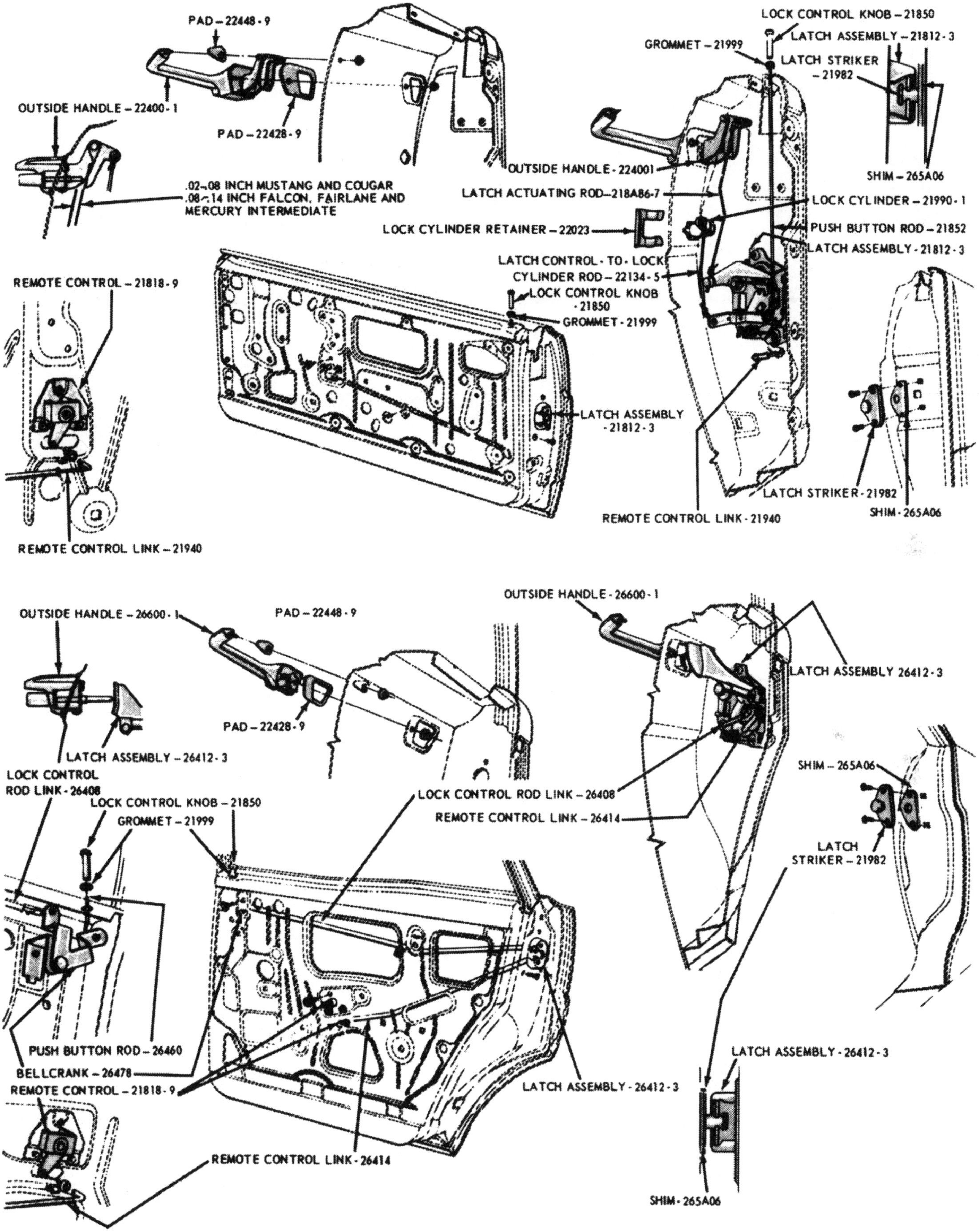

FIG 13:5 Door handle and latch installation, typical

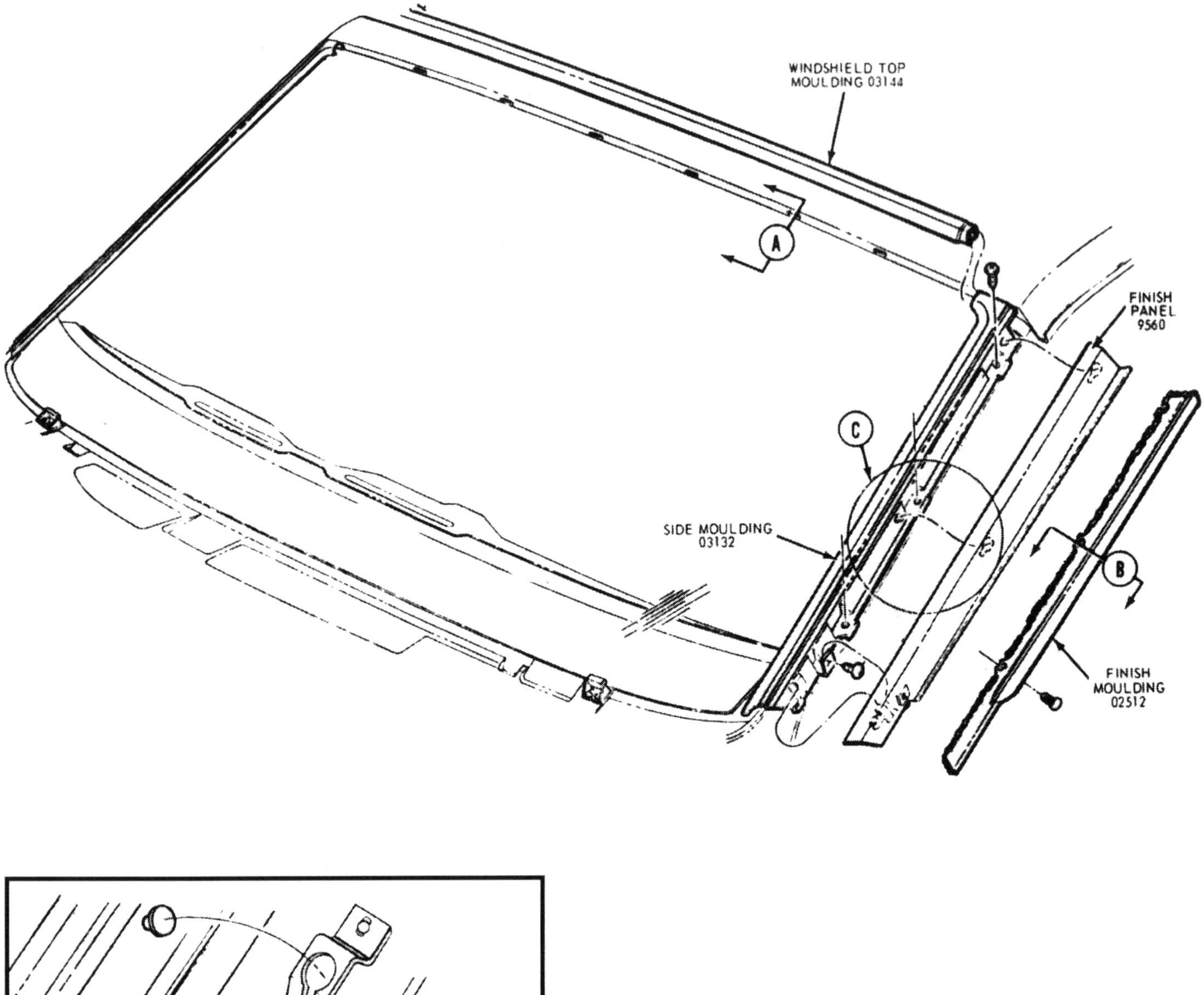

FIG 13:6 Late model windshield installation

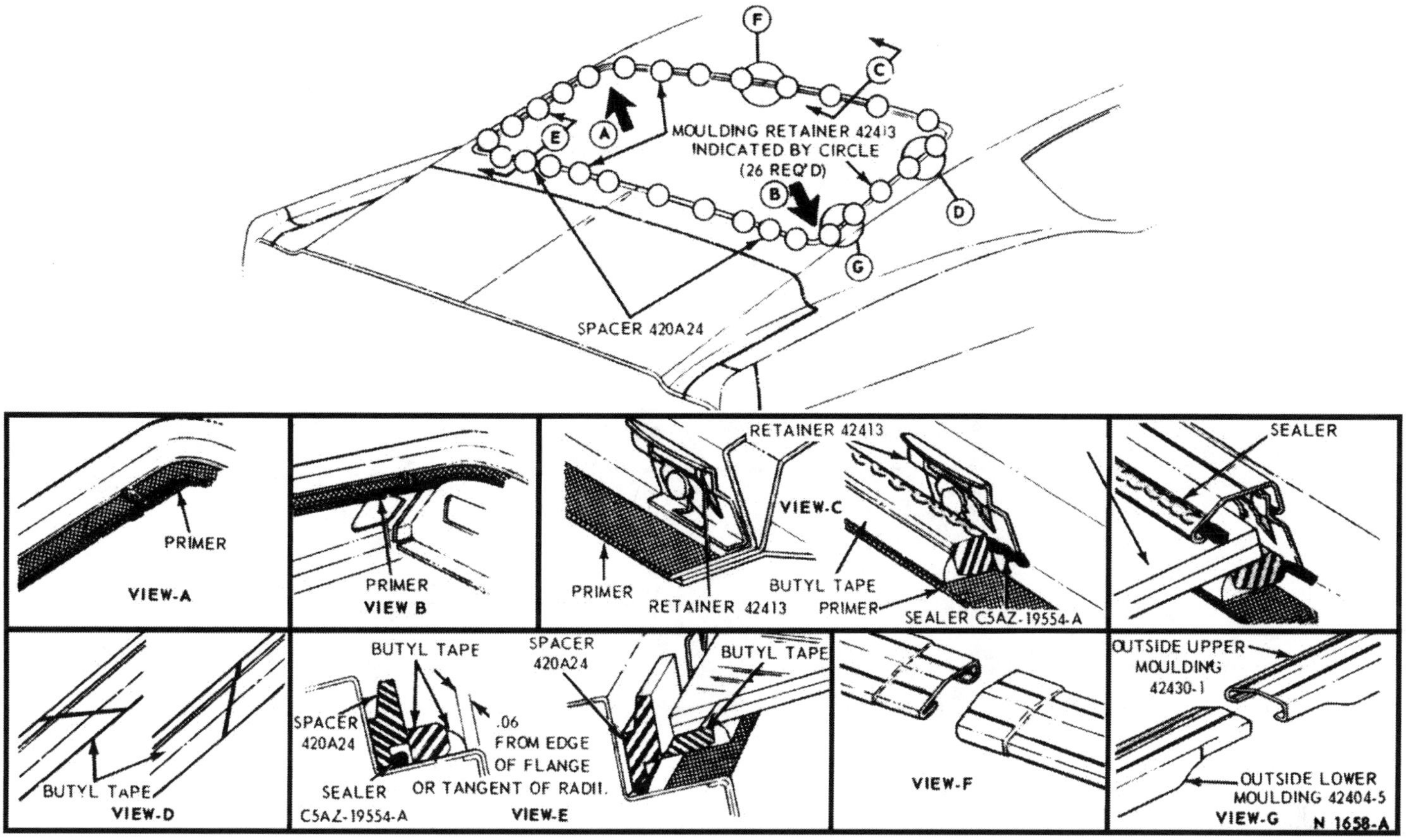

FIG 13:7 Late model back window installation

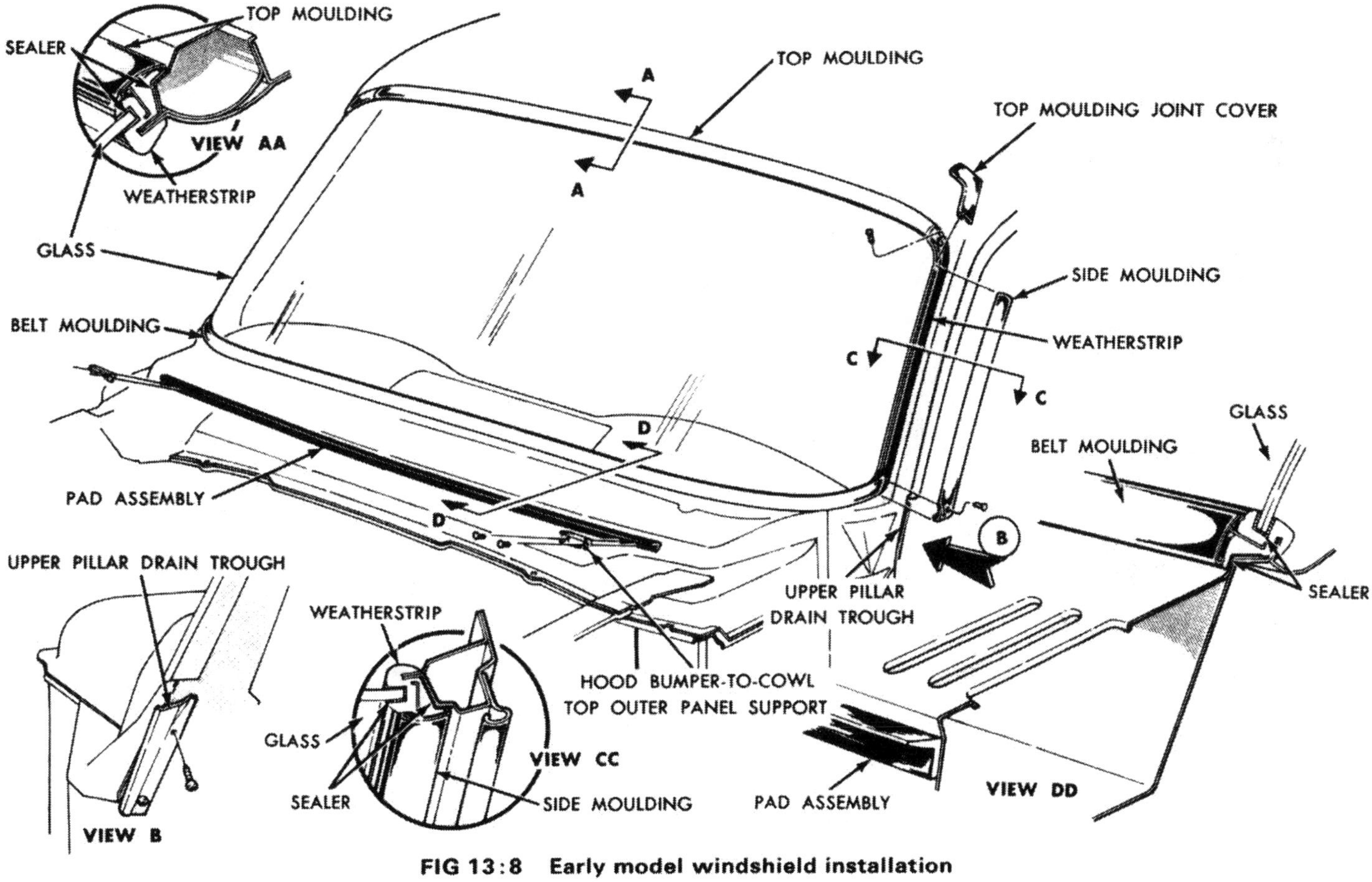

FIG 13:8 Early model windshield installation

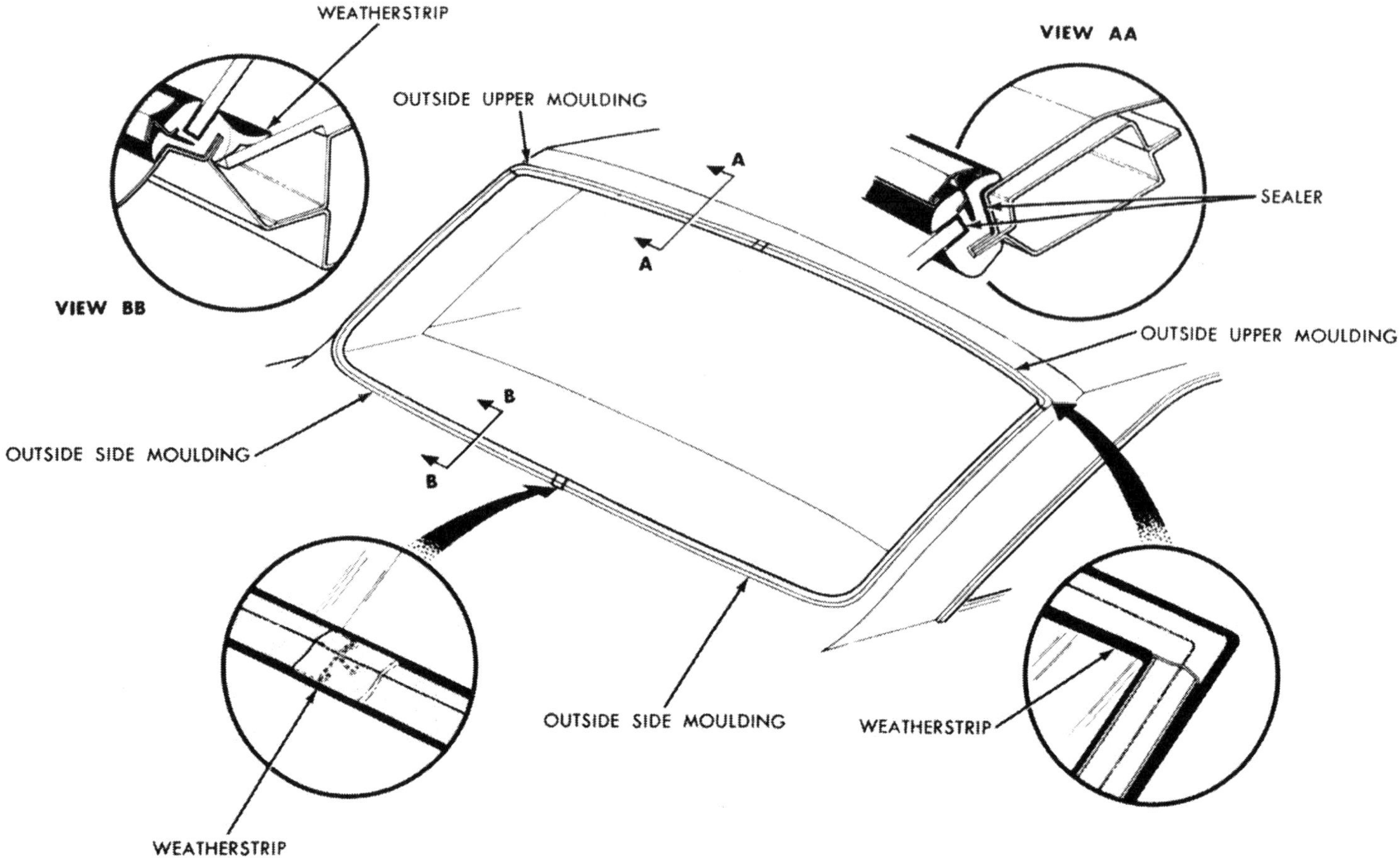

FIG 13:9 Early model back window installation

Installation using weatherstrip:

Inject sealant around the weatherstrip glass opening groove. Position the weatherstrip around the glass. Insert a length of cord into the body aperture flange groove in the weatherstrip making sure that there is sufficient cord to leave the ends overlapped and hanging free. Tape the ends of the cord to the inside of the glass, as shown in **FIG 13:10.** Apply a bead of sealer around the body aperture flange.

With the aid of an assistant, position the glass—from the outside of the vehicle—against the aperture. While the assistant applies firm pressure to hold the glass in place, pull the cord up and out of the groove, parallel to and towards the centre of the glass, so that as the cord comes free it lifts the lip of the weatherstrip up and over the flange in the aperture.

On models fitted with heated back windows, position a piece of tape over the flange where the wires for the glass pass. Install the mouldings and any parts removed. Surplus sealer should be removed with a cloth moistened with solvent. **Do not use excessive solvent as it may creep into the weatherstrip and remove the sealer that should be there.**

Rear view mirror:

On some models the rear view mirror is bonded directly to the glass, using a vinyl pad. Heat from a hot air gun is used to soften the vinyl pad for removal or installation. Insulate any chips or scratches with wet rags or putty. Blow hot air over the windshield and after approximately 8 to 10 minutes the rear view mirror may be removed by levering carefully on the mirror. Gently reduce the heat and then allow the glass to cool naturally.

Preheat the glass in the area before installing the mirror and then heat the button. Press the button firmly into place and keep heating until vinyl can just be seen extruding between glass and mirror support. Pressure must be applied to the mirror mount while bonding it in place. Maintain pressure for at least 2 minutes while reducing the heat and then allow the glass to cool naturally. Do not remove insulation while the glass is hot.

13:4 The hood

The hood hinge on an early model is shown in **FIG 13:11** and on a late model in **FIG 13:12.** The arrows indicate the direction in which the parts can be moved to align the hood accurately with the body.

Before removing the hood, lightly mark around the hinge with pencil to ensure that the hood will be installed back in its original position. Lay pads of cloth onto the paintwork to protect it from damage and have enough assistants to support the hood as it is freed. Open the hood and remove the bolts that secure the hinge to the hood.

Install the hood in the reverse order of removal. If the hood is not in accurate alignment with the body, slacken the attaching bolts sufficiently to hold the weight of the hood but allow it to be moved under pressure. Move the hood into alignment and tighten the attaching bolts fully.

FIG 13:10 **The draw cord for the weatherstrip installation**

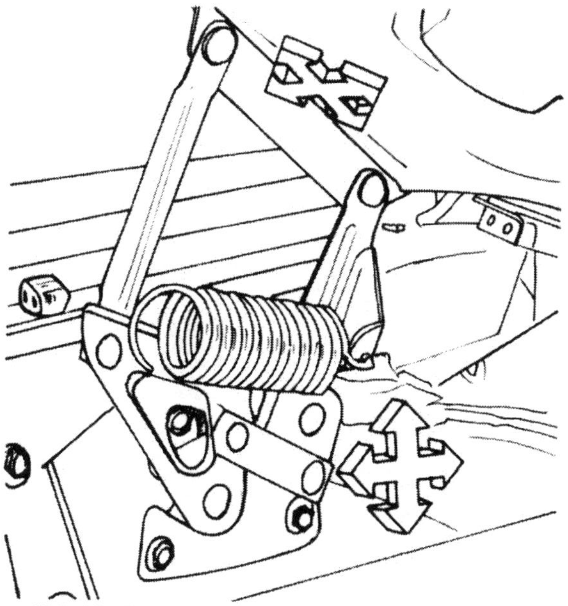

FIG 13:11 **Early hood hinge adjustments**

An early type hood lock mechanism is shown in **FIG 13:13.** Adjust the lock until the lock dowel enters it centrally and the safety catch engages on the striker. The lock dowel can be screwed up or down to adjust the tension on the hood when it is closed. If the hood is loose and rattles when closed screw the dowel upwards, while if it is stiff and difficult to close screw the dowel slightly downwards.

13:5 The luggage locker door

A typical late model hinge attachment is shown in **FIG 13:14.** Removal and installation of the door is very similar to the method used for the hood (see previous section) though any shims fitted should be carefully collected and installed back in their original positions.

Torsion bars are used to counterbalance the weight of the door. The method of altering the torsion bar tension is shown in **FIG 13:15.**

13:6 Air conditioning

A typical air conditioning system is shown in **FIG 13:16.** The system is charged with a non-toxic refrigerant (Refrigerant 12). The compressor is driven by a belt from the engine through a magnetic clutch. The clutch only operates when the system is selected and the temperature in the car rises above the selected temperature. With the system operating, the compressor draws the gas from the evaporator and pressurizes it. The operating pressure from the compressor varies between 80 to 300 lb/sq inch depending on the ambient air temperature and compressor operating speed. The high pressure gas, hot because of compression, passes from the compressor to the condenser where it is cooled by the passage of air over the condensor. The refrigerant changes from gas to

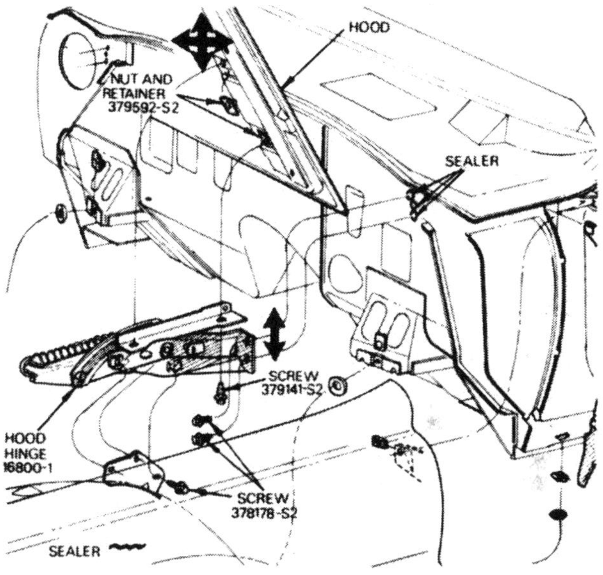

FIG 13:12 **Late hood hinge**

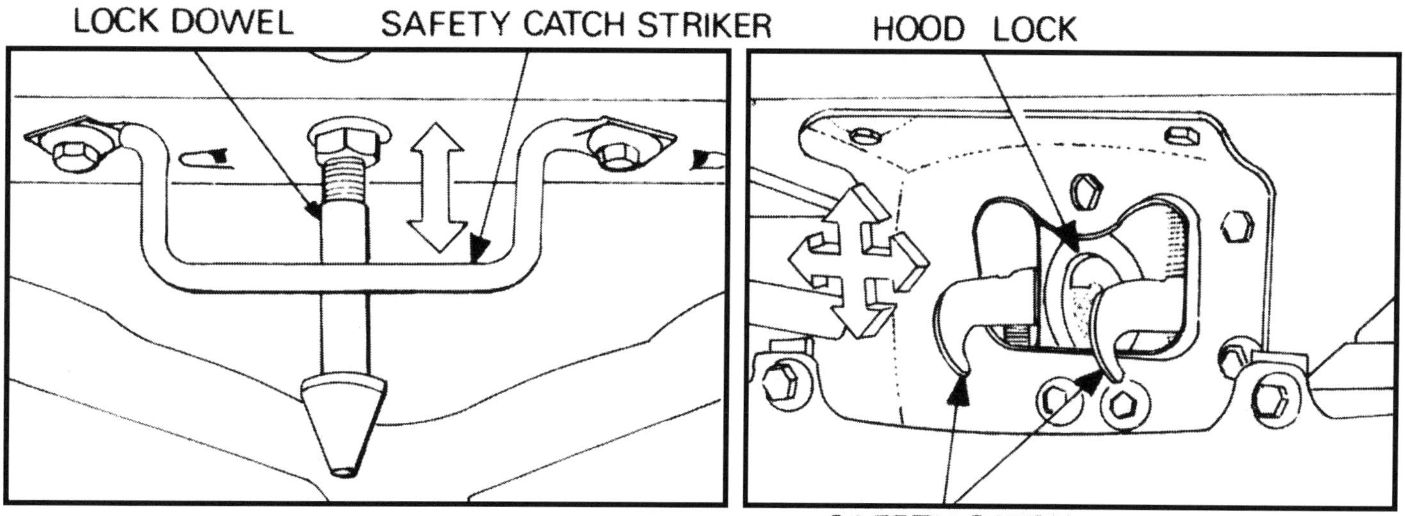

FIG 13:13 **Typical hood lock mechanism**

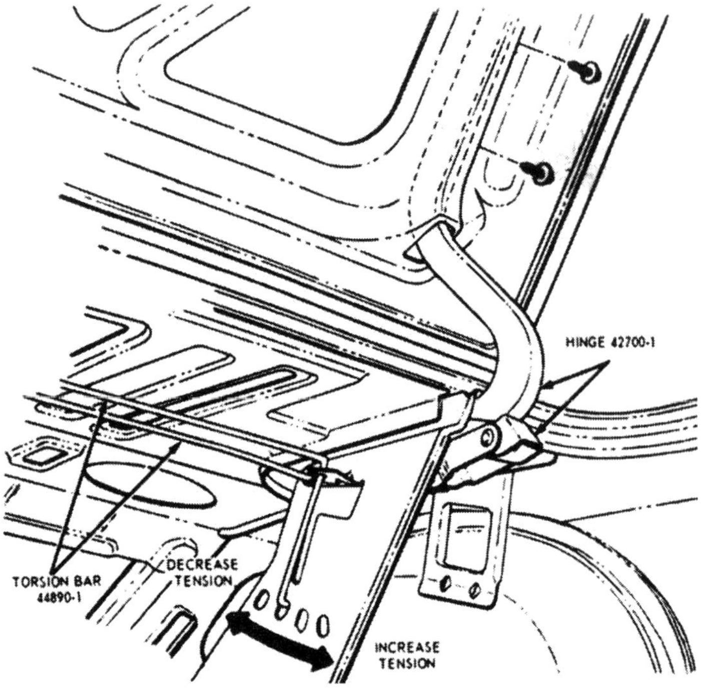

FIG 13:14 Luggage compartment lid hinge

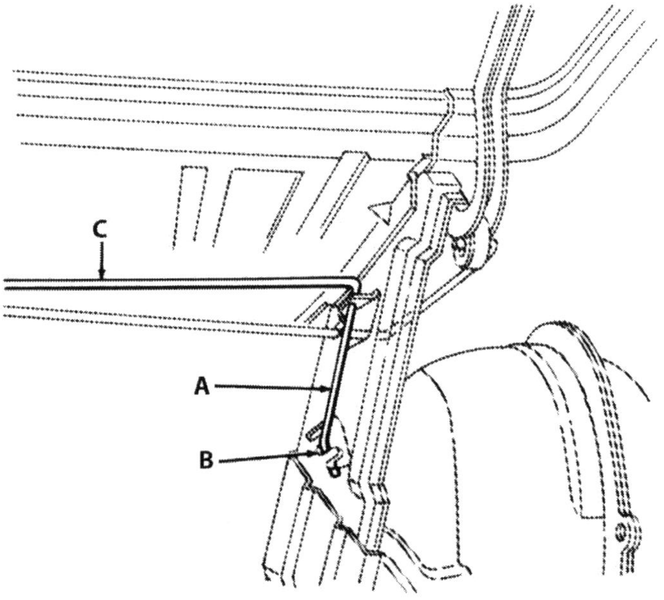

FIG 13:15 Adjusting the luggage compartment lid torsion bars

liquid in the condensor and passes onto the receiver. The receiver contains a water-absorbing material to ensure that the refrigerant is perfectly dry. From the receiver the liquid refrigerant, still at high pressure, passes through a sight glass and hoses to the expansion valve. The expansion valve controls the amount of refrigerant passing through and reduces the pressure to between 12 and 50 lb/sq inch, depending on the conditions and compressor speed. The low pressure liquid passes into the evaporator unit inside the vehicle and the heat in the air passing through it causes the liquid to change back to gas, at low pressure, and cool the air. The low pressure gas is then drawn back into the compressor. A thermostatic switch is fitted to control the compressor magnetic clutch.

Servicing of the system should only be carried out by trained personnel as though the refrigerant is non-toxic serious injury can result if the safety precautions are ignored.

Safety precautions:

At normal pressure and temperature, the refrigerant evaporates practically instantaneously, absorbing heat rapidly. For the reason spillage of the fluid should be avoided as it will cause 'burns' on the skin. Contact with the eyes is more serious, so wear safety goggles when working on the system. If the refrigerant does go into an eye, use a few drops of sterile mineral oil to wash out the eye (as the refrigerant is readily absorbed by mineral oil) and then clean the eye with a weak boric acid solution. **Seek medical aid even if the irritation has ceased.**

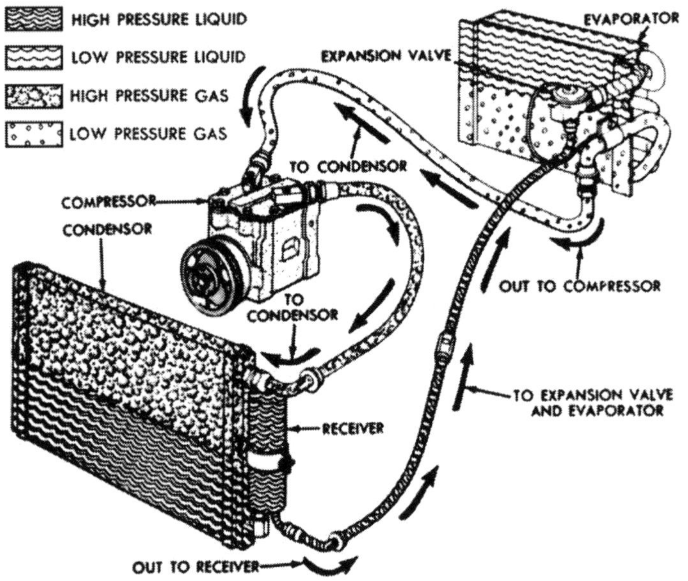

FIG 13:16 The air conditioning system

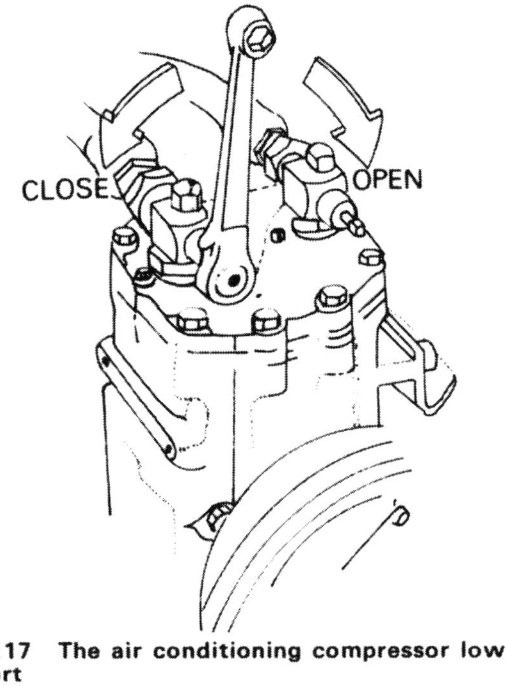

FIG 13:17 The air conditioning compressor low pressure port

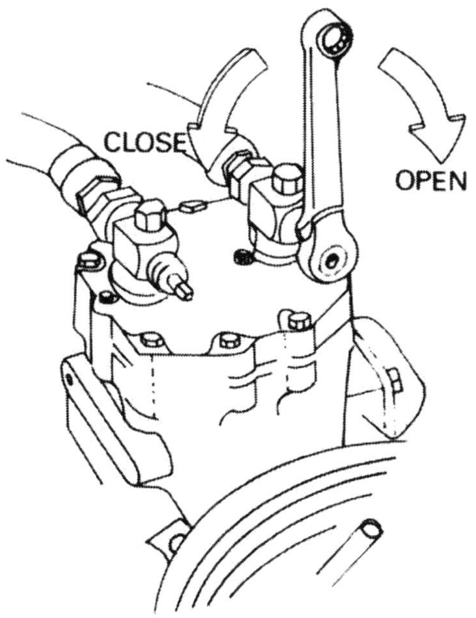

FIG 13:18 The air conditioning compressor high pressure port

If the system is discharged, do so slowly (to avoid flushing the oil out of the compressor) and discharge out directly into the open. If the refrigerant is discharged into a confined space, the rapid evaporation will cause the refrigerant to displace the air and reduce the amount of oxygen present by a serious amount. This could lead to asphyxiation.

Do not heat the system, such as when steam cleaning, soldering, or using a blow lamp in the area, as the rise in pressure could cause a component to burst explosively.

The refrigerant itself is non-toxic but it will produce a very poisonous gas if passed through a naked flame. The leak tester uses a small flame to heat a copper disc so avoid breathing the fumes from the tester when checking for leaks.

Sight glass:

Normally the glass should appear perfectly clear and free from bubbles. Check by running the engine at 1500 rev/min and cycling the magnetic clutch. Bubbles should appear when the clutch is off. If there are no bubbles then the system is empty.

Continuous bubbles indicate a low charge in the system. Under conditions of high temperatures, bubbles or foam may appear in the sight glass.

Compressor removal:

This is the only servicing that should be carried out by the owner.

Turn both the high and low pressure valves fully clockwise. The valves are shown in **FIGS 13:17** and **13:18.** Loosen the cap on the high pressure valve, by a small amount only, and allow the refrigerant to escape from the compressor. Disconnect the service valves from the compressor. Undo the compressor attachments and remove it from the engine.

The compressor is installed in the reverse order of removal. Before opening the service valves, evacuate the compressor using a vacuum pump. After installation, the owner will probably find it most convenient to leave the service valves closed and drive the vehicle to a service agent, without using the air conditioner, so that the compressor can be evacuated.

Inches	Decimals	Millimetres	Inches to Millimetres: Inches	Inches to Millimetres: mm	Millimetres to Inches: mm	Millimetres to Inches: Inches
1/64	.015625	.3969	.001	.0254	.01	.00039
1/32	.03125	.7937	.002	.0508	.02	.00079
3/64	.046875	1.1906	.003	.0762	.03	.00118
1/16	.0625	1.5875	.004	.1016	.04	.00157
5/64	.078125	1.9844	.005	.1270	.05	.00197
3/32	.09375	2.3812	.006	.1524	.06	.00236
7/64	.109375	2.7781	.007	.1778	.07	.00276
1/8	.125	3.1750	.008	.2032	.08	.00315
9/64	.140625	3.5719	.009	.2286	.09	.00354
5/32	.15625	3.9687	.01	.254	.1	.00394
11/64	.171875	4.3656	.02	.508	.2	.00787
3/16	.1875	4.7625	.03	.762	.3	.01181
13/64	.203125	5·1594	.04	1.016	.4	.01575
7/32	.21875	5.5562	.05	1.270	.5	.01969
15/64	.234375	5.9531	.06	1.524	.6	.02362
1/4	.25	6.3500	.07	1.778	.7	.02756
17/64	.265625	6.7469	.08	2.032	.8	.03150
9/32	.28125	7.1437	.09	2.286	.9	.03543
19/64	.296875	7.5406	.1	2.54	1	.03937
5/16	.3125	7.9375	.2	5.08	2	.07874
21/64	.328125	8.3344	.3	7.62	3	.11811
11/32	.34375	8.7312	.4	10.16	4	.15748
23/64	.359375	9.1281	.5	12.70	5	.19685
3/8	.375	9.5250	.6	15.24	6	.23622
25/64	.390625	9.9219	.7	17.78	7	.27559
13/32	.40625	10.3187	.8	20.32	8	.31496
27/64	.421875	10.7156	.9	22.86	9	.35433
7/16	.4375	11.1125	1	25.4	10	.39370
29/64	.453125	11.5094	2	50.8	11	.43307
15/32	.46875	11.9062	3	76.2	12	.47244
31/64	.484375	12.3031	4	101.6	13	.51181
1/2	.5	12.7000	5	127.0	14	.55118
33/64	.515625	13.0969	6	152.4	15	.59055
17/32	.53125	13.4937	7	177.8	16	.62992
35/64	.546875	13.8906	8	203.2	17	.66929
9/16	.5625	14.2875	9	228.6	18	.70866
37/64	.578125	14.6844	10	254.0	19	.74803
19/32	.59375	15.0812	11	279.4	20	.78740
39/64	.609375	15.4781	12	304.8	21	.82677
5/8	.625	15.8750	13	330.2	22	.86614
41/64	.640625	16.2719	14	355.6	23	.90551
21/32	.65625	16.6687	15	381.0	24	.94488
43/64	.671875	17.0656	16	406.4	25	.98425
11/16	.6875	17.4625	17	431.8	26	1.02362
45/64	.703125	17.8594	18	457.2	27	1.06299
23/32	.71875	18.2562	19	482.6	28	1.10236
47/64	.734375	18.6531	20	508.0	29	1.14173
3/4	.75	19.0500	21	533.4	30	1.18110
49/64	.765625	19.4469	22	558.8	31	1.22047
25/32	.78125	19.8437	23	584.2	32	1.25984
51/64	.796875	20.2406	24	609.6	33	1.29921
13/16	.8125	20.6375	25	635.0	34	1.33858
53/64	.828125	21.0344	26	660.4	35	1.37795
27/32	.84375	21.4312	27	685.8	36	1.41732
55/64	.859375	21.8281	28	711.2	37	1.4567
7/8	.875	22.2250	29	736.6	38	1.4961
57/64	.890625	22.6219	30	762.0	39	1.5354
29/32	.90625	23.0187	31	787.4	40	1.5748
59/64	.921875	23.4156	32	812.8	41	1.6142
15/16	.9375	23.8125	33	838.2	42	1.6535
61/64	.953125	24.2094	34	863.6	43	1.6929
31/32	.96875	24.6062	35	889.0	44	1.7323
63/64	.984375	25.0031	36	914.4	45	1.7717

UNITS	Pints to Litres	Gallons to Litres	Litres to Pints	Litres to Gallons	Miles to Kilometres	Kilometres to Miles	Lbs. per sq. In. to Kg. per sq. Cm.	Kg. per sq. Cm. to Lbs. per sq. In.
1	.57	4.55	1.76	.22	1.61	.62	.07	14.22
2	1.14	9.09	3.52	.44	3.22	1.24	.14	28.50
3	1.70	13.64	5.28	.66	4.83	1.86	.21	42.67
4	2.27	18.18	7.04	.88	6.44	2.49	.28	56.89
5	2.84	22.73	8.80	1.10	8.05	3.11	.35	71.12
6	3.41	27.28	10.56	1.32	9.66	3.73	.42	85.34
7	3.98	31.82	12.32	1.54	11.27	4.35	.49	99.56
8	4.55	36.37	14.08	1.76	12.88	4.97	.56	113.79
9		40.91	15.84	1.98	14.48	5.59	.63	128.00
10		45.46	17.60	2.20	16.09	6.21	.70	142.23
20				4.40	32.19	12.43	1.41	284.47
30				6.60	48.28	18.64	2.11	426.70
40				8.80	64.37	24.85		
50					80.47	31.07		
60					96.56	37.28		
70					112.65	43.50		
80					128.75	49.71		
90					144.84	55.92		
100					160.93	62.14		

UNITS	Lb ft to kgm	Kgm to lb ft	UNITS	Lb ft to kgm	Kgm to lb ft
1	.138	7.233	7	.967	50.631
2	.276	14.466	8	1.106	57.864
3	.414	21.699	9	1.244	65.097
4	.553	28.932	10	1.382	72.330
5	.691	36.165	20	2.765	144.660
6	.829	43.398	30	4.147	216.990

APPENDIX

TECHNICAL DATA

(Dimensions are in inches unless otherwise stated)

ENGINE

Type	Straight six, water cooled
Capacity	170, 200 or 250 cu inch

Engines will be referred to by their capacity

Bore:	
170	3.50
200	3.68
250	3.68
Stroke:	
170	2.94
200	3.13
250	3.91
Compression ratios:	
170	9.1:1
200	8.1:1
250	9.1:1
Firing order	1-5-3-6-2-4
Crankshaft:	
Main bearings 170	4—thrust taken at No. 3
200, 250	7—thrust taken at No. 5
End play	.004 to .008 (limit .012)
Journals diameter 170, 200	2.2486 to 2.2490 (Red)
	2.2482 to 2.2486 (Blue)
250	2.3982 to 2.3990
clearance	.0005 to .0015 (limit .0022)
Crankpins diameter	2.1236 to 2.1240 (Red)
	2.1232 to 2.1236 (Blue)
clearance 170, 200	.0008 to .0024
250	.001 to .0015
Connecting rods:	
Length, centre to centre 170, 200	4.7135 to 4.7165
250	5.8185 to 5.8815
Side clearance	.0035 to .0105
Small-end diameter	.9107 to .9112
Big-end diameter	2.2390 to 2.2398
Piston pins:	
Length	3.010 to 3.040
Clearance to piston	.0003 to .0005 (limit .0008)
Diameter	.9119 to .9124
Oversizes	.001, .002
Pistons:	
Diameter	At piston pin bore centre line at 90 deg. to bore
170 (Red)	3.4982 to 3.4987
(Blue)	3.4993 to 3.4999
(.003 OS)	3.5001 to 3.5005
200, 250 (Red)	3.6778 to 3.6784
(Blue)	3.6790 to 3.6796
(.003 OS)	3.6802 to 3.6808
Ring groove width:	
Compression rings	.080 to .081
Oil ring	.188 to .189

Piston rings:	
Width:	
Compression rings	.077 to .078
Oil ring	.188 to .189
Clearance	.002 to .004
Gap:	
Compression rings	.010 to .020
Oil ring	.015 to .055
Camshaft:	
Lobe lift	.232 (.245 or .250 cu inch)
Maximum loss of lift	.005
Theoretical valve lift	.3480
End play	.001 to .007 (wear limit .012)
Journal to bearing clearance	.001 to .002 (wear limit .006)
Cylinder block	
Cylinder bore (standard):	
170	3.5000 to 3.5024
200, 250	3.6800 to 3.6824
Maximum bore out of round:	
Machined	.001
Wear limit	.005
Cylinder head:	
Valve seat width:	
Intake	.060 to .080
Exhaust	.070 to .090
Valve seat angle	45 deg.
Surface flatness	.003 inch in any 6 inches, .007 overall
Valve springs:	
Free length	1.79
Fitted height	$1\frac{9}{16}$ to $1\frac{5}{8}$
Test length	1.59 inch at 51 to 57 lb 1.222 inch at 142 to 158 lb
Wear limit	1.59 inch at 46 lb 1.222 inch at 128 lb
Valves:	
Stem diameter:	
Intake	.3100 to .3107
Exhaust	.3098 to .3105
Oversizes	+.003, +.015, +.030
Clearance to valve guide:	
Intake	.0008 to .0085
Exhaust	.0010 to .0027
Seat angle	44 deg.
Oil pump:	
Relief valve springs	1.078 inch at 9 to 10.1 lb
Drive shaft to housing clearance	.0015 to .0029
Relief valve clearance	.0015 to .0029
Rotor assembly end clearance	.0011 to .0041
Outer rotor to housing radial clearance	.006 to .013
Valve clearances:	
1964 to 1967:	
Hydraulic lifters	.066 to .216
Mechanical	.018 (hot or cold)
1968 to 1970:	
Mechanical	.018
170	.066 to .166 (.117 desired)
200, 250	.095 to .195 (.145 desired)

FUEL SYSTEM

Fuel pump:

Static pressure	4.0 to 6.0 lb/sq inch
Delivery	1 pint in 30 to 36 seconds

Carburetter settings:

	T.P.C.	*C.P.*	*F.L.*	*F.1.*	*C.1.*
C6DF-R ...		.140-.160	$1\frac{3}{32}$	1500	500
C6DF-S ...		.100-.120	$1\frac{3}{32}$	1300	575
C6OF-AD ...		.130-.150	$1\frac{3}{32}$	1300	575
C6OF-AC ...		.140-.160	$1\frac{3}{32}$	1500	500
C7DF-J ...		.100-.120	$1\frac{3}{32}$	1400	700
C7DF-K ...		.140-.160	$1\frac{3}{32}$	1500	550
C7DF-T ...	.063	.250	.218	—	550
C7OF-N ...		.100-.120	$1\frac{3}{32}$	1400	700
C7OF-R ...		.140-.160	$1\frac{3}{32}$	1500	500
C8DF-A ...	.035		$1\frac{3}{32}$	1400	700
C8DF-B ...	.046	.150	$1\frac{3}{32}$	1400	750
C8DF-G ...	.046	.265	$\frac{7}{32}$	—	550
C8DF-H ...	.040	.265	$\frac{7}{32}$	—	750
C8OF-A ...		.150	$1\frac{3}{32}$	1400	700
C8OF-B ...	.040	.130	$1\frac{3}{32}$	1500	550
D0DF-L ...	.036	.265	$\frac{3}{8}$	—	600
D0DF-M ...	.031	.265	$\frac{3}{8}$	—	750
D0DF-T ...	.031	.265	$\frac{3}{8}$	—	750
DODF-V ...	.036	.265	$\frac{3}{8}$	—	600

T.P.C. Throttle plate clearance
C.P. Choke plate pull-down (with .036 inch gauge when applicable)
F.L. Float level
F.1. Fast idle speed
C.1. Curb idle speed

Anti-stall dashpot settings:

Autolite	1100 Manual	3 turns in $\pm \frac{1}{4}$ turn
	1100 Automatic	2 turns in $\pm \frac{1}{4}$ turn
	1101	.08
Carter	YF I-V	.125

IGNITION SYSTEM

Firing order	1-5-3-6-2-4
Spark plug gap	.032 to .036
Spark plug type	BF 82 (Autolite), UF-14Y (Champion)

Distributor contact gap and dwell angle:

	Gap Inches	*Dwell angle Degrees*
1964	.024 to .026	35 to 38
1965	.024 to .026	35 to 38
1966	.025	37 to 42
1967	.025	37 to 42
1968	.027	35 to 40
1969	.027	35 to 40
1970	.027	35 to 40

Initial static setting: *Given in degrees BTDC*

	Manual shift	*Automatic*
1964 and 1965	6	12
1966	6 Standard TDC with Thermactor	12

1967:		
Standard	6	12
With emission control	5	5
1968 to 1970	6	6

COOLING SYSTEM

Antifreeze ... ... Ethylene Glycol, should always be used to give protection to 0°F

Thermostat:

	Starts opening	*Fully open*
Low temperature	155 to 162°F	182°F
High temperature...	188 to 195°F	210 to 212°F

CLUTCH

Type	Single dry-plate
Operation	Mechanical linkage
Adjustment	By release lever to give minimum ½ inch free pedal travel at 3000 rev/min

MANUAL SHIFT TRANSMISSION

	First	*Second*	*Third*	*Reverse*
Ratios:				
Up to 1965	2.79:1	1.70:1	1:1	2.87:1
1966 onwards	2.99:1	1.75:1	1:1	3.17:1

SUSPENSION

	Min.	*Max.*	*Desired*	*Max. variation between wheels*
Castor:	*Given in degrees*			
1964	0	+1	+½	½
1965	—1¼	+¾	—¼	½
1966	—1	+1	0	½
1967	—1½	+½	—1	½
1968	—1½	+½	—½	½
1969 and 1970	—1¾	+¼	—¾	½
Camber:	*Given in degrees*			
1964	0	+1	+½	½
1965	—¼	+1¼	+½	½
1966 on	—½	+1	+¼	½
Toe:in:		*Given in inches*		
1964		$\frac{1}{4}$	$\frac{5}{16}$	—
1965		$\frac{5}{32}$	$\frac{11}{32}$	$\frac{9}{32}$
1966		$\frac{1}{8}$	$\frac{3}{8}$	$\frac{1}{4}$
1967		$\frac{1}{8}$	$\frac{1}{4}$	$\frac{3}{16}$
1968		$\frac{1}{8}$	$\frac{3}{8}$	$\frac{1}{4}$
1969 and 1970		$\frac{1}{8}$	$\frac{3}{8}$	$\frac{3}{16}$

STEERING

Type	Recirculating ball. Power steering as optional extra	
Power steering fluid	Ford Automatic Transmission Fluid	
Steering unit preloads	Given in inch lb	
	Worm only	*Total*
Manual	4 to 5	9 to 10
Power	3 to 4	8 to 9
Turns lock to lock:		
	Manual	*Power*
1964	$4\frac{5}{8}$	$4\frac{5}{8}$
1965	$4\frac{5}{8}$	$3\frac{3}{4}$
1966 and 1967	5	4
1968 to 1970	5.5	4

BRAKING SYSTEM

Type	Hydraulically operated service brakes. Cable operated park brakes
Fluid	Ford Super Heavy Duty
Adjustments	Service brakes self-adjusting Park brakes linkage adjusted at cable
Linings	Use only Ford-approved shoes

ELECTRICAL SYSTEM

Polarity	Negative ground
Voltage	12 Volts
Battery	Lead/acid
Charging	Alternator, except 1964 models which use standard generator
Starter motor:	
No load current draw	70 amps
Brush length:	
New	.5 inch
Minimum	.25 inch
Brush tension	40 oz
Operating current:	
1964 to 1967	100 to 150 amps
1968 on	150 to 200 amps
Stall current	
1964 to 1967	460 amps at 9 lb ft on 5 volts
1968 on	460 amps at 9 lb ft on 5 volts

CAPACITIES*

Automatic transmission	8 quarts refill
Cooling system	Including 1 quart for heater
1964	15 quarts
1965	14 quarts
1966 and 1967	15 quarts
1968 on	13.6 quarts
Manual transmission	$3\frac{1}{2}$ pints
Steering:	
1964 and 1965	8 oz
1966	11 oz
1967 onwards	.87 lb

*Multiply by .83 to obtain Imperial measures

TORQUE WRENCH SETTINGS

Given in lb ft unless otherwise stated

Engine:	
Cylinder head bolts:	
Stage 1	55 to 65
Stage 2	70 to 75
Exhaust manifold to head	13 to 18
Flywheel bolts	75 to 85
Oil pan attachments	7 to 9
Connecting rod nuts	19 to 24
Main bearing bolts	60 to 70
Camshaft sprocket	35 to 45
Camshaft thrust plate	12 to 15
Valve rocker shaft attachments	30 to 35
Pulley to damper:	
UBS bolts	25 to 35
Place bolts	35 to 45
Front bracket to insulator	18 to 25
Front insulator to support bracket	24 to 34
Support bracket to crossmember	20 to 30
Rear insulator to transmission	30 to 60
Clutch:	
Flywheel housing to block	40 to 50
Cover to flywheel	12 to 20
Manual-shift transmission:	*Threespeed*
Extension housing bolts	42 to 50
Input shaft retainer	30 to 36
Speedometer cable retainer	3 to 4.5
Transmission to flywheel housing	37 to 42
Automatic transmission:	
Converter to flywheel	20 to 30
Converter housing to case	28 to 40
Rear servo cover	12 to 20
Intermediate servo cover	16 to 22
Cooler line fittings	80 to 120 lb inch
Converter drain plug	20 to 30
Transmission to engine	40 to 50
Extension housing	28 to 40
Governor body to collector	80 to 120 lb inch
Suspension:	
Shock absorber upper attachment	20 to 28
Shock absorber bracket, upper	15 to 20
Brake backing plate	28 to 42
Suspension ball joints	60 to 90
Strut to lower arm	75 to 105
Stabilizer mounting bracket	17 to 25
Strut to underbody	60 to 80
Stabilizer link	5 to 10
Caliper:	
Upper bolt	110 to 120
Lower bolt	60 to 70
Spring shackle bars	15 to 22
Rear shock absorber, upper and lower	15 to 25
Spring to axle	30 to 45
Spring front hanger bolt	90 to 120
Spring front hanger nut	40 to 60
Wheel attachments:	
4-lug	55 to 85
5-lug	70 to 115

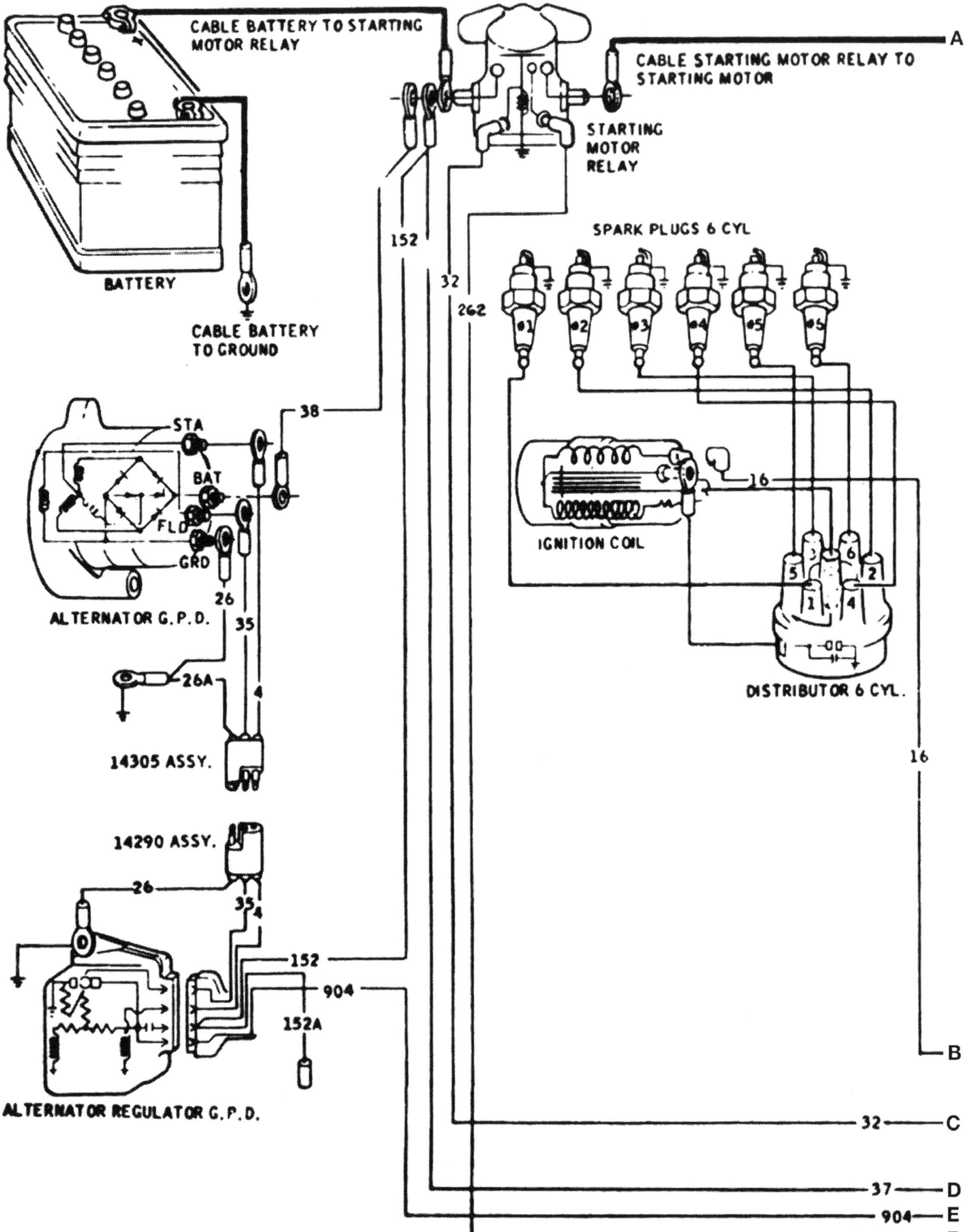

FIG 14:1 The engine wiring system

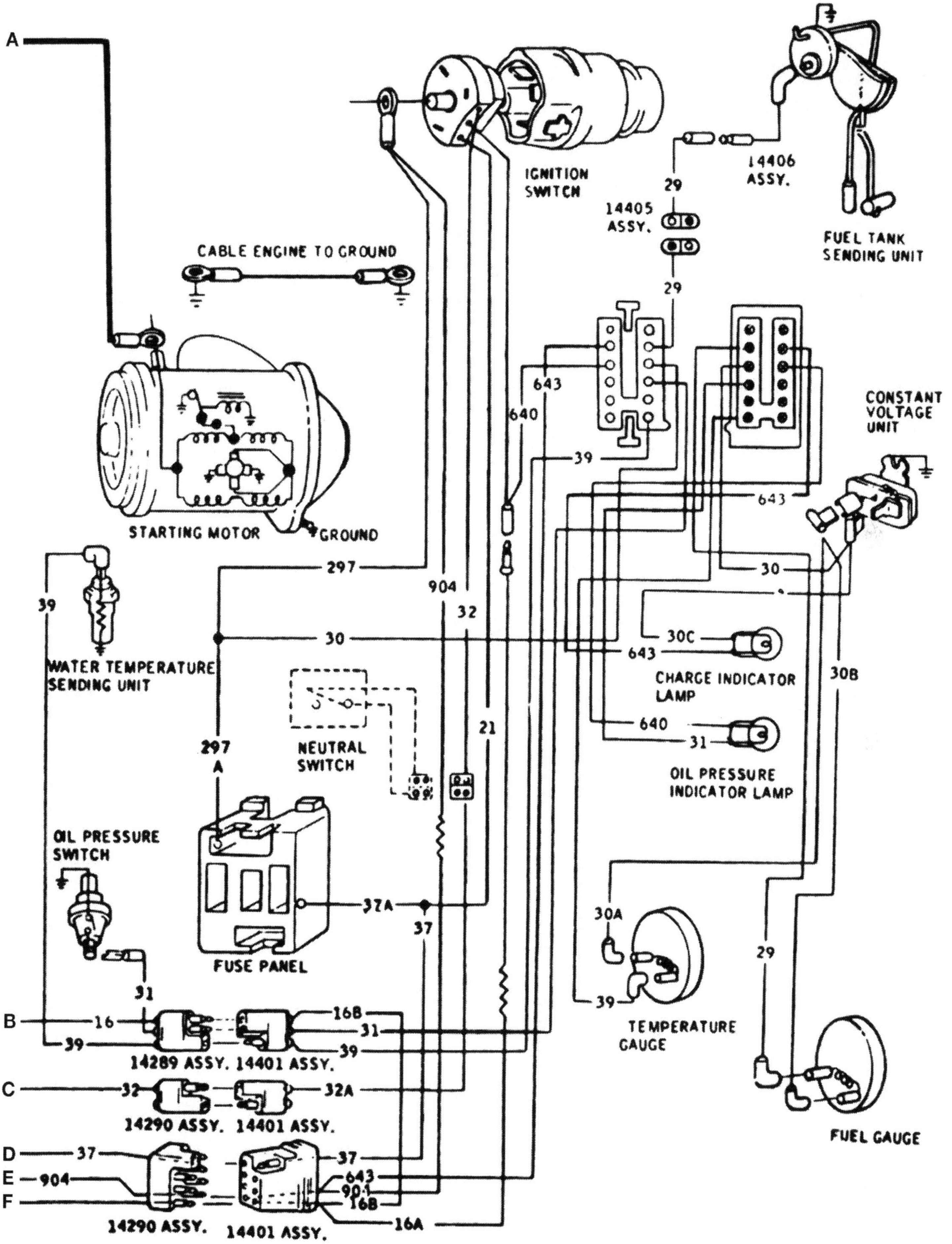

A
IGNITION SWITCH
14406 ASSY.
29
14405 ASSY.
FUEL TANK SENDING UNIT
CABLE ENGINE TO GROUND
29
643
640
CONSTANT VOLTAGE UNIT
39
STARTING MOTOR
GROUND
643
297
904
32
30
39
30
30C
WATER TEMPERATURE SENDING UNIT
643
CHARGE INDICATOR LAMP
30B
21
640
297
A
NEUTRAL SWITCH
31
OIL PRESSURE INDICATOR LAMP
OIL PRESSURE SWITCH
32A
37
30A
FUSE PANEL
29
31
39
TEMPERATURE GAUGE
B
16
16B
31
39
39
14289 ASSY.
14401 ASSY.
C
32
32A
14290 ASSY.
14401 ASSY.
FUEL GAUGE
D
37
37
E
904
643
904
F
16B
16A
14290 ASSY.
14401 ASSY.

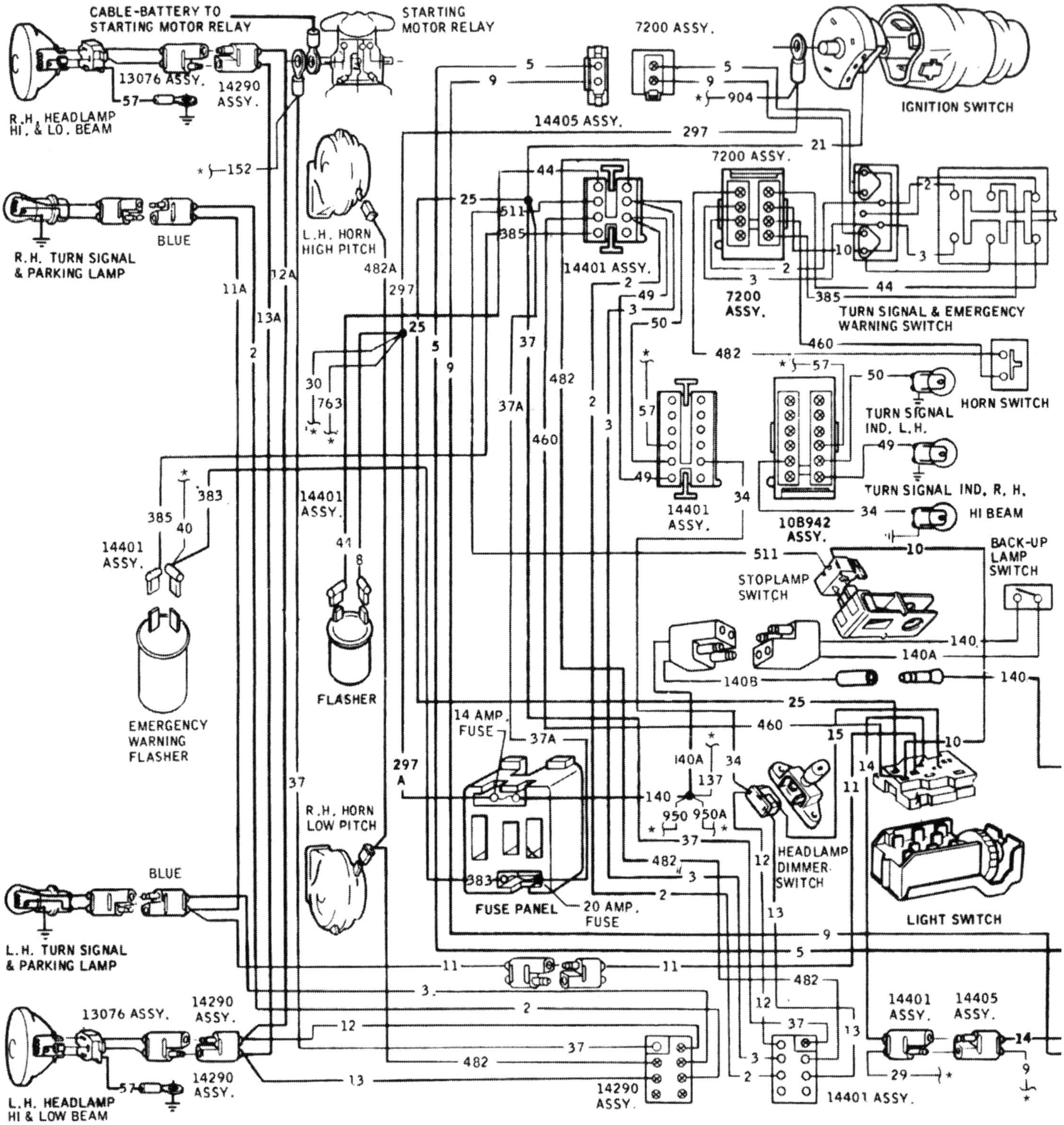

FIG 14 : 2 Exterior lighting 1

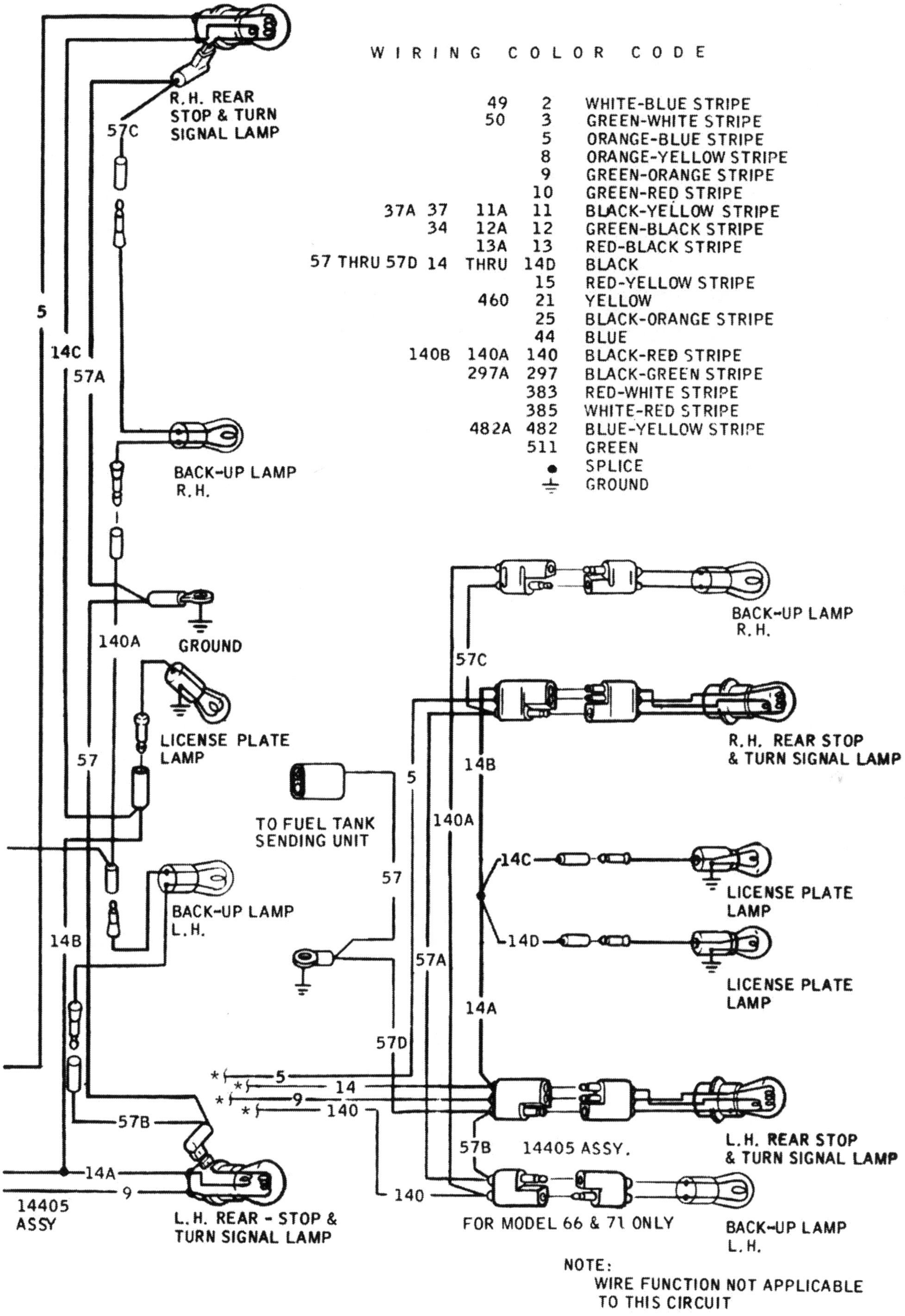

FIG 14:3 Exterior lighting 2

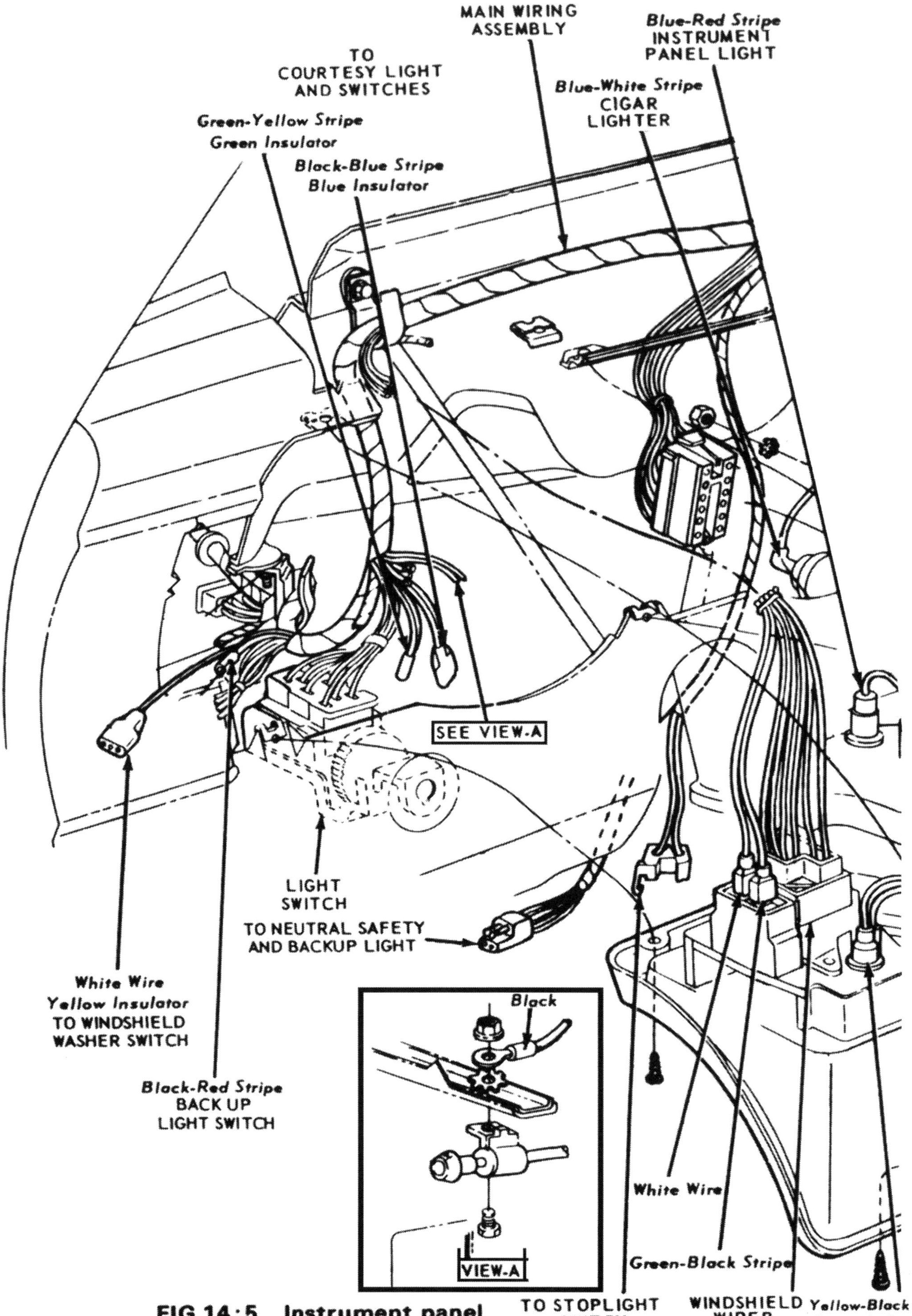

FIG 14:5 Instrument panel

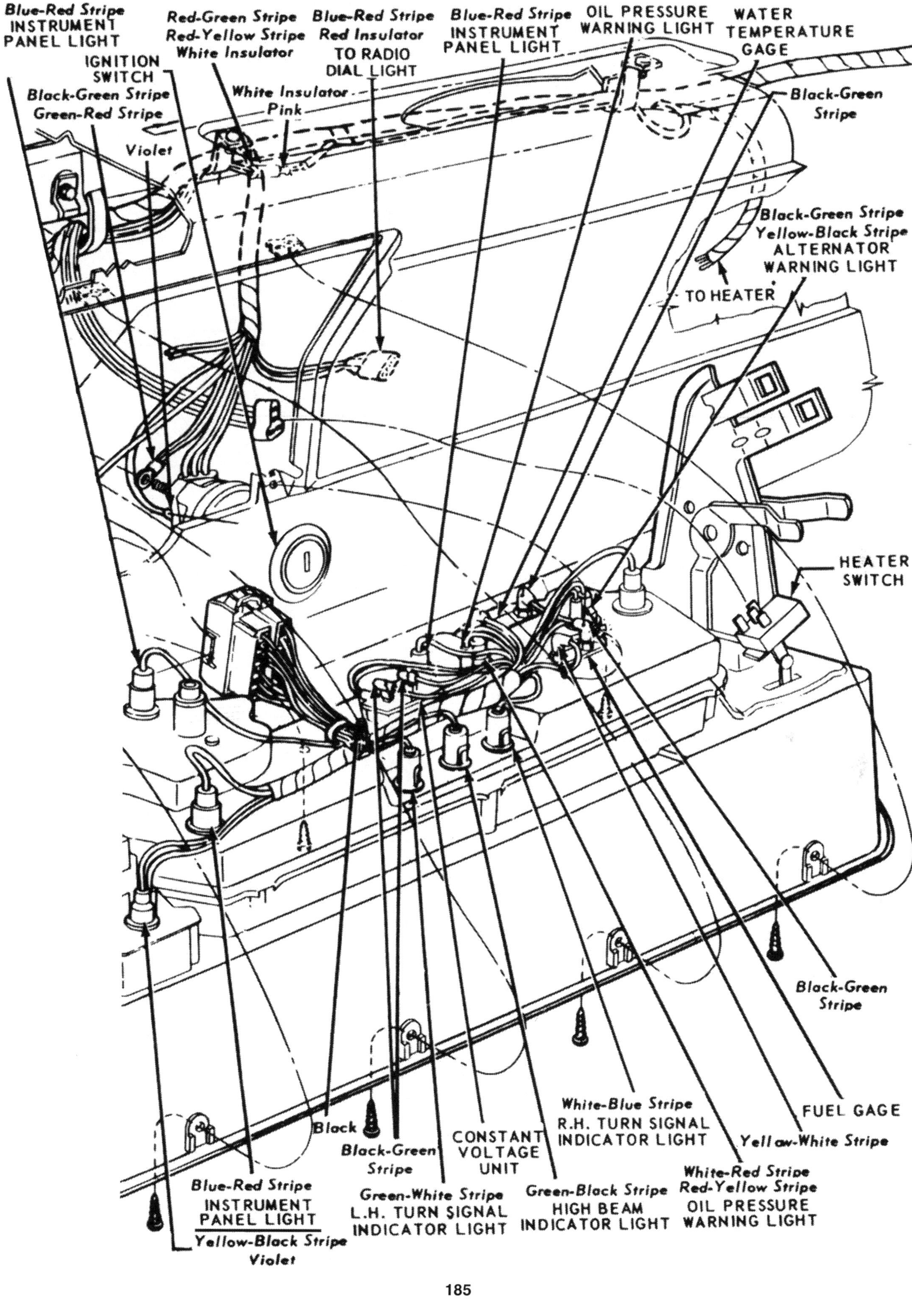
Blue-Red Stripe
INSTRUMENT PANEL LIGHT
IGNITION SWITCH
Black-Green Stripe
Green-Red Stripe
Violet
Red-Green Stripe
Red-Yellow Stripe
White Insulator
White Insulator
Pink
Blue-Red Stripe
Red Insulator
TO RADIO DIAL LIGHT
Blue-Red Stripe
INSTRUMENT PANEL LIGHT
OIL PRESSURE WARNING LIGHT
WATER TEMPERATURE GAGE
Black-Green Stripe
Black-Green Stripe
Yellow-Black Stripe
ALTERNATOR WARNING LIGHT
TO HEATER
HEATER SWITCH
Black-Green Stripe
FUEL GAGE
Yellaw-White Stripe
White-Blue Stripe
R.H. TURN SIGNAL INDICATOR LIGHT
White-Red Stripe
Red-Yellow Stripe
OIL PRESSURE WARNING LIGHT
Green-Black Stripe
HIGH BEAM INDICATOR LIGHT
CONSTANT VOLTAGE UNIT
Green-White Stripe
L.H. TURN SIGNAL INDICATOR LIGHT
Black
Black-Green Stripe
Blue-Red Stripe
INSTRUMENT PANEL LIGHT
Yellow-Black Stripe
Violet

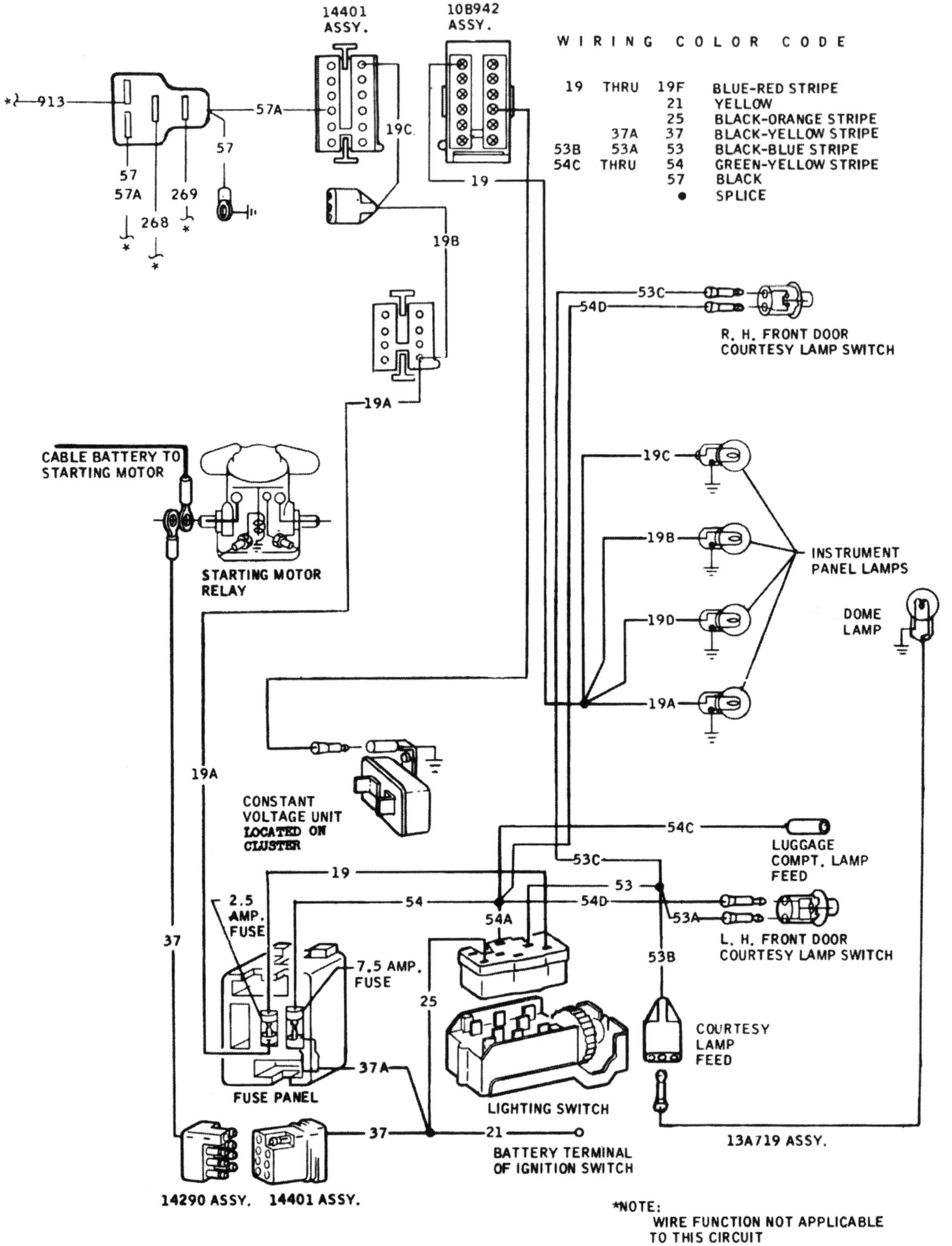

FIG 14:5 Interior lighting

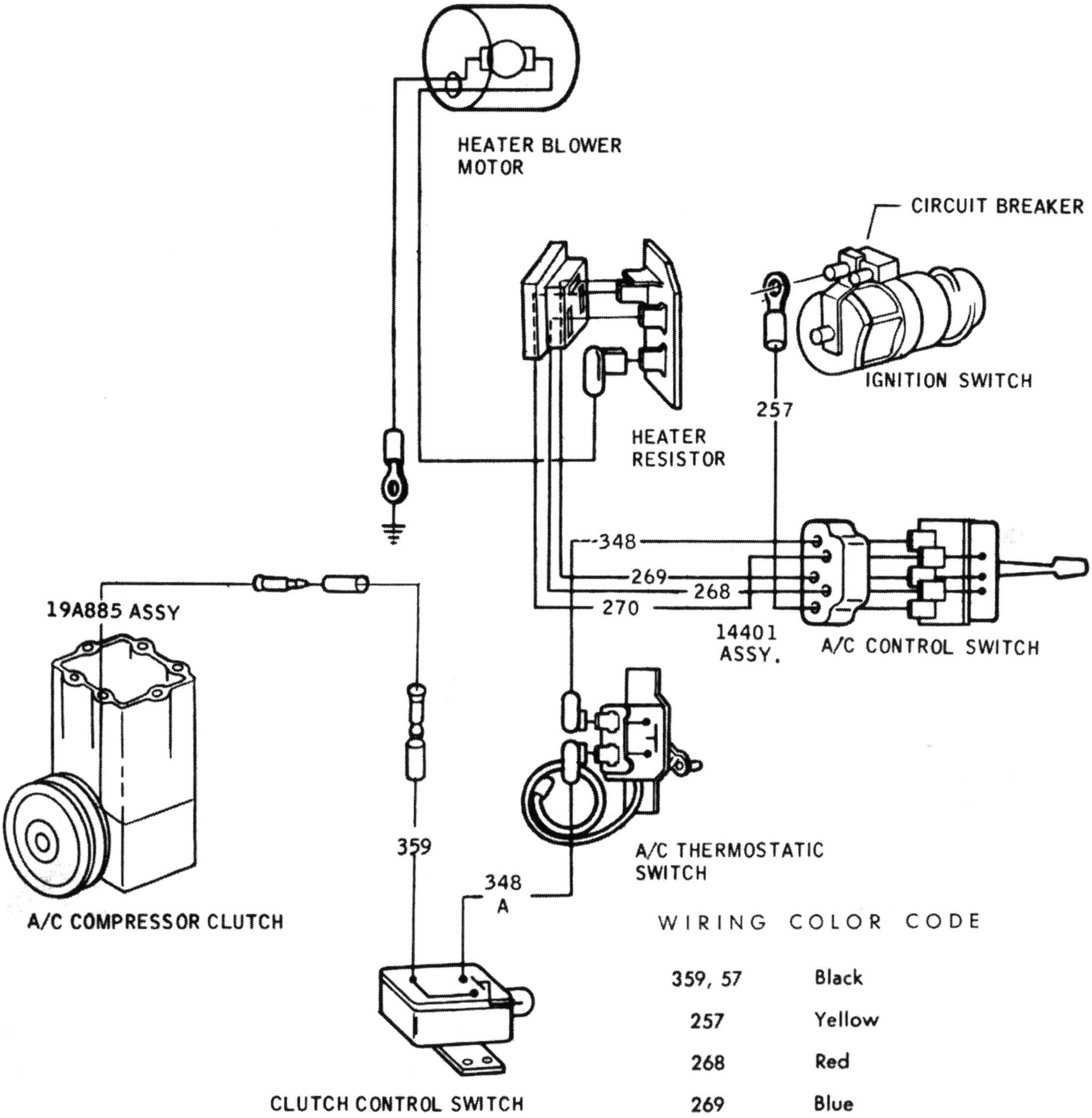

WIRING COLOR CODE

359, 57	Black
257	Yellow
268	Red
269	Blue
270	Black—Yellow Stripe
348, 348A	Green

FIG 14:6 Air conditioning

Glossary of Terms

Allen key — Cranked wrench of hexagonal section for use with socket head screws.

Alternator — Electrical generator producing alternating current. Rectified to direct current for battery charging.

Ambient temperature — Surrounding atmospheric temperature.

Annulus — Used in engineering to indicate the outer ring gear of an epicyclic gear train.

Armature — The shaft carrying the windings, which rotates in the magnetic field of a generator or starter motor. That part of a solenoid or relay which is activated by the magnetic field.

Axial — In line with, or pertaining to, an axis.

Backlash — Play in meshing gears.

Balance lever — A bar where force applied at the centre is equally divided between connections at the ends.

Banjo axle — Axle casing with large diameter housing for the crownwheel and differential.

Bendix pinion — A self-engaging and self-disengaging drive on a starter motor shaft.

Bevel pinion — A conical shaped gearwheel, designed to mesh with a similar gear with an axis usually at 90 degree to its own.

bhp — Brake horse power, measured on a dynamometer.

bmep — Brake mean effective pressure. Average pressure on a piston during the working stroke.

Brake cylinder — Cylinder with hydraulically operated piston(s) acting on brake shoes or pad(s).

Brake regulator — Control valve fitted in hydraulic braking system which limits brake pressure to rear brakes during heavy braking to prevent rear wheel locking.

Camber — Angle at which a wheel is tilted from the vertical.

Capacitor — Modern term for an electrical condenser. Part of distributor assembly, connected across contact breaker points, acts as an interference suppressor.

Castellated — Top face of a nut, slotted across the flats, to take a locking splitpin.

Castor — Angle at which the kingpin or swivel pin is tilted when viewed from the side.

cc — Cubic centimetres. Engine capacity is arrived at by multiplying the area of the bore in sq. cm by the stroke in cm by the number of cylinders.

Clevis — U-shaped forked connector used with a clevis pin, usually at handbrake connections.

Collet — A type of collar, usually split and located in a groove in a shaft, and held in place by a retainer. The arrangement used to retain the spring(s) on a valve stem in most cases.

Commutator — Rotating segmented current distributor between armature windings and brushes in generator or motor.

Compression ratio — The ratio, or quantitative relation, of the total volume (piston at bottom of stroke) to the unswept volume (piston at top of stroke) in an engine cylinder.

Condenser — See 'Capacitor'.

Core plug — Plug for blanking off a manufacturing hole in a casting.

Crownwheel — Large bevel gear in rear axle, driven by a bevel pinion attached to the propeller shaft. Sometimes called a 'ring gear'.

'C'-spanner — Like a 'C' with a handle. For use on screwed collars without flats, but with slots or holes.

Damper — Modern term for shock absorber, used in vehicle suspension systems to damp out spring oscillations.

Depression — The lowering of atmospheric pressure as in the inlet manifold and carburetter.

Dowel — Close tolerance pin, peg, tube or bolt which accurately locates mating parts.

Drag link — Rod connecting steering box drop arm (pitman arm) to nearest front wheel steering arm in certain types of steering systems.

Dry liner — Thinwall tube pressed into cylinder bore.

Dry sump — Lubrication system where all oil is scavenged from the sump and returned to a separate tank.

Dynamo — See 'Generator'

Electrode — Terminal part of an electrical component such as the points or 'Electrodes' of a sparking plug.

Electrolyte — In lead-acid car batteries a solution of sulphuric acid and distilled water.

End float — The axial movement between associated parts, end play.

EP — Extreme pressure. In lubricants, special grades for heavily loaded bearing surfaces, such as gear teeth in a gearbox, or crown wheel and pinion in a rear axle.

Fade — Of brakes. Reduced efficiency due to overheating.

Field coils — Windings on the polepieces of motors and generators.

Fillets — Narrow finishing strips usually applied to interior bodywork.

First motion shaft — Input shaft from clutch to gearbox.

Fullflow filter — Filters in which all the oil is pumped to the engine. If the element becomes clogged, a bypass valve operates to pass unfiltered oil to the engine.

FWD — Front wheel drive.

Gear pump — Two meshing gears in a close fitting casing. Oil is carried from the inlet round the outside of both gears in the spaces between the gear teeth and casing to the outlet. The meshing gear teeth prevent oil passing back to the inlet and the oil is forced through the outlet port.

Generator — Modern term for 'Dynamo'. When rotated produces electrical current.

Grommet	A ring of protective or sealing material. Can be used to protect pipes or leads passing through bulkheads.
Grubscrew	Fully threaded headless screw with screwdriver slot. Used for locking or alignment purposes.
Gudgeon pin	Shaft which connects a piston to its connecting rod. Sometimes called 'wrist pin' or 'piston pin'.
Halfshaft	One of a pair transmitting drive from the differential.
Helical	In spiral form. The teeth of helical gears are cut at a spiral angle to the side faces of the gearwheel.
Hot spot	Hot area that assists vapourisation of fuel on its way to cylinders. Often provided by close contact between inlet and exhaust manifolds.
HT	High Tension. Applied to electrical current produced by the ignition coil for the sparking plugs.
Hydrometer	A device for checking specific gravity of liquids. Used to check specific gravity of electrolyte.
Hypoid bevel gears	A form of bevel gear used in the rear axle drive gears. The bevel pinion meshes below the centre line of the crownwheel, giving a lower propeller shaft line.
Idler	A device for passing on movement. A free running gear between driving and driven gears. A lever transmitting track rod movement to a side rod in steering gear.
Impeller	A centrifugal pumping element. Used in water pumps to stimulate flow.
Journals	Those parts of a shaft that are in contact with the bearings.
Kingpin	The main vertical pin which carries the front wheel spindle and permits steering movement. May be called 'steering pin' or 'swivel pin'.
Layshaft	The shaft which carries the laygear in the gearbox. The laygear is driven by the first motion shaft and drives the third motion shaft according to the gear selected. Called the 'countershaft' or 'second motion shaft'.
lb ft	A measure of twist or torque. A pull of 10lb at a radius of 1 ft is a torque of 10 lb ft.
lb/sq in	Pounds per square inch.
Little-end	The small, or piston end of a connecting rod. Sometimes called the 'small-end'.
LT	Low Tension. The current output from the battery.
Mandrel	Accurately manufactured bar or rod used for test or centring purposes.
Manifold	A pipe, duct, or chamber, with several branches.
Needle rollers	Bearing rollers with a length many times their diameter.
Oil bath	Reservoir which lubricates parts by immersion. In air filters, a separate oil supply for wetting a wire mesh element to hold the dust.
Oil wetted	In air filters, a wire mesh element lightly oiled to trap and hold airborne dust.
Overlap	Period during which inlet and exhaust valves are open together.
Panhard rod	Bar connected between fixed point on chassis and another on axle to control sideways movement.
Pawl	Pivoted catch which engages in the teeth of a ratchet to permit movement in one direction only.
Peg spanner	Tool with pegs, or pins, to engage in holes or slots in the part to be turned.
Pendant pedals	Pedals with levers that are pivoted at the top end.
Phillips screwdriver	The cross-point screwdriver for using with the cross-slotted heads of Phillips screws.
Pinion	A small gear in relation to another gear.
Piston-type damper	Shock absorber in which damping is controlled controlled by a piston working in a closed oil-filled cylinder.
Preloading	Preset static pressure on ball or roller bearings not due to working loads.
Radial	Radiating from a centre, like the spokes of a wheel.
Radius rod	Pivoted arm confining movement of a part to an arc of fixed radius.
Ratchet	Toothed wheel or rack which can move in one direction only, movement in the other being prevented by a pawl.
Ring gear	A gear tooth ring attached to outer periphery of flywheel. Starter pinion engages with it during starting.
Runout	Amount by which rotating part is out of true.
Semi-floating axle	Outer end of rear axle halfshaft is carried on bearing inside axle casing. Wheel hub is secured to end of shaft.
Servo	A hydraulic or pneumatic system for assisting, or, augmenting a physical effort. See 'Vacuum Servo'.
Setscrew	One which is threaded for the full length of the shank.
Shackle	A coupling link, used in the form of two parallel pins connected by side plates to secure the end of the master suspension spring and absorb the effects of deflection.
Shell bearing	Thinwalled steel shell lined with anti-friction metal. Usually semi-circular and used in pairs for main and big-end bearings.
Shock absorber	See 'Damper'.
Silentbloc	Rubber bush bonded to inner and outer metal sleeves.
Socket-head screw	Screw with hexagonal socket for an Allen key.
Solenoid	A coil of wire creating a magnetic field when electric current passes through it. Used with a soft iron core to operate contacts or a mechanical device.
Spur gear	A gear with teeth cut axially across the periphery.

Stub axle Short axle fixed at one end only.

Tachometer An instrument for accurate measurement of rotating speed. Usually indicates in revolutions per minute.

TDC Top Dead Centre. The highest point reached by a piston in a cylinder, with the crank and connecting rod in line.

Thermostat Automatic device for regulating temperature. Used in vehicle coolant systems to open a valve which restricts circulation at low temperature.

Third motion shaft Output shaft of gearbox.

Three-quarter floating axle Outer end of rear axle half-shaft flanged and bolted to wheel hub,which runs on bearing mounted on outside of axle casing. Vehicle weight is not carried by the axle shaft.

Thrust bearing or washer Used to reduce friction in rotating parts subject to axial loads.

Torque Turning or twisting effort. See 'lb ft'.

Track rod The bar(s) across the vehicle which connect the steering arms and maintain the front wheels in their correct alignment.

UJ Universal joint. A coupling between shafts which permits angular movement.

UNF Unified National Fine screw thread.

Vacuum servo Device used in brake system, using difference between atmospheric pressure and inlet manifold depression to operate a piston which acts to augment brake pressure as required. See 'Servo'.

Venturi A restriction or 'choke' in a tube, as in a carburetter, used to increase velocity to obtain a reduction in pressure.

Vernier A sliding scale for obtaining fractional readings of the graduations of an adjacent scale.

Welch plug A domed thin metal disc which is partially flattened to lock in a recess. Used to plug core holes in castings.

Wet liner Removable cylinder barrel, sealed against coolant leakage, where the coolant is in direct contact with the outer surface.

Wet sump A reservoir attached to the crankcase to hold the lubricating oil.

Made in United States
Orlando, FL
04 November 2026

90200813R00105